STERLING
Test Prep

AF694846

College Physics

Practice Questions

Vol. 2

5th edition

Customer Satisfaction Guarantee

Your feedback is important because we strive to provide the highest quality prep materials. Email us comments or suggestions.

info@sterling–prep.com

We reply to emails – check your spam folder

5 4 3 2 1

ISBN-13: 978-1-9547257-7-5

Sterling Test Prep materials are available at quantity discounts.

Contact info@sterling–prep.com

Sterling Test Prep
6 Liberty Square #11
Boston, MA 02109

Published by Sterling Test Prep

Printed in the U.S.A.

Thousands of students use our study aids to achieve higher grades!

To achieve a high grade in college physics, you need to do well on your tests and final exam. This book helps you develop and apply knowledge to quickly choose the correct answers to questions typically tested in college physics courses. Solving targeted practice questions builds your understanding of fundamental concepts and is a more effective strategy than merely memorizing terms.

This book ha 690 practice questions covering college physics topics. Physics instructors with years of teaching and applied physics experience prepared this practice material to build your knowledge and skills crucial for success in a college physics course. Our editorial team reviewed and systematized the content for targeted preparation.

The detailed explanations describe why an answer is correct and – more important for your learning – why another attractive choice is wrong. They provide step-by-step solutions for quantitative questions and teach the scientific foundations and details of essential physics topics needed to answer conceptual questions. Read the explanations carefully to understand how they apply to the question and learn important principles and the relationships between them. With the practice material contained in this book, you will significantly improve your understanding, test scores, and your grade.

We wish you great success in your academic achievements and look forward to being an important part of your preparation!

211124gdx

College study aids by Sterling Test Prep

Cell and Molecular Biology Review

Organismal Biology Review

Cell and Molecular Biology Practice Questions

Organismal Biology Practice Questions

Physics Review (Paert 1 and 2)

Physics Practice Questions (Vol. 1 and 2)

Organic Chemistry Practice Questions

United States History 101

American Government and Politics 101

Environmental Science 101

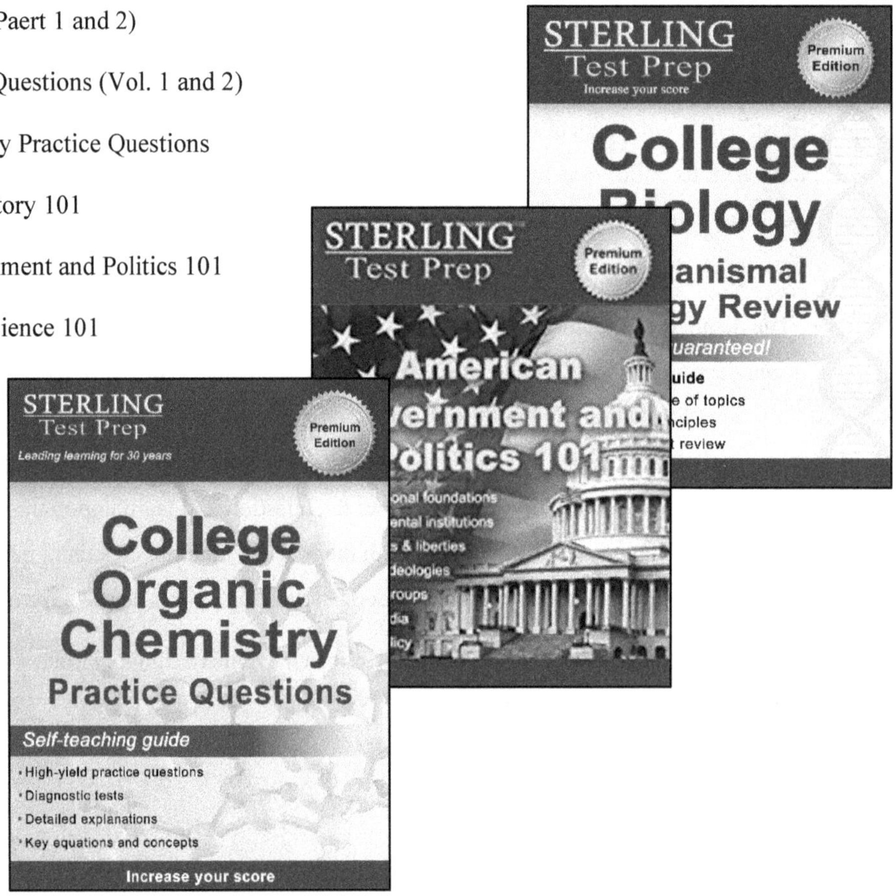

Visit our Amazon store

Table of Contents

Common Physics Equations and Conversions 7

Topical Practice Questions **15**

Fluids and Solids 17

Heat and Thermodynamics 29

Electric Circuits 41

Electrostatics and Electromagnetism 53

Atomic and Nuclear Structure 65

Particle Physics 77

Quantum Mechanics 83

Special Relativity 93

Diagnostic Tests **103**

Diagnostic Test 1 105

Diagnostic Test 2 129

Diagnostic Test 3 151

Topical Practice Questions: Answer Keys & Detailed Explanations **173**

Answer Keys 175

Fluids and Solids 179

Heat and Thermodynamics 197

Electric Circuits 215

Electrostatics and Electromagnetism 231

Atomic and Nuclear Structure 247

Particle Physics 259

Quantum Mechanics 267

Special Relativity 283

Diagnostic Tests: Answer Keys & Detailed Explanations **305**

Diagnostic Test 1 307

Diagnostic Test 2 341

Diagnostic Test 3 371

Glossary of Physics Terms **401**

College Level Examination Program (CLEP)

Biology Review

Biology Practice Questions

Chemistry Review

Chemistry Practice Questions

Introductory Business Law Review

College Algebra Practice Questions

College Mathematics Practice Questions

History of the United States I Review

History of the United States II Review

Western Civilization I Review

Western Civilization II Review

Social Sciences and History Review

American Government Review

Introductory Psychology Review

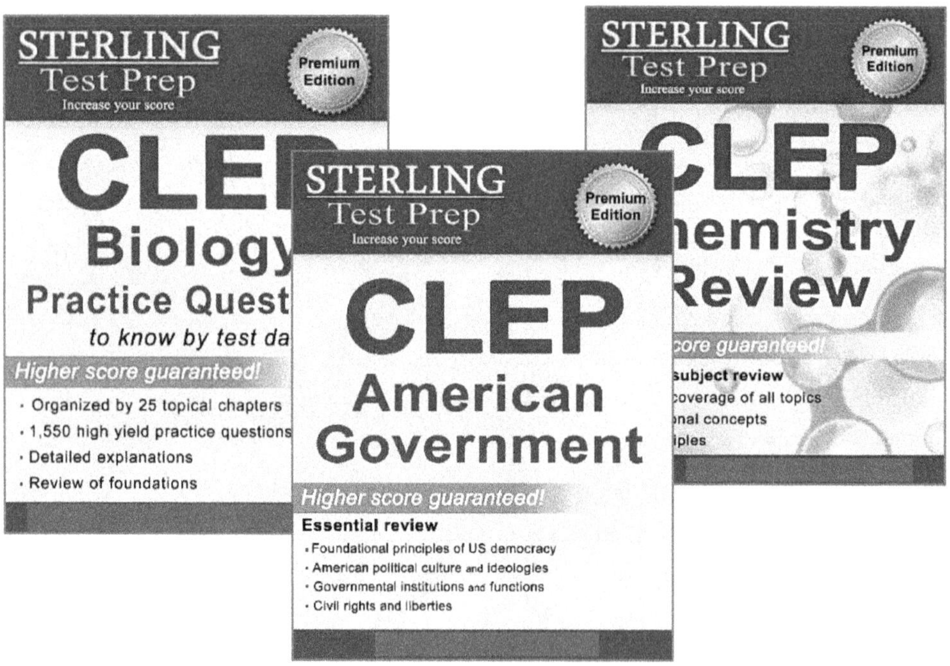

Visit our Amazon store

Common Physics Equations and Conversions

Constants and Conversion Factors

1 unified atomic mass unit	$1 \text{ u} = 1.66 \times 10^{-27}$ kg $1 \text{ u} = 931 \text{ MeV}/c^2$
Proton mass	$m_p = 1.67 \times 10^{-27}$ kg
Neutron mass	$m_n = 1.67 \times 10^{-27}$ kg
Electron mass	$m_e = 9.11 \times 10^{-31}$ kg
Electron charge magnitude	$e = 1.60 \times 10^{-19}$ C
Avogadro's number	$N_0 = 6.02 \times 10^{23} \text{ mol}^{-1}$
Universal gas constant	$R = 8.31$ J/(mol·K)
Boltzmann's constant	$k_B = 1.38 \times 10^{-23}$ J/K
Speed of light	$c = 3.00 \times 10^8$ m/s
Planck's constant	$h = 6.63 \times 10^{-34}$ J·s $h = 4.14 \times 10^{-15}$ eV·s $hc = 1.99 \times 10^{-25}$ J·m $hc = 1.24 \times 10^3$ eV·nm
Vacuum permittivity	$\varepsilon_0 = 8.85 \times 10^{-12} \text{ C}^2/\text{N}\cdot\text{m}^2$
Coulomb's law constant	$k = 1/4\pi\varepsilon_0 = 9.0 \times 10^9 \text{ N}\cdot\text{m}^2/\text{C}^2$
Vacuum permeability	$\mu_0 = 4\pi \times 10^{-7}$ (T·m)/A
Magnetic constant	$k' = \mu_0/4\pi = 10^{-7}$ (T·m)/A
Universal gravitational constant	$G = 6.67 \times 10^{-11} \text{ m}^3/\text{kg}\cdot\text{s}^2$
Acceleration due to gravity at Earth's surface	$g = 9.8 \text{ m/s}^2$
1 atmosphere pressure	$1 \text{ atm} = 1.0 \times 10^5 \text{ N/m}^2$ $1 \text{ atm} = 1.0 \times 10^5$ Pa
1 electron volt	$1 \text{ eV} = 1.60 \times 10^{-19}$ J
Balmer constant	$B = 3.645 \times 10^{-7}$ m
Rydberg constant	$R = 1.097 \times 10^7 \text{ m}^{-1}$
Stefan constant	$\sigma = 5.67 \times 10^{-8} \text{ W/m}^2\text{K}^4$

Units

Name	Symbol
meter	m
kilogram	kg
second	s
ampere	A
kelvin	K
mole	mol
hertz	Hz
newton	N
pascal	Pa
joule	J
watt	W
coulomb	C
volt	V
ohm	Ω
henry	H
farad	F
tesla	T
degree Celsius	°C
electronvolt	eV

Prefixes

Factor	Prefix	Symbol
10^{12}	tera	T
10^{9}	giga	G
10^{6}	mega	M
10^{3}	kilo	k
10^{-2}	centi	c
10^{-3}	mili	m
10^{-6}	micro	μ
10^{-9}	nano	n
10^{-12}	pico	p

Electrostatics, Magnetism, Circuits

Electric Field	$\vec{E} = \frac{\vec{F}_E}{q}$
Electric Field Strength	$\lvert\vec{E}\rvert = \frac{1}{4\pi\varepsilon_0}\frac{\lvert q\rvert}{r^2}$
Electric Field Strength	$\lvert\vec{E}\rvert = \frac{\lvert\Delta V\rvert}{\lvert\Delta r\rvert}$
Electrostatic Force Between Charged Particles	$\lvert\vec{F}_E\rvert = \frac{1}{4\pi\varepsilon_0}\frac{\lvert q_1 q_2\rvert}{r^2}$
Electric Potential Energy	$\Delta U_E = q\Delta V$
Electrostatic Potential due to a Charge	$V = \frac{1}{4\pi\varepsilon_0}\frac{q}{r}$
Capacitor Voltage	$V = \frac{Q}{C}$
Capacitance of Parallel Plate Capacitor	$C = \kappa\varepsilon_0\frac{A}{d}$
Electric Field Inside a Parallel Plate Capacitor	$E = \frac{Q}{\varepsilon_0 A}$
Capacitor Potential Energy	$U_C = \frac{1}{2}Q\Delta V = \frac{1}{2}C(\Delta V)^2$
Current	$I = \frac{\Delta Q}{\Delta t}$
Resistance	$R = \frac{\rho l}{A}$
Power	$P = I\Delta V$
Current	$I = \frac{\Delta V}{R}$
Resistors in Series	$R_s = \sum_i R_i$
Resistors in Parallel	$\frac{1}{R_p} = \sum_i \frac{1}{R_i}$

A = area
B = magnetic field
C = capacitance
d = distance
E = electric field
ϵ = emf
F = force
I = current
l = length
P = power
Q = charge
q = point charge
R = resistance
r = separation
t = time
U = potential energy
V = electric potential
v = speed
κ = dielectric constant
ρ = resistivity
θ = angle
Φ = flux

Capacitors in Parallel

$$C_p = \sum_i C_i$$

Capacitors in Series

$$\frac{1}{C_s} = \sum_i \frac{1}{C_i}$$

Magnetic Field Strength (from a long straight current-carrying wire)

$$B = \frac{\mu_0 I}{2\pi r}$$

Magnetic Force

$$\vec{F}_M = q\vec{v} \times \vec{B}$$

$$\vec{F}_M = |q\vec{v}||\sin\theta||\vec{B}|$$

$$\vec{F}_M = I\vec{l} \times \vec{B}$$

$$\vec{F}_M = |I\vec{l}||\sin\theta||\vec{B}|$$

Magnetic Flux

$$\Phi_B = \vec{B} \cdot \vec{A}$$

$$\Phi_B = |\vec{B}|\cos\theta\,|\vec{A}|$$

Electromagnetic Induction

$$\epsilon = \frac{-\Delta\Phi_B}{\Delta t}$$

$$\epsilon = Blv$$

Fluid Mechanics and Thermal Physics

Density	$\rho = \frac{m}{V}$	
Pressure	$P = \frac{F}{A}$	
Absolute Pressure	$P = P_0 + \rho g h$	
Buoyant Force	$F_b = \rho V g$	
Fluid Continuity Equation	$A_1 v_1 = A_2 v_2$	
Bernoulli's Equation	$P_1 + \rho g y_1 + \frac{1}{2}\rho v_1^2 = P_2 + \rho g y_2 + \frac{1}{2}\rho v_2^2$	
Heat Conduction	$\frac{Q}{\Delta t} = \frac{kA\Delta T}{d}$	
Thermal Radiation	$P = e\sigma A(T^4 - T_C^4)$	
Ideal Gas Law	$PV = nRT = Nk_BT$	
Average Energy	$K = \frac{3}{2}k_BT$	
Work	$W = -P\Delta V$	
Conservation of Energy	$\Delta E = Q + W$	
Linear Expansion	$\Delta l = \alpha l_o \Delta T$	
Heat Engine Efficiency	$n_c = \lvert W/Q_H \rvert$	
Carnot Heat Engine Efficiency	$n_c = \frac{T_H - T_C}{T_H}$	
Energy of Temperature Change	$Q = mc\Delta T$	
Energy of Phase Change	$Q = mL$	

A = area
c = specific heat
d = thickness
e = emissivity
F = force
h = depth
k = thermal conductivity
K = kinetic energy
l = length
L = latent heat
m = mass
n = number of moles
n_c= efficiency
N = number of molecules
P = pressure or power
Q = energy transferred to system by heating
T = temperature
t = time
E = internal energy
V = volume
v = speed
W = work done on a system
y = height
σ = Stefan constant
α = coefficient of linear expansion
ρ = density

Modern Physics

Photon Energy	$E = hf$	B = Balmer constant
		c = speed of light
Photoelectric Electron Energy	$K_{max} = hf - \phi$	E = energy
Electron Wavelength	$\lambda = \frac{h}{p}$	f = frequency
		K = kinetic energy
Energy Mass Relationship	$E = mc^2$	m = mass
		p = momentum
Rydberg Formula	$\frac{1}{\lambda} = R(\frac{1}{n_f^2} - \frac{1}{n_i^2})$	R = Rydberg constant
		v = velocity
Balmer Formula	$\lambda = B(\frac{n^2}{n^2 - 2^2})$	λ = wavelength
		$ø$ = work function
Lorentz Factor	$\gamma = \frac{1}{\sqrt{1 - \frac{v^2}{c^2}}}$	γ = Lorentz factor

Geometry and Trigonometry

Rectangle	$A = bh$
Triangle	$A = \frac{1}{2}bh$
Circle	$A = \pi r^2$ $C = 2\pi r$
Rectangular Solid	$V = lwh$
Cylinder	$V = \pi r^2 l$ $S = 2\pi rl + 2\pi r^2$
Sphere	$V = \frac{4}{3}\pi r^3$ $S = 4\pi r^2$
Right Triangle	$a^2 + b^2 = c^2$ $\sin\theta = \frac{a}{c}$ $\cos\theta = \frac{b}{c}$ $\tan\theta = \frac{a}{b}$

A = area

C = circumference

V = volume

S = surface area

b = base

h = height

l = length

w = width

r = radius

θ = angle

Trigonometric Functions for Common Angles

$\boldsymbol{\theta}$	sin $\boldsymbol{\theta}$	cos $\boldsymbol{\theta}$	tan $\boldsymbol{\theta}$
0°	0	1	0
30°	1/2	$\sqrt{3}/2$	$\sqrt{3}/3$
37°	3/5	4/5	3/4
45°	$\sqrt{2}/2$	$\sqrt{2}/2$	1
53°	4/5	3/5	4/3
60°	$\sqrt{3}/2$	1/2	$\sqrt{3}$
90°	1	0	∞

Notes for active learning

Topical Practice Questions

Fluids and Solids

Questions **1-3** are based on the following:

A container has a vertical tube with an inner radius of 20 mm connected to the container at its side. An unknown liquid reaches level A in the container and level B in the tube. Level A is 5 cm higher than level B. The liquid supports a 20 cm high column of oil between levels B and C with a density of 850 kg/m^3. (Use the acceleration due to gravity g = 9.8 m/s^2)

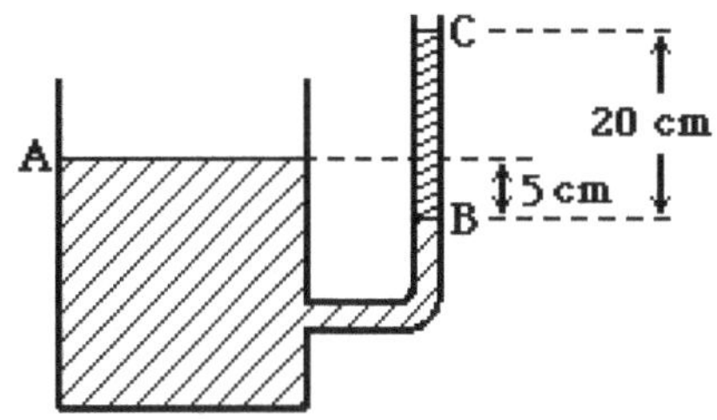

1. What is the density of the unknown liquid?

A. 2,800 kg/m^3
B. 2,100 kg/m^3
C. 3,400 kg/m^3
D. 3,850 kg/m^3
E. 1,800 kg/m^3

2. The gauge pressure at level B is closest to:

A. 1,250 Pa
B. 1,830 Pa
C. 340 Pa
D. 1,666 Pa
E. 920 Pa

3. What is the mass of the oil?

A. 210 g
B. 453 g
C. 620 g
D. 847 g
E. 344 g

4. A cubical block of stone is lowered at a steady rate into the ocean by a crane, keeping the top and bottom faces horizontal. Which of the following graphs best describes the gauge pressure P on the bottom of this block as a function of time t if the block just enters the water at time $t = 0$ s?

A.
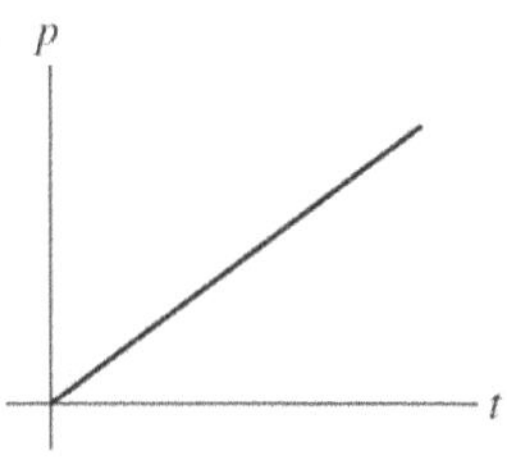

C.
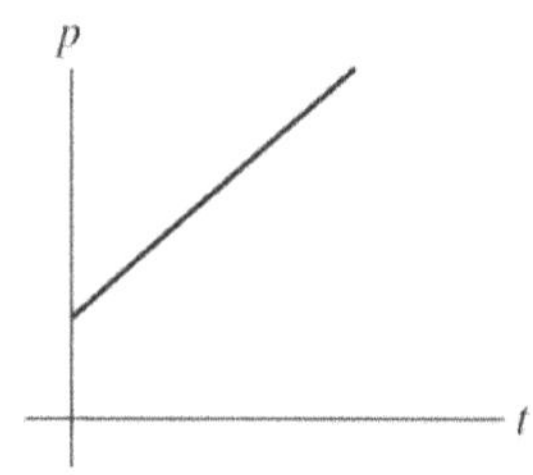

B.
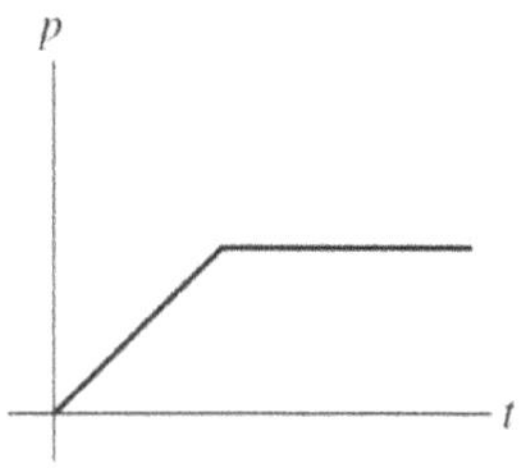

D.
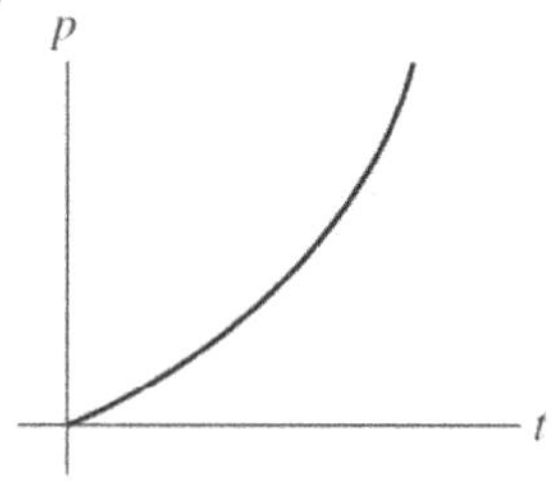

E.
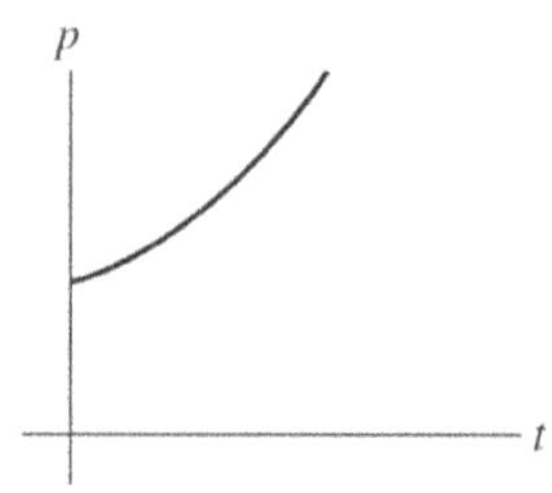

5. Consider a very small hole in the bottom of a tank that is 19 cm in diameter and is filled with water to a height of 80 cm. What is the speed at which the water exits the tank through the hole? (Use the acceleration due to gravity $g = 9.8$ m/s^2)

A. 8.6 m/s
B. 12 m/s
C. 14.8 m/s
D. 8.4 m/s
E. 4 m/s

6. An ideal gas at standard temperature and pressure is compressed until its volume is half the initial volume, and then it is allowed to expand until its pressure is half the initial pressure. This is achieved while holding the temperature constant. If the initial internal energy of the gas is U, the final internal energy of the gas is:

A. U/2
B. U/3
C. U
D. 2U
E. U/4

7. When 9.5 kg mass is suspended from a 4.5 m long wire with 1.2 mm diameter, the wire stretches by 4 mm. What is the effective spring constant for the wire? (Use the acceleration due to gravity $g = 9.8$ m/s^2)

A. 5.6×10^4 N/m
B. 2.3×10^4 N/m
C. 3.7×10^5 N/m
D. 7.2×10^3 N/m
E. 9.2×10^4 N/m

8. An object is sinking in a fluid. What is the weight of the fluid displaced by the sinking object when the object is completely submerged?

A. Dependent on the viscosity of the liquid
B. Equal to the weight of the object
C. Greater than the weight of the object
D. Zero
E. Less than the weight of the object

9. A submarine in neutral buoyancy is 100 m below the surface of the water. How much air pressure must be supplied to remove water from the ballast tanks for the submarine to the surface? (Use the acceleration due to gravity $g = 9.8$ m/s^2 and density of water $\rho = 10^3$ kg/m^3)

A. 9.8×10^5 N/m^2
B. 4.7×10^5 N/m^2
C. 7.6×10^5 N/m^2
D. 5.6×10^5 N/m^2
E. Requires the cross-sectional area of the tank

10. When atmospheric pressure increases, what happens to the absolute pressure at the bottom of a pool?

A. It does not change
B. It increases by double the amount
C. It increases by the same amount
D. It increases by half the amount
E. It depends on the depth of the pool

11. When soup gets cold, it often tastes greasy because oil spreads out on the surface of the soup instead of staying in small globules. This is explained in terms of the:

A. increase in the surface tension of water with a decreasing temperature
B. Archimedes' principle
C. decrease in the surface tension of water with a decreasing temperature
D. Joule-Thomson effect
E. Braggs' law

12. For a wire of diameter d and fixed length L, the weight that causes the wire to stretch a given distance is:

A. directly proportional to d^2 of the wire
B. inversely proportional to d^2 of the wire
C. proportional to d of the wire
D. independent of d of the wire
E. independent of L of the wire

13. An object whose weight is 60 N is floating at the surface of a container of water. How much of the object's volume is submerged? (Use acceleration due to gravity $g = 10\ \text{m/s}^2$)

A. $0.006\ \text{m}^3$
B. $0.06\ \text{m}^3$
C. $0.6\ \text{m}^3$
D. 6%
E. 60%

14. What is the volume flow rate of fluid if it flows at 2.5 m/s through a pipe of diameter 3 cm?

A. $0.9\ \text{m}^3/\text{s}$
B. $4.7\ \text{m}^3/\text{s}$
C. $5.7 \times 10^{-4}\ \text{m}^3/\text{s}$
D. $5.7 \times 10^{-3}\ \text{m}^3/\text{s}$
E. $1.8 \times 10^{-3}\ \text{m}^3/\text{s}$

15. What is the specific gravity of a cork that floats with three-quarters of its volume in and one-quarter of its volume out of the water?

A. 0.25
B. 0.5
C. 0.75
D. 2
E. 1

16. What volume does 600 g of cottonseed oil occupy if the density of cottonseed oil is $0.93\ \text{g/cm}^3$?

A. $255\ \text{cm}^3$
B. $360\ \text{cm}^3$
C. $470\ \text{cm}^3$
D. $645\ \text{cm}^3$
E. $690\ \text{cm}^3$

17. The kinetic theory of a monatomic gas suggests the average kinetic energy per molecule is:

A. $1/3\ k_B\text{T}$
B. $2\ k_B\text{T}$
C. $3/2\ k_B\text{T}$
D. $2/3\ k_B\text{T}$
E. $\sqrt{2}k_B\text{T}$

18. An object is weighed in air and while totally submerged in water. If it weighs 150 N less when submerged, find the volume of the object. (Use acceleration due to gravity $g = 10\ \text{m/s}^2$ and the density of water $\rho = 1{,}000\ \text{kg/m}^3$)

A. $0.0015\ \text{m}^3$
B. $0.015\ \text{m}^3$
C. $0.15\ \text{m}^3$
D. $1\ \text{m}^3$
E. $1.5\ \text{m}^3$

19. When a container of water is on a laboratory scale, the scale reads 140 g. Now a 30 g piece of copper is suspended from a thread and lowered into the water without contacting the bottom of the container. What does the scale read? (Use acceleration due to gravity $g = 9.8\ m/s^2$, density of water $\rho = 1\ g/cm^3$ and density of copper $\rho = 8.9\ g/cm^3$)

A. 122 g
B. 168 g
C. 143 g
D. 110 g
E. 94 g

20. An ideal, incompressible fluid flows through a 6 cm diameter pipe at 1 m/s. There is a 3 cm decrease in diameter within the pipe. What is the speed of the fluid in this constriction?

A. 3 m/s
B. 1.5 m/s
C. 8 m/s
D. 4 m/s
E. 2.5 m/s

21. Two blocks are submerged in a fluid. Block A has dimensions 2 cm high × 3 cm wide × 4 cm long, and Block B is 2 cm × 3 cm × 8 cm. Both blocks are submerged with their large faces pointing up and down (i.e., the blocks are horizontal), and they are submerged to the same depth. Compared to the fluid pressure on the bottom of Block A, the bottom of Block B experiences:

A. pressure that depends on the density of the objects
B. greater fluid pressure
C. exactly double the fluid pressure
D. less fluid pressure
E. equal fluid pressure

22. A piece of thread of diameter d is in the shape of a rectangle (length l, width w) and is lying on the surface of the water in a beaker. If A is the surface tension of the water, what is the maximum weight that the thread can have without sinking?

A. $Ad(l + w) / \pi$
B. $A(l + w)$
C. $4A(l + w)$
D. $A(l + w) / 2$
E. $Ad(l + w)$

23. What is the difference between the pressure inside and outside a tire called?

A. Absolute pressure
B. Fluid pressure
C. Atmospheric pressure
D. Gauge pressure
E. N/m^2

24. Which of the following is NOT a unit of pressure?

A. atm
B. psi
C. inches of mercury
D. Pascal
E. N·m^2

25. Which of the following is a dimensionless number?

I. Reynolds number II. specific gravity III. shear stress

A. I only
B. II only
C. III only
D. I and II only
E. I and III only

26. Water flows out of a large reservoir through a 5 cm diameter pipe. The pipe connects to a 3 cm diameter pipe that is open to the atmosphere, as shown. What is the speed of the water in the 5 cm pipe? Treat the water as an ideal incompressible fluid. (Use the acceleration due to gravity $g = 9.8$ m/s^2)

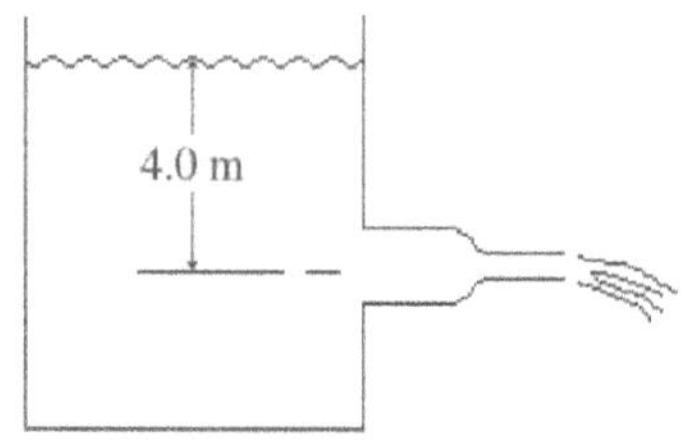

A. 2.6 m/s
B. 3.2 m/s
C. 4.8 m/s
D. 8.9 m/s
E. 6.7 m/s

27. If the pressure acting on an ideal gas at constant temperature is tripled, what is the resulting volume of the ideal gas?

A. Increased by a factor of two
B. Remains the same
C. Reduced to one-half
D. Increased by a factor of three
E. Reduced to one-third

28. A circular plate with an area of 1 m^2 covers a drain-hole at the bottom of a tank of water 1 m deep. Approximately how much force is required to lift the cover if it weighs 1,500 N? (Use the acceleration due to gravity $g = 10$ m/s^2)

A. 4,250 N
B. 9,550 N
C. 16,000 N
D. 11,500 N
E. 14,000 N

29. A bowling ball that weighs 80 N is dropped into a swimming pool filled with water. If the buoyant force on the bowling ball is 20 N when the ball is 1 m below the surface (and sinking), what is the normal force exerted by the bottom of the pool on the ball when it comes to rest there 4 m below the surface?

A. 0 N
B. 60 N
C. 50 N
D. 70 N
E. 40 N

30. A block of unknown material is floating in a fluid, half-submerged. If the specific gravity of the fluid is 1.6, what is the block's density? (Use the specific gravity = $\rho_{\text{fluid}} / \rho_{\text{water}}$ and the density of water ρ = 1,000 kg/m^3)

A. 350 kg/m^3
B. 800 kg/m^3
C. 900 kg/m^3
D. 1,250 kg/m^3
E. 1,600 kg/m^3

31. Three equal mass Styrofoam balls of radii R, 2R, and 3R, are released simultaneously from a tall tower. Which reaches the ground last?

A. The smallest Styrofoam ball
B. The middle size Styrofoam ball
C. The largest Styrofoam ball
D. All three reach the ground simultaneously
E. The result will depend on the atmospheric pressure

32. In a closed container of fluid, object A is submerged at 6 m from the bottom, and object B is submerged at 12 m from the bottom. Compared to object A, object B experiences:

A. less fluid pressure
B. double the fluid pressure
C. equal fluid pressure
D. triple the fluid pressure
E. one fifth the fluid pressure

33. Ideal, incompressible water flows at 14 m/s in a horizontal pipe with a pressure of 3.5×10^4 Pa. If the pipe widens to twice its original radius, what is the pressure in the wider section? (Use the density of water ρ = 1,000 kg/m^3)

A. 7.6×10^4 Pa
B. 12.7×10^4 Pa
C. 2×10^5 Pa
D. 11.1×10^3 Pa
E. 6.3×10^4 Pa

34. Two kilometers above the surface of the Earth, the atmospheric pressure is:

A. unrelated to the atmospheric pressure at the surface
B. twice the atmospheric pressure at the surface
C. triple the atmospheric pressure at the surface
D. less than the atmospheric pressure at the surface
E. requires more information

35. An 80 kg man would weigh 784 N if there were no atmosphere. By how much does the buoyancy due to air reduce the man's weight? (Use the density of the man = 1 g/cm^3, the density of the air = 1.2×10^{-3} g/cm^3, the mass m = 80 kg and the acceleration due to gravity g = 9.8 m/s^2)

A. 0.58 N
B. 0.32 N
C. 0.94 N
D. 2.8 N
E. 1.2 N

36. Diffusion is described by which law?

A. Dulong's
B. Faraday's
C. Kepler's
D. Graham's
E. Archimedes'

37. As a metal rod is stretched, which condition is reached first?

A. Elastic limit
B. Proportional limit
C. Breaking point
D. Fracture point
E. Elastic modulus

38. A pump uses a piston 12 cm in diameter that moves 3 cm/s. What is the fluid velocity in a tube that is 2 mm in diameter?

A. 218 cm/s
B. 88 cm/s
C. 136 cm/s
D. 108 m/s
E. 52 m/s

39. What is the specific gravity of an object floating with one-tenth of its volume out of the water?

A. 0.3
B. 0.9
C. 1.3
D. 2.1
E. 2.9

40. If each factor listed below were changed by 15%, which would have the greatest effect on the flow rate?

A. Fluid density
B. Pressure difference
C. Length of the pipe
D. Fluid viscosity
E. Radius of the pipe

41. A 680 g steel hammer (m_h) is tied to a string hanging from a force meter. A 5 kg container of water (m_w) sits on a scale. The hammer is lowered completely into the water but above the bottom. What does the force meter read? (Use the density of steel $\rho = 7.9$ g/cm^3, density of water $\rho = 1$ g/cm^3 and acceleration due to gravity $g = 10$ m/s^2)

A. 5.9 N
B. 8.4 N
C. 10.7 N
D. 5.2 N
E. 11.2 N

42. An external pressure applied to an enclosed fluid is transmitted unchanged to every point within the fluid by:

A. Torricelli's law
B. Bernoulli's principle
C. Archimedes' principle
D. Pascal's principle
E. Fermat's principle

43. A submarine rests on the bottom of the sea. What is the normal force exerted upon the submarine by the seafloor equal to?

A. weight of the submarine
B. weight of the submarine minus the weight of the displaced water
C. buoyant force minus the atmospheric pressure acting on the submarine
D. weight of the submarine plus the weight of the displaced water
E. buoyant force plus the atmospheric pressure acting on the submarine

44. Consider a brick totally immersed in water, with the long edge of the brick vertical. Which statement describes the pressure on the brick?

A. Highest on the sides of the brick
B. Highest on the top of the brick
C. Lowest on the sides with the largest area
D. Same on all surfaces of the brick
E. Highest on the bottom of the brick

45. Water is flowing in a drainage channel of a rectangular cross-section. The width of the channel is 14 m, the depth of water is 7 m, and the flow speed is 3 m/s. What is the mass flow rate of the water? (Use the density of water $\rho = 1{,}000$ kg/m^3)

A. 2.9×10^5 kg/s
B. 4.8×10^4 kg/s
C. 6.2×10^5 kg/s
D. 9.3×10^4 kg/s
E. 4.3×10^2 kg/s

46. What is the magnitude of the buoyant force if a 3 kg object floats motionlessly in a fluid of specific gravity 0.8? (Use acceleration due to gravity $g = 10$ m/s^2)

A. 15 N
B. 7.5 N
C. 30 N
D. 45 N
E. 0 N

47. What is the pressure 6 m below the surface of the ocean? (Use density of water $\rho = 10^3$ kg/m^3, atmospheric pressure $P_{atm} = 1.01 \times 10^5$ Pa and acceleration due to gravity $g = 10$ m/s^2)

A. 1.6×10^5 Pa
B. 0.8×10^5 Pa
C. 2.7×10^4 Pa
D. 3.3×10^4 Pa
E. 4.8×10^4 Pa

48. Density is:

A. inversely proportional to mass and volume
B. proportional to mass and inversely proportional to the volume
C. inversely proportional to mass and proportional to the volume
D. proportional to mass and volume
E. inversely proportional to mass and independent of volume

49. Which of the following would be expected to have the smallest bulk modulus?

A. Solid plutonium
B. Liquid water
C. Solid lead
D. Liquid mercury
E. Helium vapor

50. When an 8 kg object is suspended from a steel wire that is 2.7 m long and 0.8 mm in diameter, how much does the wire stretch? (Use the acceleration due to gravity $g = 9.8$ m/s^2 and Young's modulus = 20×10^{10} N/m^2)

A. 2.8 mm
B. 3.2 mm
C. 0.8 mm
D. 2.1 mm
E. 1.1 mm

51. For the wire and the object from the previous question, if the diameter of the wire is doubled, what would be the stretch? (Use the acceleration due to gravity $g = 9.8$ m/s^2 and Young's modulus = 20×10^{10} N/m^2)

A. 0.5 mm
B. 2.1 mm
C. 1.2 mm
D. 0.2 mm
E. 0.9 mm

52. If atmospheric pressure increases by an amount ΔP, which of the following statements about the pressure in a large pond is true?

A. The gauge pressure increases by ΔP
B. The absolute pressure increases by ΔP
C. The absolute pressure increases, but by an amount less than ΔP
D. The absolute pressure does not change
E. The gauge pressure decreases by ΔP

53. Strain is defined as the:

A. stress per unit area
B. applied force per unit area
C. ratio of stress to the elastic modulus
D. ratio of the change in length to the original length
E. ratio of stress to change in length

54. Fluid flows through a 19 cm long tube with a radius of 2.1 mm at an average speed of 1.8 m/s. What is the viscosity of the fluid if the pressure drop is 970 Pa?

A. 0.036 $N{\cdot}s/m^2$
B. 0.013 $N{\cdot}s/m^2$
C. 0.0044 $N{\cdot}s/m^2$
D. 0.0016 $N{\cdot}s/m^2$
E. 0.0022 $N{\cdot}s/m^2$

55. The pressure differential across the cross-section of a condor's wing due to the difference in airflow is explained by:

A. Torricelli's law
B. Poiseuille's law
C. Bernoulli's equation
D. Newton's First Law
E. Pascal's principle

Notes for active learning

Heat and Thermodynamics

1. Compared to the initial value, what is the resulting pressure for an ideal gas compressed isothermally to one-third of its initial volume?

A. Equal
B. Three times larger
C. Larger, but less than three times larger
D. More than three times larger
E. Requires more information

2. A uniform hole in a brass plate has a diameter of 1.2 cm at 25 °C. What is the diameter of the hole when the plate is heated to 225 °C? (Use the coefficient of linear thermal expansion for brass = 19×10^{-6} K^{-1})

A. 2.2 cm
B. 2.8 cm
C. 1.2 cm
D. 1.6 cm
E. 0.8 cm

3. A student heats 90 g of water using 50 W of power, with 100% efficiency. How long does it take to raise the temperature of the water from 10 °C to 30 °C? (Use the specific heat of water $c = 4.186$ J/g·°C)

A. 232 s
B. 81 s
C. 59 s
D. 151 s
E. 102 s

4. A runner generates 1,260 W of thermal energy. If her heat is to be dissipated only by evaporation, how much water does she shed in 15 minutes of running? (Use latent heat of vaporization of water $L_v = 22.6 \times 10^5$ J/kg)

A. 500 g
B. 35 g
C. 350 g
D. 50 g
E. 40 g

5. Phase changes occur as temperature:

I. decreases
II. increases
III. remains the same

A. I only
B. II only
C. III only
D. I and II only
E. I and III only

6. How much heat is needed to melt a 55 kg sample of ice at 0 °C? (Use the latent heat of fusion for water L_f = 334 kJ/kg and the latent heat of vaporization L_v = 2,257 kJ/kg)

A. 0 kJ
B. 2.6×10^5 kJ
C. 3×10^5 kJ
D. 4.6×10^6 kJ
E. 1.8×10^4 kJ

7. Metals are excellent heat conductors and electrical conductors because of the:

A. relatively high densities of metals
B. high elasticity of metals
C. ductility of metals
D. looseness of outer electrons in metal atoms
E. crystal structure of metals

8. Solar houses are designed to retain the heat absorbed during the day so that the stored heat can be released during the night. A botanist produces steam at 100 °C during the day and then allows the steam to cool to 0 °C and freeze during the night. How many kilograms of water are needed to store 200 kJ of energy for this process? (Use the latent heat of vaporization of water $L_v = 22.6 \times 10^5$ J/kg, the latent heat of fusion of water L_f = 33.5×10^4 J/kg, and the specific heat capacity of water c = 4,186 J/kg·K)

A. 0.066 kg
B. 0.103 kg
C. 0.482 kg
D. 1.18 kg
E. 3.66 kg

9. The heat required to change a substance from the solid to the liquid state is referred to as the heat of:

A. condensation
B. freezing
C. fusion
D. vaporization
E. sublimation

10. A rigid container holds 0.2 kg of hydrogen gas. How much heat is needed to change the temperature of the gas from 250 K to 280 K? (Use specific heat of hydrogen gas = 14.3 J/g·K)

A. 46 kJ
B. 72 kJ
C. 56 kJ
D. 35 kJ
E. 86 kJ

11. A 500 g aluminum electric tea kettle is heated with a 500 W heating coil. How many minutes are required to heat 1 kg of water from 18 °C to 98 °C in the tea kettle? (Use specific heat of aluminum = 900 J/kg·K and specific heat of water = 4,186 J/kg·K)

A. 16 min
B. 12 min
C. 8 min
D. 4 min
E. 10 min

12. Heat is added at a constant rate to a pure substance in a closed container. The temperature of the substance as a function of time is shown in the graph. If $L_f =$ the latent heat of fusion and L_v = latent heat of vaporization, what is the value of the ratio L_v / L_f for this substance?

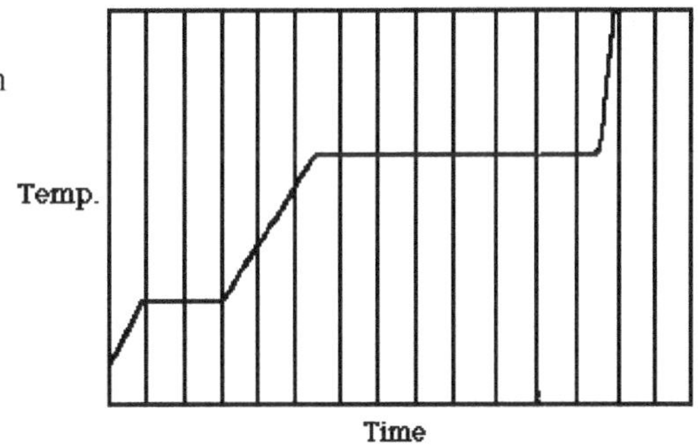

A. 3.5
B. 7.2
C. 4.5
D. 5.0
E. 1.5

13. The moderate temperatures of islands throughout the world have much to do with water's:

A. high evaporation rate
B. high specific heat capacity
C. vast supply of thermal energy
D. poor conductivity
E. absorption of solar energy

14. A 4.5 g lead BB moving at 46 m/s penetrates a woodblock and comes to rest inside the block. If the BB absorbs half of the kinetic energy, what is the change in the temperature of the BB? (Use the specific heat of lead = 128 J/kg·K)

A. 2.8 K
B. 3.6 K
C. 1.6 K
D. 0.8 K
E. 4.1 K

15. The heat required to change a substance from the liquid to the vapor state is referred to as the heat of:

A. melting
B. condensation
C. vaporization
D. fusion
E. sublimation

16. A Carnot engine operating between a reservoir of liquid mercury at its melting point and a colder reservoir extracts 18 J of heat from the mercury. It does 5 J of work during each cycle. What is the temperature of the colder reservoir? (Use melting temperature of mercury = 233 K)

A. 168 K
B. 66 K
C. 57 K
D. 82 K
E. 94 K

17. A 920 g empty iron pan is on a stove. How much heat must the iron pan absorb in joules to raise its temperature from 18 °C to 96 °C? (Use the specific heat for iron = 113 cal/kg·°C and the conversion of 1 cal = 4.186 J)

A. 50,180 J
B. 81,010 J
C. 63,420 J
D. 33,940 J
E. 26,500 J

18. When a solid melts, what change occurs in the substance?

A. Heat energy dissipates
B. Heat energy enters
C. Temperature increases
D. Temperature decreases
E. Kinetic energy increases

19. Which of the following is an accurate statement about the work done for a cyclic process carried out in a gas? (Use P for pressure and V for volume on the graph)

A. It is equal to the area under *ab* minus the area under *dc*
B. It is equal to the area under the curve *adc*
C. It is equal to the area under the curve *abc*
D. It equals zero
E. It is equal to the area enclosed by the cyclic process

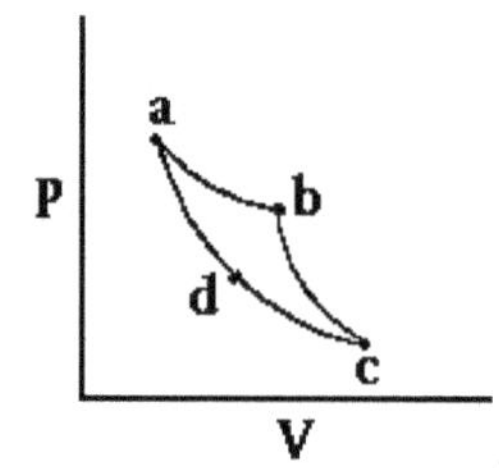

20. Substance A has higher specific heat than substance B. With other factors equal, which substance requires more energy to be heated to the same temperature?

A. Substance A
B. Substance B
C. Both require the same amount of heat
D. Depends on the density of each substance
E. Depends on the volume of each substance

21. A 6.5 g meteor hits the Earth at a speed of 300 m/s. If the meteor's kinetic energy is entirely converted to heat, by how much does its temperature rise? (Use the specific heat of the meteor = 120 cal/kg·°C and the conversion of 1 cal = 4.186 J)

A. 134 °C
B. 68 °C
C. 120 °C
D. 90 °C
E. 44 °C

22. When a liquid freezes, what change occurs in the substance?

A. Heat energy dissipates
B. Heat energy enters
C. Temperature increases
D. Temperature decreases
E. Kinetic energy increases

23. A monatomic ideal gas ($C_V = 3/2\ R$) undergoes an isothermal expansion at 300 K as the volume increases from 0.05 m^3 to 0.2 m^3. The final pressure is 130 kPa. What is the heat transfer of the gas? (Use the ideal gas constant R = 8.314 J/mol·K)

A. –14 kJ
B. 36 kJ
C. 14 kJ
D. –21 kJ
E. 0 kJ

24. What is the maximum temperature rise expected for 1 kg of water falling from a waterfall with a vertical drop of 30 m? (Use the acceleration due to gravity $g = 9.8$ m/s^2 and the specific heat of water = 4,186 J/kg·K)

A. 0.1 °C
B. 0.06 °C
C. 0.15 °C
D. 0.07 °C
E. 0.03 °C

25. When 0.75 kg of water at 0 °C freezes, what is the change in entropy of the water? (Use the latent heat of fusion of water $L_f = 33{,}400$ J/kg)

A. –92 J/K
B. –18 J/K
C. 44 J/K
D. 80 J/K
E. –60 J/K

26. When a bimetallic bar made of a copper and iron strip is heated, the copper part of the bar bends toward the iron strip. The reason for this is:

A. copper expands more than iron
B. iron expands more than copper
C. iron gets hotter before copper
D. copper gets hotter before iron
E. copper and iron expand at the same rate

27. In a flask, 110 g of water is heated using 60 W of power, with perfect efficiency. How long does it take to raise the temperature of the water from 20 °C to 30 °C? (Use the specific heat of water $c = 4{,}186$ J/kg·K)

A. 132 s
B. 57 s
C. 9.6 s
D. 77 s
E. 41 s

28. When a liquid evaporates, what change occurs in the substance?

A. Heat energy dissipates
B. Heat energy enters
C. Temperature increases
D. Temperature decreases
E. Kinetic energy increases

29. A flask of liquid nitrogen is at –243 °C. If the nitrogen is heated until the average energy of the particles is doubled, what is the new temperature?

A. 356 °C
B. –356 °C
C. –134 °C
D. 134 °C
E. –213 °C

30. If a researcher attempts to determine how much the temperature of a particular piece of material would rise when a known amount of heat is added to it, knowing which of the following quantities would be most helpful?

A. density
B. coefficient of linear expansion
C. initial temperature
D. specific heat
E. thermal conductivity

31. A substance has a density of 1,800 kg/m^3 in the liquid state. At atmospheric pressure, the substance has a boiling point of 170 °C. The vapor has a density of 6 kg/m^3 at the boiling point at atmospheric pressure. What is the change in the internal energy of 1 kg of the substance as it vaporizes at atmospheric pressure? (Use the heat of vaporization L_v= 1.7×10^5 J/kg)

A. 180 kJ
B. 170 kJ
C. 6 kJ
D. 12 kJ
E. 200 kJ

32. If an aluminum rod that is at 5 °C is heated until it has twice the thermal energy, its temperature is:

A. 108 °C
B. 56 °C
C. 278 °C
D. 283 °C
E. 556 °C

33. A thermally isolated system comprises a hot piece of aluminum and a cold piece of copper, with the aluminum and the copper in thermal contact. The specific heat capacity of aluminum is more than double that of copper. Which object experiences the greater temperature change during the time the system takes to reach thermal equilibrium?

A. Both experience the same magnitude of temperature change
B. The volume of each is required
C. The copper
D. The aluminum
E. The mass of each is required

34. In liquid water of a given temperature, the water molecules move randomly at different speeds. Electrostatic forces of cohesion tend to hold them together. However, occasionally one molecule gains enough energy through multiple collisions to pull away and escape from the liquid. Which of the following is an illustration of this phenomenon?

A. When a large steel suspension bridge is built, gaps are left between the girders
B. When a body gets too warm, it produces sweat to cool itself down
C. Increasing the atmospheric pressure over a liquid causes the boiling temperature to decrease
D. If the snow begins to fall when Mary is skiing, she feels colder than before it started to snow
E. A hot water bottle is more effective in keeping a person warm than would a rock of the same mass heated to the same temperature

35. A 2,200 kg water sample at 0 °C is cooled to –30 °C and freezes in the process. Approximately how much heat is liberated during this process? (Use heat of fusion for water L_f = 334 kJ/kg, heat of vaporization L_v = 2,257 kJ/kg and specific heat for ice = 2,050 J/kg·K)

A. 328,600 kJ
B. 190,040 kJ
C. 637,200 kJ
D. 870,100 kJ
E. 768,200 kJ

36. Object 1 has three times the specific heat capacity and four times the mass of Object 2. The same amount of heat is transferred to the two objects. If Object 1 changes by an amount of ΔT, what is the change in temperature of Object 2?

A. (4/3)ΔT
B. 3ΔT
C. ΔT
D. (3/4)ΔT
E. 12ΔT

37. The process whereby heat flows through molecular collisions is:

A. radiation
B. inversion
C. conduction
D. convection
E. evaporation

38. The graph shows a PV diagram for 5.1 g of oxygen gas in a sealed container. The temperature of T_1 is 20 °C. What are the values for temperatures of T_3 and T_4, respectively? (Use the gas constant R= 8.314 J/mol·K)

A. –53 °C and 387 °C
B. –14 °C and 34 °C
C. 210 °C and 640 °C
D. 12 °C and 58 °C
E. 29 °C and 171 °C

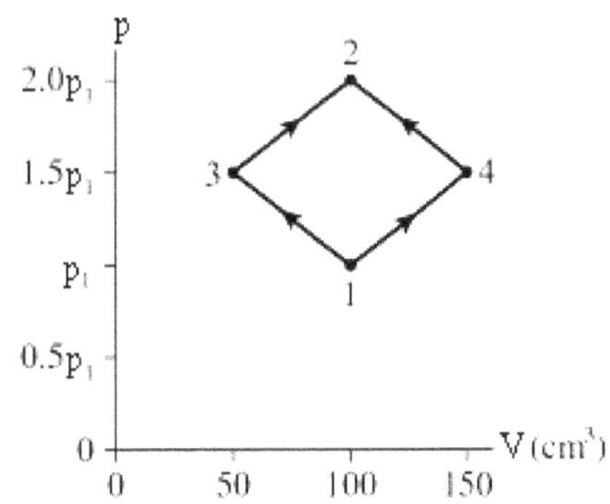

39. On a cold day, a piece of steel feels much colder to the touch than a piece of plastic. This is due to the difference in which one of the following physical properties of these materials?

A. Emissivity
B. Thermal conductivity
C. Density
D. Specific heat
E. Mass

40. What is the term for a process when a gas is allowed to expand as heat is added to it at constant pressure?

A. Isochoric
B. Isobaric
C. Adiabatic
D. Isothermal
E. Isentropic

41. A Carnot engine is used as an air conditioner to cool a house in the summer. The air conditioner removes 20 kJ of heat per second from the house and maintains the inside is at 293 K, while the outside temperature is 307 K. What is the power required for the air conditioner?

A. 2.3 kW
B. 3.22 kW
C. 1.6 kW
D. 4.88 kW
E. 0.96 kW

42. Heat energy is measured in units of:

I. Joules　　II. calories　　III. work

A. I only
B. II only
C. I and II only
D. III only
E. I, II and III

43. The process in which heat flows by the mass movement of molecules from one place to another is:

I. conduction　　II. convection　　III. radiation

A. I only
B. II only
C. III only
D. I and II only
E. I and III only

44. The process whereby heat flows in the absence of a medium is:

A. radiation
B. inversion
C. conduction
D. convection
E. evaporation

45. The figure shows 0.008 mol of gas that undergoes the process $1 \rightarrow 2 \rightarrow 3$. What is the volume of V_3?

(Use the ideal gas constant R = 8.314 J/mol·K and the conversion of 1 atm = 101,325 Pa)

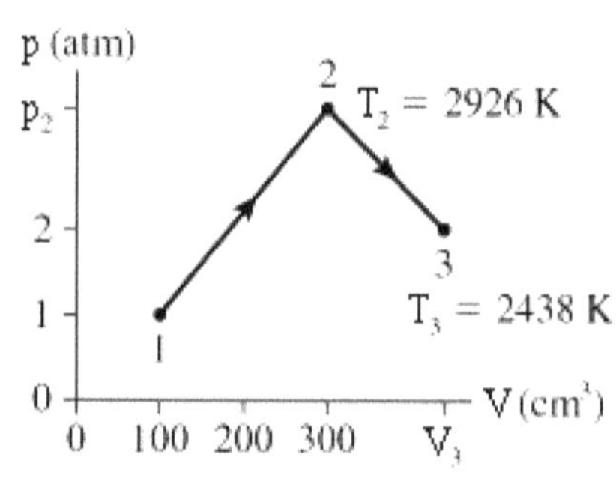

A. 435 cm^3
B. 568 cm^3
C. 656 cm^3
D. 800 cm^3
E. 940 cm^3

46. When a gas expands adiabatically:

A. it does no work
B. work is done on the gas
C. the internal (thermal) energy of the gas decreases
D. the internal (thermal) energy of the gas increases
E. the temperature of the gas remains constant

47. Why is it that when a swimmer gets out of a swimming pool and stands in a breeze dripping wet, he feels much colder than when he dries off?

A. This is a physiological effect resulting from the skin's sensory nerves
B. The water on his skin is colder than the surrounding air
C. The moisture on his skin has good thermal conductivity
D. Water has a relatively small specific heat
E. To evaporate a gram of water from his skin requires heat and most of this heat flows out of his body

48. Which method of heat flow requires the movement of energy through solid matter to a new location?

I. Conduction　　II. Convection　　III. Radiation

A. I only
B. II only
C. III only
D. I and II only
E. I and III only

49. An ideal gas is compressed via an isobaric process to one-third of its initial volume. Compared to the initial pressure, the resulting pressure is:

A. more than three times greater
B. nine times greater
C. three times greater
D. the same
E. requires more information

50. Which of the following would be the best radiator of thermal energy?

A. A metallic surface
B. A black surface
C. A white surface
D. A shiny surface
E. Styrofoam

51. A brass rod is 59.1 cm long, and an aluminum rod is 39.3 cm long when both rods are at an initial 0 °C. The rods have 1.1 cm between them. The distance between the far ends of the rods is maintained at 99.5 cm. The temperature is raised until the two rods are barely in contact. In the figure, what is the temperature at which contact of the rods barely occurs? (Use coefficient of linear expansion of brass = 2×10^{-5} K^{-1} and coefficient of linear expansion of aluminum = 2.4×10^{-5} K^{-1})

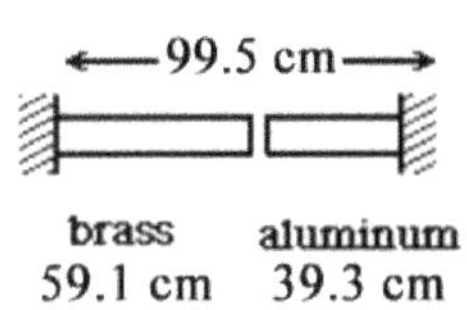

A. 424 °C
B. 588 °C
C. 483 °C
D. 363 °C
E. 518 °C

52. At room temperature, a person loses energy to the surroundings at the rate of 60 W. If an equivalent food intake compensates for this energy loss, how many kilocalories does he need to consume every 24 hours? (Use the conversion of 1 cal = 4.186 J)

A. 1,240 kcal
B. 1,660 kcal
C. 600 kcal
D. 880 kcal
E. 1,920 kcal

53. What primary heat transfer mechanism does one end of an iron bar become hot when the other end is placed in a flame?

A. Convection
B. Forced convection
C. Radiation
D. Conduction
E. Diffusion

Questions **54-55** are based on the following:

Two experiments were performed to determine the calorimetric properties of alcohol with a melting point of –10 °C. In the first trial, a 220 g cube of frozen alcohol, at the melting point, is added to 350 g of water at 26 °C in a Styrofoam container. When thermal equilibrium is reached, the alcohol-water solution is at 5 °C. In the second trial, an identical cube of alcohol is added to 400 g of water at 30 °C, and the temperature at thermal equilibrium is 10 °C. (Use the specific heat of water = 4,190 J/kg·K and assume no heat exchange between the Styrofoam container and the surroundings).

54. What is the specific heat capacity of the alcohol?

A. 2,150 J/kg·K
B. 2,475 J/kg·K
C. 1,175 J/kg·K
D. 1,820 J/kg·K
E. 2,730 J/kg·K

55. What is the heat of fusion of the alcohol?

A. 7.2×10^3 J/kg
B. 1.9×10^5 J/kg
C. 5.2×10^4 J/kg
D. 10.3×10^4 J/kg
E. 3.3×10^4 J/kg

Notes for active learning

Notes for active learning

Electric Circuits

1. What is the new resistance of a wire if the length of a particular wire is doubled and its radius is doubled?

A. It is √2 times as large
B. It is ½ as large
C. It stays the same
D. It is 2 times as large
E. It is ¼ as large

2. A 6 Ω resistor is connected across the terminals of a 12 V battery. If 0.6 A of current flows, what is the internal resistance of the battery?

A. 2 Ω
B. 26 Ω
C. 20 Ω
D. 14 Ω
E. 3.6 Ω

3. Three 8 V batteries are connected in series to power light bulbs A and B. The resistance of light bulb A is 60 Ω and the resistance of light bulb B is 30 Ω. How does the current through light bulb A compare with the current through light bulb B?

A. The current through light bulb A is less
B. The current through light bulb A is greater
C. The current through light bulb A is the same
D. The current through light bulb A is exactly doubled that through light bulb B
E. None are true

4. A sphere with a radius 2 mm carries a 1 μC charge. What is the potential difference, $V_B - V_A$, between point B 3.5 m from the center of the sphere and point A 8 m from the sphere's center? (Use Coulomb's constant $k = 9 \times 10^9$ N·m²/C²)

A. –485 V
B. 1,140 V
C. –140 V
D. 1,446 V
E. 2,457 V

5. Which of the following effect (s) capacitance of capacitors?

I. material between the conductors
II. distance between the conductors
III. geometry of the conductors

A. I only
B. II only
C. III only
D. I and III only
E. I, II and III

6. A proton with an initial speed of 1.5×10^5 m/s falls through a potential difference of 100 volts, gaining speed. What is the speed reached? (Use the mass of a proton = 1.67×10^{-27} kg and the charge of a proton = 1.6×10^{-19} C)

A. 2×10^5 m/s
B. 4×10^5 m/s
C. 8.6×10^5 m/s
D. 7.6×10^5 m/s
E. 6.6×10^5 m/s

7. The current flowing through a circuit of constant resistance is doubled. What is the effect on the power dissipated by that circuit?

A. Decreases to one-half its original value
B. Decreases to one-fourth its original value
C. Quadruples its original value
D. Doubles its original value
E. Decreases to one-eighth its original value

8. A positively-charged particle is at rest in an unknown medium. What is the magnitude of the magnetic field generated by this particle?

A. Constant everywhere and dependent only on the mass of the medium
B. Lower at points near to the particle compared to a distant point
C. Higher at points near to the particle compared to a distant point
D. Equal to zero
E. Constant everywhere and dependent only on the density of the medium

9. The heating element of a toaster is a long wire of some metal, often a metal alloy, which heats up when a 120 V potential difference is applied across it. Consider a 300 W toaster connected to a wall outlet. Which statement would result in an increase in the rate by which heat is produced?

A. Use a longer wire
B. Use a thicker wire
C. Use a thicker and longer wire
D. Use a thinner and longer wire
E. Use a thinner wire of the same length

10. A 4 μC point charge and an 8 μC point charge are initially infinitely far apart. How much work is required to bring the 4 μC point charge to (x = 2 mm, y = 0 mm), and the 8 μC point charge to (x = –2 mm, y = 0 mm)? (Use Coulomb's constant $k = 9 \times 10^9$ N·m^2/C^2)

A. 32.6 J
B. 9.8 J
C. 47 J
D. 81 J
E. 72 J

11. What current flows when a 400 Ω resistor is connected across a 220 V circuit?

A. 0.55 A
B. 1.8 A
C. 5.5 A
D. 0.18 A
E. 18 A

12. Which statement is accurate for when different resistors are connected in parallel across an ideal battery?

A. Power dissipated in each is the same
B. Their equivalent resistance is higher than the resistance of any individual resistor
C. Current flowing in each is the same
D. Their equivalent resistance is equal to the average of the individual resistances
E. Potential difference across each is the same

13. An electron was accelerated from rest through a potential difference of 990 V. What is its speed? (Use the mass of an electron = 9.11×10^{-31} kg, the mass of a proton = 1.67×10^{-27} kg, and charge of a proton = 1.6×10^{-19} C)

A. 0.8×10^7 m/s
B. 3.7×10^7 m/s
C. 7.4×10^7 m/s
D. 1.9×10^7 m/s
E. 6.9×10^7 m/s

14. A circular conducting loop with a radius of 0.5 m and a small gap filled with a 12 Ω resistor oriented in the *xy*-plane. If a magnetic field of 1 T, making an angle of 30° with the *z*-axis, increases to 12 T in 5 s, what is the magnitude of the current flowing in the conductor?

A. 0.33 A
B. 0.13 A
C. 0.88 A
D. 1.5 A
E. 4.5 A

15. For an electric motor with a resistance of 35 Ω that draws 10 A of current, what is the voltage drop?

A. 3.5 V
B. 25 V
C. 350 V
D. 3,500 V
E. 0.3 V

16. A charged parallel-plate capacitor has an electric field E_0 between its plates. The bare nuclei of a stationary ^{1}H and ^{4}He are between the plates. Ignoring gravity, how does the magnitude of the acceleration of the hydrogen nucleus a_H compare with the magnitude of the acceleration of the helium nucleus a_{He}? (Use the mass of an electron = 9×10^{-31} kg, mass of a proton = 1.67×10^{-27} kg, mass of a neutron = 1.67×10^{-27} kg and charge of a proton = 1.6×10^{-19} C)

A. $a_H = 2a_{He}$
B. $a_H = 4a_{He}$
C. $a_H = \frac{1}{4}a_{He}$
D. $a_H = a_{He}$
E. $a_H = \frac{1}{2}a_{He}$

17. Identical light bulbs are attached to identical batteries in three different ways (A, B, or C), as shown in the figure. What is the ranking (from lowest to highest) of the total power produced by the battery?

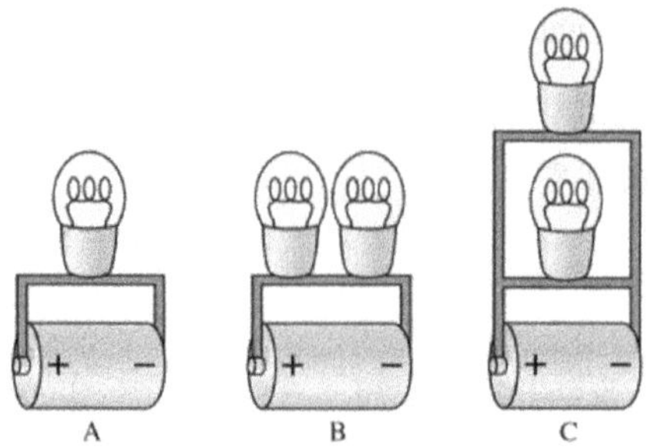

A. C, B, A
B. B, A, C
C. A, C, B
D. A, B, C
E. C, A, B

18. A parallel-plate capacitor consists of two parallel, square plates with dimensions 1 cm by 1 cm. What is the capacitance if the plates are separated by 1 mm, and the space between them is filled with Teflon? (Use the dielectric constant k for Teflon = 2.1 and the electric permittivity $\mathcal{E}_0 = 8.854 \times 10^{-12}$ F/m)

A. 0.83 pF
B. 2.2 pF
C. 0.46 pF
D. 1.9 pF
E. 0.11 pF

19. The resistor R has a variable resistance. Which statement is true when R is decreased? (Neglect the minimal internal resistance r of the battery)

A. I_1 decreases, I_2 increases
B. I_1 increases, I_2 remains the same
C. I_1 remains the same, I_2 increases
D. I_1 remains the same, I_2 decreases
E. I_1 increases, I_2 increases

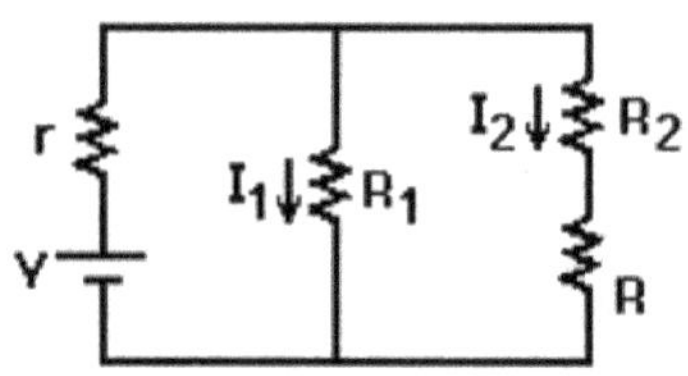

20. What physical quantity does the slope of the graph represent?

A. 1 / Current
B. Voltage
C. Current
D. Resistivity
E. 1 / Voltage

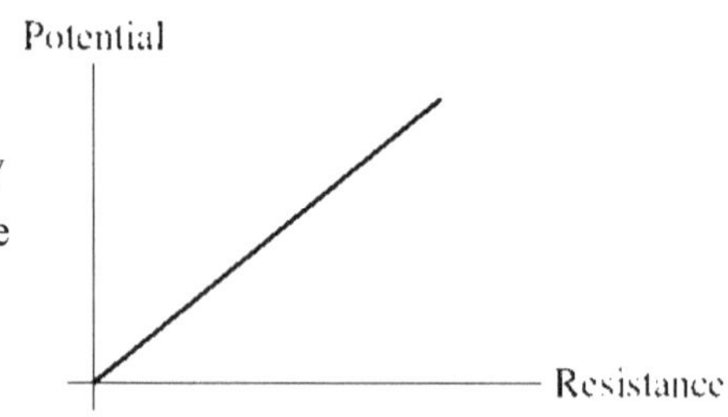

21. An alternating current is supplied to an electronic component with a rating used only for voltages below 12 V. What is the highest V_{rms} that can be supplied to this component while staying below the voltage limit?

A. 6 V
B. 12 V
C. $3\sqrt{2}$ V
D. $12\sqrt{2}$ V
E. $6\sqrt{2}$ V

22. A generator produces alternating current electricity with a frequency of 40 cycles per second. What is the maximum potential difference created by the generator if the RMS voltage is 150 V?

A. 54 V
B. 91 V
C. 212 V
D. 141 V
E. 223 V

23. Kirchhoff's junction rule is a statement of:

A. Law of conservation of energy
B. Law of conservation of angular momentum
C. Law of conservation of momentum
D. Law of conservation of charge
E. Newton's Second Law

24. Four identical capacitors are connected in parallel to a battery. If a total charge of Q flows from the battery, how much charge does each capacitor carry?

A. $Q/4$
B. Q
C. $4Q$
D. $16Q$
E. $Q/16$

25. Which statement is correct for two conductors joined by a long copper wire?

A. The electric field at the surface of each conductor is the same
B. Each conductor must be at the same potential
C. Each conductor must have the same resistivity
D. A free charge must be present on either conductor
E. The potential of the wire is the average of the potential of each conductor

26. Electromagnetic induction occurs in a coil when there is a change in the:

A. coil's charge
B. current in the coil
C. magnetic field intensity in the coil
D. electric field intensity in the coil
E. electromagnetic polarity

27. When unequal resistors are connected in series across an ideal battery, the:

A. current flowing in each is the same
B. equivalent resistance of the circuit is less than that of the greatest resistor
C. power dissipated in each turn is the same
D. potential difference across each is the same
E. current is still unequal in the two resistors

28. Electric current flows from the point of:

A. equal potential
B. high pressure to the point of lower pressure
C. low pressure to the point of higher pressure
D. high potential to the point of lower potential
E. low potential to the point of higher potential

29. Consider the group of charges in this figure. All three charges have $Q = 3.8$ nC. What is their electric potential energy?

(Use Coulomb's constant $k = 9.0 \times 10^9\ \mathrm{N{\cdot}m^2/C^2}$)

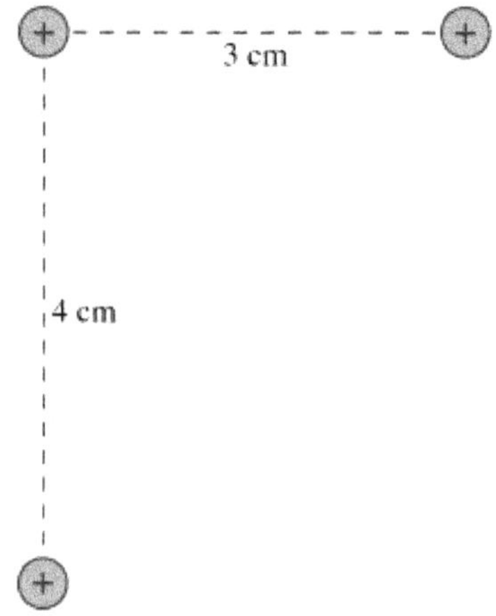

A. 1.9×10^{-6} J
B. 7.4×10^{-5} J
C. 8.8×10^{-6} J
D. 9.7×10^{-6} J
E. 1.0×10^{-5} J

30. A positively-charged and negatively-charged particles are traveling on the same path perpendicular to a constant magnetic field. How do the forces experienced by the two particles differ if the magnitudes of the charges are equal?

A. Differ in direction, but not in magnitude
B. Differ in magnitude, but not in the direction
C. No difference in magnitude or direction
D. Differ in magnitude and direction
E. Cannot be predicted

31. An electron moves in a direction opposite to an electric field. The potential energy of the system:

A. decreases, and the electron moves toward a region of lower potential
B. increases, and the electron moves toward a region of higher potential
C. decreases, and the electron moves toward a region of higher potential
D. remains constant, and the electron moves toward a region of higher potential
E. remains constant, and the electron moves toward a region of lower potential

32. What is the name of a device that transforms electrical energy into mechanical energy?

A. Magnet
B. Transformer
C. Turbine
D. Generator
E. Motor

33. A hydrogen atom consists of a proton and an electron. If the orbital radius of the electron increases, the absolute magnitude of the potential energy of the electron:

A. remains the same
B. decreases
C. increases
D. depends on the potential of the electron
E. is independent of the orbital radius

34. Copper wire A has a length L and a radius r. Copper wire B has a length $2L$ and a radius $2r$. Which of the following is true regarding the resistances across the ends of the wires?

A. The resistance of wire A is one-half that of wire B
B. The resistance of wire A is four times higher than that of wire B
C. The resistance of wire A is twice as high as that of wire B
D. The resistance of wire A is equal to that of wire B
E. The resistance of wire A is one-fourth that of wire B

35. When a negative charge is free, it tries to move:

A. toward infinity
B. away from infinity
C. from high potential to low potential
D. from low potential to high potential
E. in the direction of the electric field

36. Four 6 V batteries (in a linear sequence of A → B → C → D) are connected in series to power lights A and B. The resistance of light A is 50 Ω and the resistance of light B is 25 Ω. What is the potential difference at a point between battery C and battery D? (Assume that the potential at the start of the sequence is zero)

A. 4 volts
B. 12 volts
C. 18 volts
D. 26 volts
E. 30 volts

37. By what factor does the dielectric constant change when a material is introduced between the plates of a parallel-plate capacitor if the capacitance increases by a factor of 4?

A. ½
B. 4
C. 0.4
D. ¼
E. 2

38. Two isolated copper plates, each of area 0.4 m^2, carry opposite charges of magnitude 6.8×10^{-10} C. They are opposite each other in parallel alignment. What is the potential difference between the plates when their spacing is 4 cm? (Use the dielectric constant $k = 1$ in air and electric permittivity $\varepsilon_0 = 8.854 \times 10^{-12}$ F/m)

A. 1.4 V
B. 4.1 V
C. 5.8 V
D. 3.2 V
E. 7.7 V

39. The force on an electron moving in a magnetic field is largest when its direction is:

A. perpendicular to the magnetic field direction
B. at an angle higher than 90° to the magnetic field direction
C. at an angle lower than 90° to the magnetic field direction
D. exactly opposite to the magnetic field direction
E. parallel to the magnetic field direction

40. What is the quantity that is calculated in units of A·s?

A. Passivity
B. Capacitance
C. Potential
D. Current
E. Charge

41. A proton with a speed of 1.7×10^5 m/s falls through a potential difference V and increases its speed to 3.2×10^5 m/s. Through what potential difference did the proton fall? (Use the mass of a proton = 1.67×10^{-27} kg and the charge of a proton = 1.6×10^{-19} C)

A. 880 V
B. 1,020 V
C. 384 V
D. 430 V
E. 130 V

42. Three capacitors are connected to a battery, as shown. The capacitances are: $C_1 = 2C_2$ and $C_1 = 3C_3$. Which of the three capacitors stores the smallest amount of charge?

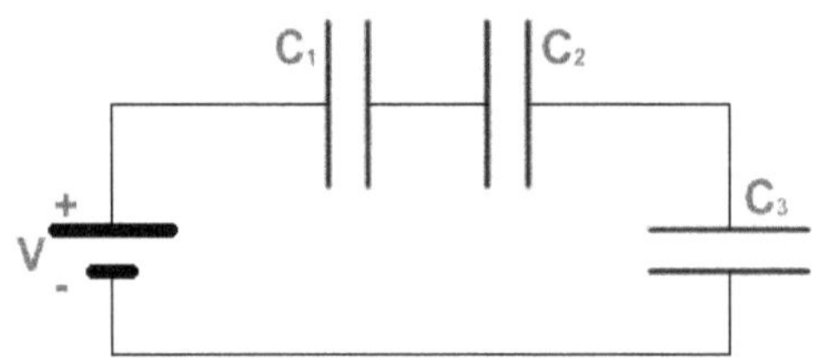

A. C_1
B. C_1 or C_3
C. C_2
D. C_3
E. The charge amount is the same in the three capacitors

43. Two isolated copper plates, each of area 0.6 m^2, carry opposite charges of magnitude 7.08×10^{-10} C. They are placed opposite each other in parallel alignment, with a spacing of 2 mm. What will be the potential difference between the plates when their spacing is increased to 6 cm? (Use the dielectric constant $k = 1$ in air and electric permittivity $\varepsilon_0 = 8.854 \times 10^{-12}$ F/m)

A. 8.0 V
B. 3.1 V
C. 4.3 V
D. 7.2 V
E. 0.9 V

44. Electric current can flow:

A. in a region of negligible resistance
B. through a potential difference
C. in a perfect conductor
D. in the absence of resistance
E. in a semi-perfect conductor

45. The metal detectors used to screen passengers at airports operate via:

A. Newton's Laws
B. Bragg's Law
C. Faraday's Law
D. Ohm's Law
E. Ampere's Law

46. A 7 μC negative charge is attracted to a large, well-anchored, positive charge. How much kinetic energy does the negatively-charged object gain if the potential difference through which it moves is 3.5 mV?

A. 0.86 J
B. 6.7 μJ
C. 36.7 μJ
D. 0.5 kJ
E. 24.5 nJ

47. A wire of resistivity ρ is replaced in a circuit by a wire of the same material but four times as long. If the total resistance remains the same, the diameter of the new wire must be:

A. one-fourth the original diameter
B. two times the original diameter
C. the same as the original diameter
D. one-half the original diameter
E. four times the original diameter

48. The addition of resistors in series to a resistor in an existing circuit, while voltage remains constant, would result in [ ] in the original resistor.

A. an increase in current
B. a decrease in resistance
C. an increase in resistance
D. a decrease in current
E. no change

49. In an experiment, a battery is connected to a variable resistor R, adjusting the resistance by turning a knob. The potential difference across the resistor and the current through it are recorded for different settings of the resistor knob. The battery is an ideal potential source in series with an internal resistor. The emf of the potential source is 9 V, and the internal resistance is 0.1 Ω. What is the current if the variable resistor is set at 0.5 Ω?

A. 15 A
B. 0.9 A
C. 4.5 A
D. 45 A
E. 9 A

50. Two parallel plates initially uncharged are separated by 1.6 mm. What charge must be transferred from one plate to the other if 10 kJ of energy is stored in the plates? The area of each plate is 24 mm^2. (Use the dielectric constant $k = 1$ in air and the electric permittivity $\mathcal{E}_0 = 8.854 \times 10^{-12}$ F/m)

A. 78 μC
B. 15 mC
C. 52 μC
D. 29 μC
E. 66 mC

51. When a proton moves in the direction of the electric field, the potential energy of the system [ ] and electric potential [ ]. (Use the dielectric constant $k = 1$ in air and the electric permittivity $\mathcal{E}_0 = 8.854 \times 10^{-12}$ F/m)

A. increases ... increases
B. decreases ... decreases
C. increases ... decreases
D. decreases ... increases
E. remains the same ... decreases

52. Each plate of a parallel-plate air capacitor has an area of 0.004 m^2, and the separation of the plates is 0.02 mm. An electric field of 8.6×10^6 V/m is present between the plates. What is the energy density between the plates? (Use the electric permittivity $\mathcal{E}_0 = 8.854 \times 10^{-12}$ F/m)

A. 100 J/m^3
B. 400 J/m^3
C. 220 J/m^3
D. 330 J/m^3
E. 510 J/m^3

53. What is the quantity calculated with units of kg·m^2/(s·C^2)?

A. Resistance
B. Capacitance
C. Potential
D. Resistivity
E. Current

54. At a constant voltage, an increase in the resistance of a circuit results in:

A. no change in I or V
B. an increase in I
C. an increase in power
D. constant power
E. a decrease in I

55. A charge $+Q$ is located at one of the corners of a square. The absolute potential at the center of a square is 3 V. If a second charge $-Q$ is at one of the other three corners, what is the absolute potential at the square's center?

A. –6 V
B. 12 V
C. 6 V
D. 0 V
E. –12 V

Notes for active learning

Notes for active learning

Electrostatics and Electromagnetism

1. How many excess electrons are present for an object that has a charge of –1 Coulomb? (Use Coulomb's constant $k = 9 \times 10^9$ N·m²/C² and charge of an electron $e = -1.6 \times 10^{-19}$ C)

A. 3.1×10^{19} electrons
B. 6.3×10^{19} electrons
C. 6.3×10^{18} electrons
D. 1.6×10^{19} electrons
E. 6.5×10^{17} electrons

2. A flat disk 1 m in diameter is oriented so that the area vector of the disk makes an angle of $\pi/6$ radians with a uniform electric field. What is the electric flux through the surface if the field strength is 740 N/C?

A. 196π N·m²/C
B. $250/\pi$ N·m²/C
C. 644π N·m²/C
D. 160π N·m²/C
E. $10/\pi$ N·m²/C

3. A positive charge $Q = 1.3 \times 10^{-9}$ C is located along the x-axis at $x = -10^{-3}$ m, and a negative charge of the same magnitude is located at the origin. What is the magnitude and direction of the electric field at the point along the x-axis where $x = 10^{-3}$ m? (Use Coulomb's constant $k = 9 \times 10^9$ N·m²/C² and to the right as the positive direction)

A. 8.8×10^6 N/C to the left
B. 3.25×10^7 N/C to the right
C. 5.5×10^7 N/C to the right
D. 2.75×10^6 N/C to the right
E. 8.5×10^7 N/C to the left

4. Two charges $Q_1 = 2.4 \times 10^{-10}$ C and $Q_2 = 9.2 \times 10^{-10}$ C are near each other, and charge Q_1 exerts a force F_1 on Q_2. How does F_1 change if the distance between Q_1 and Q_2 is increased by a factor of 4?

A. Decreases by a factor of 4
B. Increases by a factor of 16
C. Decreases by a factor of 16
D. Increases by a factor of 4
E. Remains the same

5. A proton is at ($x = 1$ nm, $y = 0$ nm) and an electron is at ($x = 0$ nm, $y = 4$ nm). Find the attractive Coulomb force between them. (Use Coulomb's constant $k = 9 \times 10^9$ N·m²/C² and charge of an electron $e = -1.6 \times 10^{-19}$ C)

A. 5.3×10^8 N
B. 1.9×10^{-15} N
C. 9.3×10^4 N
D. 2.6×10^{-18} N
E. 1.4×10^{-11} N

6. Which form of electromagnetic radiation has photons with the highest energy?

A. Gamma rays
B. Visible light
C. Microwaves
D. Ultraviolet radiation
E. Infrared radiation

7. A 54,000 kg asteroid carrying a negative charge of 15 μC is 180 m from another 51,000 kg asteroid, carrying a negative charge of 11 μC. What is the net force the asteroids exert upon each other? (Use the gravitational constant $G = 6.673 \times 10^{-11}$ N·m²/kg² and Coulomb's constant $k = 9 \times 10^9$ N·m²/C²)

A. 400,000 N
B. 5,700 N
C. -4.0×10^{-5} N
D. 4.0×10^{-5} N
E. 5.7×10^{-5} N

8. Two small beads are 30 cm apart with no other charges or fields present. Bead A has 20 μC of charge, and bead B has 5 μC. Which of the following statements is true about the electric forces on these beads?

A. The force on A is 120 times the force on B
B. The force on A is exactly equal to the force on B
C. The force on B is 4 times the force on A
D. The force on A is 20 times the force on B
E. The force on B is 120 times the force on A

Questions **9-10** are based on the following:

Two parallel metal plates separated by 0.01 m are charged to create a uniform electric field of 3.5×10^4 N/C between them, which points down. A small, stationary 0.008 kg plastic ball *m* is located between the plates and has a small charge *Q*. The forces acting on it are the force of gravity and the electric field. (Use Coulomb's constant $k = 9 \times 10^9$ N·m²/C², the charge of an electron = -1.6×10^{-19} C, the charge of a proton = 1.6×10^{-19} C, the mass of a proton = 1.67×10^{-27} kg, the mass of an electron = 9.11×10^{-31} kg and the acceleration due to gravity $g = 9.8$ m/s²)

9. What is the charge on the ball?

A. −250 C
B. 250 C
C. 3.8×10^{-6} C
D. -2.2×10^{-6} C
E. 2.5×10^{-3} C

10. How would the acceleration of an electron between the plates compare to the acceleration of a proton between the plates?

A. One thousand eight hundred thirty times as large, and in the opposite direction
B. The square root times as large and in the opposite direction
C. Twice as large and in the opposite direction
D. The same magnitude, but in the opposite direction
E. One hundred times as large, and in the opposite direction

11. A positive charge $Q = 2.3 \times 10^{-11}$ C is 10^{-2} m from a negative charge of equal magnitude. Point P is located equidistant between them. What is the magnitude of the electric field at point P? (Use Coulomb's constant $k = 9 \times 10^{9}$ N·m²/C²)

A. 9.0×10^{3} N/C
B. 4.5×10^{3} N/C
C. 3.0×10^{4} N/C
D. 1.7×10^{4} N/C
E. 6.6×10^{5} N/C

12. A point charge $Q = -10$ μC. What is the number of excess electrons on charge Q? (Use the charge of an electron $e = -1.6 \times 10^{-19}$ C)

A. 4.5×10^{13} electrons
B. 1.6×10^{13} electrons
C. 9.0×10^{13} electrons
D. 8.5×10^{13} electrons
E. 6.3×10^{13} electrons

13. A distance of 3 m separates an electron and a proton. What happens to the magnitude of the force on the proton if the electron is moved 1.5 m closer to the proton?

A. It increases to twice its original value
B. It decreases to one-fourth its original value
C. It increases to four times its original value
D. It decreases to one-half its original value
E. It remains the same

14. How will the magnitude of the electrostatic force between two objects be affected if the distance between them and both of their charges are doubled?

A. It will increase by a factor of 4
B. It will increase by a factor of 2
C. It will decrease by a factor of 2
D. It will increase by a factor of $\sqrt{2}$
E. It will be unchanged

15. Which statement is true for an H nucleus, which has a charge $+e$ situated to the left of a C nucleus, which has a charge $+6e$?

A. The electrical force experienced by the H nucleus is to the right, and the magnitude is equal to the force exerted on the C nucleus
B. The electrical force experienced by the H nucleus is to the right, and the magnitude is lower than the force exerted on the C nucleus
C. The electrical force experienced by the H nucleus is to the left, and the magnitude is higher than the force exerted on the C nucleus
D. The electrical force experienced by the H nucleus is to the left, and the magnitude is equal to the force exerted on the C nucleus
E. The electrical force experienced by the H nucleus is to the right, and the magnitude is higher than the force exerted on the C nucleus

16. Two oppositely charged particles are slowly separated from each other. What happens to the force as the particles are slowly moved apart?

A. attractive and decreasing
B. repulsive and decreasing
C. attractive and increasing
D. repulsive and increasing
E. none of the above

17. Electrons move in an electrical circuit:

A. because the wires are so thin
B. by interacting with an established electric field
C. by colliding with each other
D. by being repelled by protons
E. by interacting with an established gravitational field

18. If the number of turns on the secondary coil of a transformer is less than those on the primary, the result is a:

A. 240 V transformer
B. 110 V transformer
C. step-up transformer
D. step-down transformer
E. impedance transformer

19. Two charges $Q_1 = 3 \times 10^{-8}$ C and $Q_2 = 9 \times 10^{-8}$ C are near each other and charge Q_1 exerts a force F_1 on Q_2. What is F_2, the force that charge Q_2 exerts on charge Q_1?

A. $F_1 / 3$
B. F_1
C. $3F_1$
D. $2F_1$
E. $F_1 / 2$

20. Two electrons are passing 30 mm apart. What is the repulsive electric force that they exert on each other? (Use Coulomb's constant $k = 9 \times 10^9$ N·m²/C² and the charge of an electron = -1.6×10^{-19} C)

A. 1.3×10^{-25} N
B. 3.4×10^{-27} N
C. 1.3×10^{27} N
D. 3.4×10^{10} N
E. 2.56×10^{-25} N

21. A light bulb is connected to a circuit and has a wire leading to it in a loop. What happens when a strong magnet is quickly passed through the loop?

A. The brightness of the light bulb dims or gets brighter due to an induced emf produced by the magnet
B. The light bulb's brightness remains the same although current decreases
C. The light bulb gets brighter because more energy is added to the system by the magnet inside the coil
D. The bulb gets brighter because there is an induced emf that drives more current through the light bulb
E. This does not affect the light bulb's brightness because the coil does not shift the magnitude of the current through the light bulb

22. Suppose a van de Graaff generator builds a negative static charge, and a grounded conductor is near enough to it so that an 8 μC of negative charge arcs to the conductor. What is the number of electrons transferred? (Use the charge of an electron $e = -1.6 \times 10^{-19}$ C)

A. 1.8×10^{14} electrons
B. 48 electrons
C. 5×10^{13} electrons
D. 74 electrons
E. 5×10^{20} electrons

23. Which statement must be true if two objects are electrically attracted to each other?

A. One of the objects could be electrically neutral
B. One object must be negatively charged, and the other must be positively charged
C. At least one of the objects must be positively charged
D. At least one of the objects must be negatively charged
E. None of the above statements are true

24. An amp is a unit of electrical:

A. capacity
B. current
C. potential difference
D. charge
E. pressure

25. A loop of wire is rotated about a diameter (which is perpendicular to a given magnetic field). In one revolution, the induced current in the loop reverses direction how many times?

A. 2
B. 1
C. 0
D. 4
E. 3

26. Two charges ($Q_1 = 2.3 \times 10^{-8}$ C and $Q_2 = 2.5 \times 10^{-9}$ C) are a distance 0.1 m apart. How much energy is required to bring them to a distance 0.01 m apart? (Use Coulomb's constant $k = 9 \times 10^9$ N·m²/C²)

A. 2.2×10^{-4} J
B. 8.9×10^{-5} J
C. 1.7×10^{-5} J
D. 4.7×10^{-5} J
E. 6.2×10^{-5} J

27. A solid aluminum cube rests on a wooden table in a region where a uniform external electric field is directed straight upward. What can be concluded regarding the charge on the top surface of the cube?

A. The top surface is neutral
B. The top surface is charged negatively
C. The top surface fluctuates between being charged neutral and positively
D. The top surface's charge cannot be determined without further information
E. The top surface is charged positively

28. A point charge of $+Q$ is at the center of an equilateral triangle, as shown. When the second charge of $+Q$ is at one of the triangle's vertices, an electrostatic force of 5 N acts on it. What is the magnitude of the force acting on the center charge when the third charge of $+Q$ is at one of the other vertices?

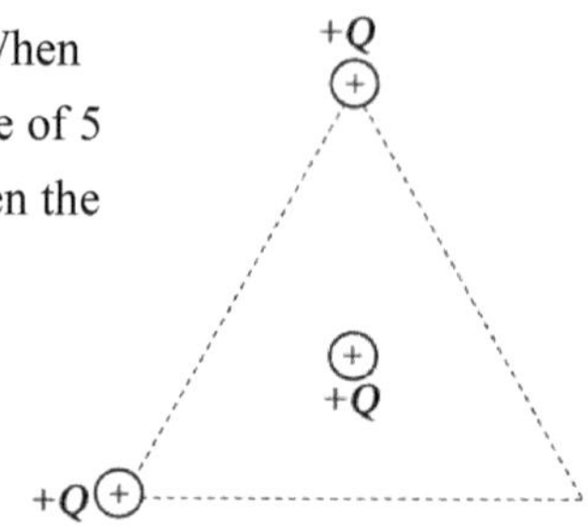

A. 0 N
B. 4 N
C. 5 N
D. 8 N
E. 12 N

29. In the figure below, the charge in the middle is fixed and $Q = -7.5$ nC. For what fixed, positive charge q_1 will non-stationary, negative charge q_2 be in static equilibrium?

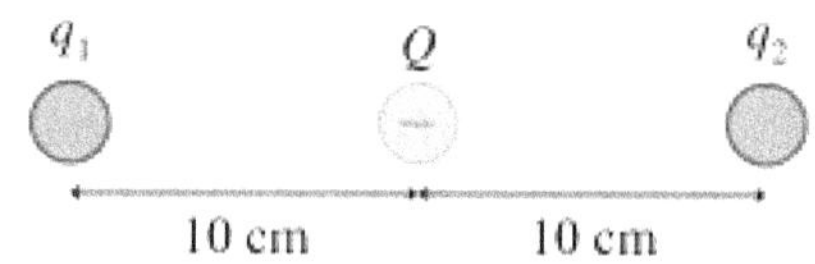

A. 53 nC
B. 7.5 nC
C. 15 nC
D. 30 nC
E. 12.8 nC

30. Which form of electromagnetic radiation has the highest frequency?

A. Gamma radiation
B. Ultraviolet radiation
C. Visible light
D. Radio waves
E. Infrared radiation

31. The following affect the electrostatic field strength at a point at a distance from a source charge, EXCEPT:

A. the sign of the source charge
B. the distance from the source charge
C. the magnitude of the source charge
D. the nature of the medium surrounding the source charge
E. the presence of nearby conducting objects

32. A charged particle is observed traveling in a circular path in a uniform magnetic field. If the particle had been traveling twice as fast, the radius of the circular path would be:

A. three times the original radius
B. twice the original radius
C. one-half of the original radius
D. four times the original radius
E. one-third of the original radius

33. Two charges separated by 1 m exert a 1 N force on each other. If the magnitude of each charge is doubled, the force on each charge is:

A. 1 N
B. 2 N
C. 4 N
D. 6 N
E. 10 N

34. In a water solution of NaCl, the NaCl dissociates into ions surrounded by water molecules. Consider a water molecule near a Na^+ ion. What tends to be the orientation of the water molecule?

A. The hydrogen atoms are nearer the Na^+ ion because of their positive charge
B. The hydrogen atoms are nearer the Na^+ ion because of their negative charge
C. The oxygen atom is nearer the Na^+ ion because of the oxygen's positive charge
D. The oxygen atom is nearer the Na^+ ion because of the oxygen's negative charge
E. The hydrogen and oxygen atoms center on either side of the Na^+ ion because an atom has no charge

35. A metal sphere is insulated electrically and is given a charge. If 30 electrons are added to the sphere giving a charge, how many Coulombs are added? (Use Coulomb's constant $k = 9 \times 10^9$ N·m²/C² and the charge of an electron $e = -1.6 \times 10^{-19}$ C)

A. –2.4 C
B. –30 C
C. -4.8×10^{-18} C
D. -4.8×10^{-16} C
E. –13 C

36. What happens to the cyclotron frequency of a charged particle if its speed doubles?

A. It is ¼ as large
B. It is ½ as large
C. It doubles
D. It is $\sqrt{2}$ times as large
E. It remains the same

37. A positive test charge q is released near a positive fixed charge Q. As q moves away from Q, it experiences:

A. increasing acceleration
B. decreasing acceleration
C. constant velocity
D. decreasing velocity
E. constant momentum

38. If a value has SI units kg m²/s²/C, this value can be:

A. electric potential difference
B. resistance
C. electric field strength
D. Newton's forces
E. electric potential energy

39. A Coulomb is a unit of electrical:

A. capacity
B. resistance
C. charge
D. potential difference
E. pressure

40. To say that electric charge is conserved means that no case has ever been found where:

A. charge has been created or destroyed
B. the total charge on an object has increased
C. the net negative charge on an object is unbalanced by a positive charge on another object
D. the total charge on an object has changed by a significant amount
E. the total charge on an object has decreased

41. Two charges $Q_1 = 1.7 \times 10^{-10}$ C and $Q_2 = 6.8 \times 10^{-10}$ C are near each other. How would F change if the charges were both doubled, but the distance between them remained the same?

A. F increases by a factor of 2
B. F increases by a factor of 4
C. F decreases by a factor of $\sqrt{2}$
D. F decreases by a factor of 4
E. F increases by a factor of $\sqrt{2}$

42. Two like charges of the same magnitude are 10 mm apart. If the force of repulsion they exert upon each other is 4 N, what is the magnitude of each charge? (Use Coulomb's constant $k = 9 \times 10^9$ N·m²/C²)

A. 6×10^{-5} C
B. 6×10^{5} C
C. 2×10^{-7} C
D. 1.5×10^{-7} C
E. 6×10^{-7} C

43. A circular loop of wire is rotated about an axis whose direction at constant angular speed can be varied. In a region where a uniform magnetic field points straight down, what orientation of the axis of the rotation guarantees that the emf will be zero (regardless of how the axis is aligned to the loop)?

A. It must be vertical
B. It must make an angle of 45° to the direction South
C. It could have any horizontal orientation
D. It must make an angle of 45° to the vertical
E. It must make an angle of 90° to the direction South

44. Two identical small charged spheres are a certain distance apart, and each initially experiences an electrostatic force of magnitude F due to the other. With time, the charge gradually diminishes on both spheres. What is the magnitude of the electrostatic force when each sphere has lost half its initial charge?

A. 1/16 F
B. 1/8 F
C. 1/4 F
D. 2 F
E. F

45. A proton, moving in a uniform magnetic field, moves in a circle perpendicular to the field lines and takes time T for each circle. If the proton's speed tripled, what would now be its time to go around each circle?

A. T/3
B. T
C. 6T
D. 3T
E. T/6

46. As measurements of the electrostatic field strength are taken at points that progressively approach a negatively-charged particle, the field vectors will point:

A. away from the particle and have a constant magnitude
B. away from the particle and have progressively decreasing magnitude
C. towards the particle and have progressively increasing magnitude
D. towards the particle and have progressively decreasing magnitude
E. towards the particle and have a constant magnitude

47. Every proton in the universe is surrounded by its own:

I. electric field   II. gravitational field   III. magnetic field

A. I only
B. II only
C. III only
D. I and III only
E. I, II and III

48. A charge $Q = 3.1 \times 10^{-5}$ C is fixed in space while another charge $q = -10^{-6}$ C is 6 m away. Charge q is slowly moved 4 m in a straight line directly toward the charge Q. How much work is required to move charge q? (Use Coulomb's constant $k = 9 \times 10^9$ N·m²/C²)

A. –0.09 J
B. –0.03 J
C. 0.16 J
D. 0.08 J
E. 0.8 J

49. A point charge $Q = -600$ nC. What is the number of excess electrons in charge Q? (Use the charge of an electron $e = -1.6 \times 10^{-19}$ C)

A. 5.6×10^{12} electrons
B. 2.1×10^{10} electrons
C. 2.8×10^{11} electrons
D. 3.8×10^{12} electrons
E. 4.3×10^{8} electrons

50. In electricity, what quantity is analogous to the acceleration of gravity, g (i.e., a force per unit mass)?

A. Electric charge
B. Electric current
C. Electric field
D. Electromagnetic force
E. Electric dipole

51. Which type of electromagnetic (EM) wave travels through space the slowest?

A. Visible light
B. Ultraviolet light
C. Gamma rays
D. Radio waves
E. EM waves travel at the same speed

52. As a proton moves in the direction of the electric field lines, it is moving from:

A. high potential to low potential and losing electric potential energy
B. high potential to low potential and gaining electric potential energy
C. low potential to high potential and gaining electric potential energy
D. low potential to high potential and retaining electric potential energy
E. high potential to low potential and retaining electric potential energy

53. Which of the following requires a measure of time?

A. Joule
B. Watt
C. Volt
D. Coulomb
E. Ohm

54. If an object is characterized as electrically polarized:

A. its internal electric field is zero
B. it is a strong insulator
C. it is electrically charged
D. its charges have been rearranged
E. it is a weak insulator

55. Two positive charges Q_1 and $Q_2 = 3.4 \times 10^{-10}$ C are located 10^{-3} m away, and point P is precisely between them. What is the magnitude of the electric field at point P?

A. 0 N/C
B. 10^{-10} N/C
C. 6.8×10^{-7} N/C
D. 1.7×10^{-5} N/C
E. 1.2×10^{-9} N/C

Notes for active learning

Notes for active learning

Atomic and Nuclear Structure

1. Which statement(s) about alpha particles is/are FALSE?

I. They are a harmless form of radiation
II. They have low penetrating power
III. They have high ionization power

A. I only
B. II only
C. III only
D. I and II only
E. I and III only

2. What is the term for nuclear radiation that is identical to an electron?

A. Positron
B. Gamma ray
C. Beta minus particle
D. Alpha particle
E. Beta plus particle

3. Protons are accelerated in a particle accelerator. When the speed of the protons is doubled, by what factor does their de Broglie wavelength change? Note: consider this situation non-relativistic.

A. Increases by $\sqrt{2}$
B. Decreases by $\sqrt{2}$
C. Increases by 2
D. Increases by 4
E. Decreases by 2

4. The Bohr model of the atom was able to explain the Balmer series because:

A. electrons were allowed to exist in specific orbits and nowhere else
B. differences between the energy levels of the orbits matched the differences between the energy levels the line spectra
C. smaller orbits require electrons to have more negative energy to match the angular momentum
D. differences between the energy levels of the orbits were precisely half the differences between the energy levels of the line spectra
E. none of the above

5. Radioactivity is the tendency for an element to:

A. become ionized easily
B. be dangerous to living things
C. emit radiation
D. emit protons
E. radiate heat

6. Which is the missing species in the nuclear equation: ${}^{100}_{44}\text{Ru} + {}^{0}_{-1}\text{e}^- \rightarrow$ ___?

A. ${}^{100}_{45}\text{Ru}$

B. ${}^{100}_{43}\text{Ru}$

C. ${}^{101}_{44}\text{Ru}$

D. ${}^{100}_{43}\text{Tc}$

E. ${}^{101}_{43}\text{Tc}$

7. The term nucleon refers to:

A. the nucleus of a specific isotope
B. protons and neutrons
C. positrons emitted from an atom that undergoes nuclear decay
D. electrons emitted from a nucleus in a nuclear reaction
E. neutrons emitted from a nucleus in a nuclear reaction

8. An isolated ^{9}Be atom spontaneously decays into two alpha particles. What can be concluded about the mass of the ^{9}Be atom?

A. The mass is less than twice the mass of the ^{4}He atom, but not equal to the mass of ^{4}He
B. No conclusions can be made about the mass
C. The mass is exactly twice the mass of the ^{4}He atom
D. The mass is equal to the mass of the ^{4}He atom
E. The mass is greater than twice the mass of the ^{4}He atom

9. Which of the following isotopes contains the most neutrons?

A. ${}^{178}_{84}\text{Po}$

B. ${}^{178}_{87}\text{Fr}$

C. ${}^{181}_{86}\text{Rn}$

D. ${}^{170}_{83}\text{Bi}$

E. ${}^{177}_{86}\text{Rn}$

10. Which of the following correctly balances this nuclear fission reaction?

$${}^{1}_{0}\text{n} + {}^{235}_{92}\text{U} \rightarrow {}^{131}_{53}\text{I} + \_\_\_ + 3\,{}^{1}_{0}\text{n}$$

A. ${}^{102}_{39}\text{Y}$

B. ${}^{102}_{36}\text{Kr}$

C. ${}^{104}_{39}\text{Y}$

D. ${}^{105}_{36}\text{Kr}$

E. ${}^{131}_{54}\text{I}$

11. What is the frequency of the light emitted by atomic hydrogen according to the Balmer formula where n = 12? (Use the Balmer series constant B = 3.6×10^{-7} m and the speed of light $c = 3 \times 10^8$ m/s)

A. 5.3×10^6 Hz
B. 9.8 Hz
C. 5.9×10^{13} Hz
D. 1.2×10^{11} Hz
E. 8.1×10^{14} Hz

12. Gamma rays require the heaviest shielding of the common nuclear radiation because gamma rays have the:

A. heaviest particles
B. lowest energy
C. most intense color
D. highest energy
E. lowest frequency

13. In transitioning from state n = 1 to state n = 2, the hydrogen atom must [ ] a photon of [ ]. (Use Planck's constant $h = 4.14 \times 10^{-15}$ eV·s, speed of light $c = 3 \times 10^8$ m/s and Rydberg constant R = 1.097×10^7 m^{-1})

A. absorb … 10.2 eV
B. absorb … 8.6 eV
C. emit … 8.6 eV
D. emit … 10.2 eV
E. absorb … 4.3 eV

14. Rubidium $^{87}_{37}$Rb is a naturally-occurring nuclide that undergoes β^- decay. What is the resultant nuclide from this decay?

A. $^{86}_{36}$Rb
B. $^{87}_{38}$Kr
C. $^{87}_{38}$Sr
D. $^{87}_{36}$Kr
E. $^{86}_{37}$Rb

15. Which of the following statements best describes the role of neutrons in the nucleus?

A. The neutrons stabilize the nucleus by attracting protons
B. The neutrons stabilize the nucleus by balancing charge
C. The neutrons stabilize the nucleus by attracting other nucleons
D. The neutrons stabilize the nucleus by repelling other nucleons
E. The neutrons stabilize the nucleus by attracting electrons

16. A Geiger–Muller counter detects radioactivity by:

A. ionizing argon gas in a chamber which produces an electrical signal
B. analyzing the mass and velocity of each particle
C. developing film which is exposed by radioactive particles
D. slowing the neutrons using a moderator and then counting the secondary charges produced
E. converting electrical charges from a chemical reaction into light

17. What percentage of the radionuclides in a given sample remains after three half-lives?

A. 25%
B. 12.5%
C. 6.25%
D. 33.3%
E. 50%

18. Electron transitions in hydrogen that form the Lyman series:

A. begin on the n = 2 shell
B. end on the n = 2 shell
C. begin on the n = 1 shell
D. end on the n = 1 shell
E. are between the n = 1 and n = 3 shells

19. Most of the volume of an atom is occupied by:

A. neutrons
B. space
C. electrons
D. protons
E. interface with protons

20. Alpha and beta minus particles are deflected in opposite directions in a magnetic field because:

I. they have opposite charges
II. alpha particles contain nucleons, and beta minus particles do not
III. they spin in opposite directions

A. I only
B. II only
C. III only
D. I and II only
E. I and III only

21. The conversion of mass to energy is measurable in:

A. chemiluminescent transformations
B. spontaneous chemical reactions
C. endothermic reactions
D. exothermic reactions
E. nuclear reactions

22. What is the term for the amount of a radioactive substance that undergoes 3.7×10^{10} disintegrations per second?

A. Rem
B. Rad
C. Curie
D. Roentgen
E. Sievert

23. An isolated ^{235}U atom spontaneously undergoes fission into two approximately equal-sized fragments. What is missing from the product side of the reaction:

$$^{235}U \rightarrow {}^{141}Ba + {}^{92}Kr + \_\_\_?$$

A. A neutron
B. Two neutrons
C. Two protons and two neutrons
D. Two protons and a neutron
E. A proton and two neutrons

24. How many protons and neutrons are in $^{34}_{16}S$?

A. 18 neutrons and 34 protons
B. 16 neutrons and 18 protons
C. 16 protons and 34 neutrons
D. 16 protons and 18 neutrons
E. 34 neutrons and 18 protons

25. The radioactive isotope Z has a half-life of 12 hours. What is the fraction of the original amount remaining after 2 days?

A. 1/16
B. 1/8
C. 1/4
D. 1/2
E. 1/3

26. Which of the following nuclear equations correctly describes alpha emission?

A. $^{238}_{92}U \rightarrow {}^{242}_{94}Pu + {}^{4}_{2}He$
B. $^{238}_{92}U \rightarrow {}^{4}_{2}He$
C. $^{238}_{92}U \rightarrow {}^{234}_{90}Th + {}^{4}_{2}He$
D. $^{238}_{92}U \rightarrow {}^{235}_{90}Th + {}^{4}_{2}He$
E. None of the above

27. A hydrogen atom transitions downward from the n = 20 state to the n = 5 state. Find the wavelength of the emitted photon. (Use Planck's constant $h = 4.14 \times 10^{-15}$ eV·s, the speed of light $c = 3 \times 10^8$ m/s and the Rydberg constant $R = 1.097 \times 10^7\ m^{-1}$)

A. 1.93 μm
B. 2.82 μm
C. 1.54 μm
D. 1.38 μm
E. 2.43 μm

28. A nuclear equation is balanced when the:

A. same elements are on both sides of the equation
B. sums of the atomic numbers of the particles and atoms are the same on both sides of the equation
C. sum of the mass numbers of the particles and the sum of atoms are the same on both sides of the equation
D. sum of the mass and atomic numbers of the particles and atoms are the same on both sides of the equation
E. charges of the particles and atoms are the same on both sides of the equation

29. A blackbody is an ideal system that:

A. absorbs 50% of the light incident upon it and emits 50% of the radiation it generates
B. absorbs 0% of the light incident upon it and emits 100% of the radiation it generates
C. absorbs 100% of the light incident upon it and emits 100% of the radiation it generates
D. emits 100% of the light it generates and absorbs 50% of the radiation incident upon it
E. absorbs 50% of the light incident upon it and emits 100% of the radiation it generates

30. Recent nuclear bomb tests have created an extra-high level of atmospheric ^{14}C. When future archaeologists date samples without knowing of these nuclear tests, will the calculated dates be correct?

A. Correct, because biological materials do not gather ^{14}C from bomb tests
B. Correct, since the ^{14}C decays within the atmosphere at the natural rate
C. Incorrect, they would appear too old
D. Incorrect, they would appear too young
E. Incorrect, since there is more ^{14}C than there should be

31. When an isotope releases gamma radiation, the atomic number:

A. and the mass number remain the same
B. and the mass number decrease by one
C. and the mass number increase by one
D. remains the same, and the mass number increases by one
E. remains the same, and the mass number decreases by one

32. If ^{14}Carbon is a beta emitter, what is the likely product of radioactive decay?

A. ^{22}Silicon
B. ^{13}Boron
C. ^{14}Nitrogen
D. ^{12}Carbon
E. None of the above

33. In a balanced nuclear equation, the:

I. sum of the mass numbers on both sides must be equal
II. sum of the atomic numbers on both sides must be equal
III. daughter nuclide appears on the right side of the arrow

A. I only
B. II only
C. III only
D. I and III only
E. I, II and III

34. What is the term for the number that characterizes an element and indicates the number of protons found in the nucleus of the atom?

A. Mass number
B. Atomic number
C. Atomic mass
D. Neutron number
E. None of the above

35. Hydrogen atoms can emit four spectral lines with visible colors from red to violet. These four visible lines emitted by hydrogen atoms are produced by electrons that:

A. end in the ground state
B. end in the n = 3 level
C. end in the n = 2 level
D. start in the ground state
E. start in the n = 3 level

36. The electron was discovered through experiments with:

A. quarks
B. foil
C. light
D. electricity
E. magnets

Questions **37-39** are based on the following:

The image shows a beam of radiation passing between two electrically-charged plates.

I. a
II. b
III. c

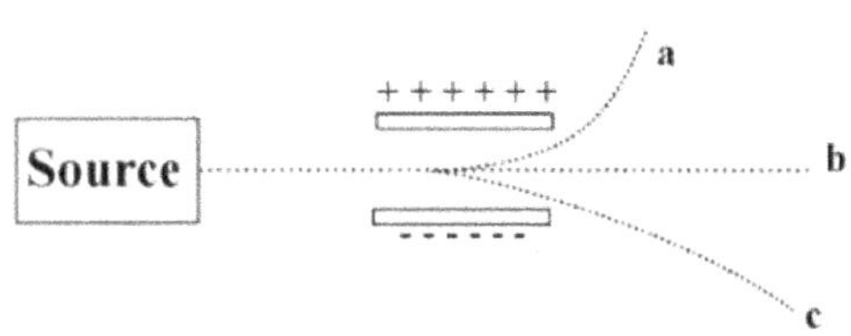

37. Which of the beams is due to an energetic light wave?

A. I only
B. II only
C. III only
D. I and II only
E. None

38. Which of the beams is/are composed of particles?

A. I only
B. II only
C. III only
D. I and III only
E. I, II and III

39. Which of the beams is due to a positively-charged helium nucleus?

A. I only
B. II only
C. III only
D. I, II and III
E. None of the above

40. Lithium atoms can absorb photons transitioning from the ground state (at –5.37 eV) to an excited state with one electron removed from the atom, corresponding to the zero-energy state. What is the wavelength of light associated with this transition? (Use Planck's constant $h = 4.14 \times 10^{-15}$ eV·s and speed of light $c = 3 \times 10^8$ m/s)

A. 6.6×10^{-6} m
B. 2.3×10^{-7} m
C. 3.6×10^{6} m
D. 4.2×10^{5} m
E. 2.6×10^{-6} m

41. The elements with atomic numbers of 84 and higher are radioactive because:

A. strong attractions between their nucleons make them unstable
B. their atomic numbers are larger than their mass numbers
C. strong repulsions between their electrons make them unstable
D. strong repulsions between their protons make their nuclei unstable
E. strong repulsions between their neutrons make their nuclei unstable

42. Which of the following statements about β particles is FALSE?

A. They have a smaller mass than α particles
B. They have high energy and a charge
C. They are created when neutrons become protons and vice versa
D. They can be positively or negatively charged
E. They are a harmless form of radioactivity

43. The Sun produces 3.85×10^{26} J each second. How much mass does it lose per second from nuclear processes alone? (Use the speed of light $c = 3 \times 10^8$ m/s)

A. 9.8×10^{1} kg
B. 2.4×10^{9} kg
C. 4.3×10^{9} kg
D. 1.1×10^{8} kg
E. 4.6×10^{6} kg

44. Which statement regarding Planck's constant is true?

A. It relates mass to the amount of energy that can be emitted
B. It sets a lower limit to the amount of energy that can be absorbed or emitted
C. It sets an upper limit to the amount of energy that can be absorbed
D. It sets an upper limit to the amount of energy that can be absorbed or emitted
E. It relates mass to the amount of energy that can be absorbed or emitted

45. The decay rate of a radioactive isotope will NOT be increased by increasing the:

I. surface area II. pressure III. temperature

A. I only
B. II only
C. III only
D. I and III only
E. I, II and III

46. Why are some smaller nuclei such as ^{14}Carbon often radioactive?

I. The attractive force of the nucleons has a limited range
II. The neutron to proton ratio is too large or too small
III. Most smaller nuclei are not stable

A. I only
B. II only
C. III only
D. II and III only
E. I and III only

47. Scandium ^{44}Sc decays by emitting a positron. What is the resulting nuclide produced by this decay?

A. $^{43}_{21}\text{Sc}$
B. $^{45}_{21}\text{Sc}$
C. $^{44}_{20}\text{Ca}$
D. $^{43}_{20}\text{Ca}$
E. $^{45}_{22}\text{Ti}$

48. A scintillation counter detects radioactivity by:

A. analyzing the mass and velocity of each electron
B. ionizing argon gas in a chamber which produces an electrical signal
C. emitting light from a NaI crystal when radioactivity passes through the crystal
D. developing film which is exposed by radioactive particles
E. slowing the neutrons using a moderator and then counting the secondary charges produced

49. Which of the following is indicated by each detection sound by a Geiger counter?

A. One half-life
B. One nucleus decaying
C. One neutron being emitted
D. One positron being emitted
E. None of the above

50. According to the Pauli Exclusion Principle, how many electrons in an atom may have a particular set of quantum numbers?

A. 1
B. 2
C. 3
D. 4
E. 5

51. The atomic number of an atom identifies the number of:

A. excited states
B. electron orbits
C. neutrons
D. protons
E. valence electrons

52. Which of the following correctly characterizes gamma radiation?

A. High penetrating power; charge = –1; mass = 0 amu
B. Low penetrating power; charge = –1; mass = 0 amu
C. Low penetrating power; charge = 0; mass = 4 amu
D. High penetrating power; charge = 0; mass = 4 amu
E. High penetrating power; charge = 0; mass = 0 amu

53. The rest mass of a proton is 1.0072764669 amu, and that of a neutron is 1.0086649156 amu. The ^{4}He nucleus weighs 4.002602 amu. What is the total binding energy of the nucleus? (Use the speed of light $c = 3 \times 10^8$ m/s and the conversion of 1 amu = 1.6606×10^{-27} kg)

A. 2.7×10^{-11} J
B. 4.4×10^{-12} J
C. 1.6×10^{-7} J
D. 2.6×10^{-12} J
E. 4.4×10^{-10} J

54. Which is the correct electron configuration for ground state boron ($Z = 5$)?

A. $1s^2 1p^2 2s$
B. $1s^2 2p^2 3s$
C. $1s^2 2p^3$
D. $1s^2 2s^2 3p^2$
E. $1s^2 2s^2 2p$

Notes for active learning

Notes for active learning

Particle Physics

1. Which of the following statements about hadrons are correct?

I. Hadrons are composed of leptons
II. All hadrons are composed of quarks
III. Protons and neutrons are hadrons, but electrons are not

A. I only
B. III only
C. II and III only
D. II and III only
E. I, II and III

2. Which of the following statements about hadrons are correct?

I. All hadrons interact by the strong nuclear force
II. Electrons, protons, and neutrons are commonly occurring hadrons
III. All hadrons are composed of quarks

A. I only
B. II only
C. I and II only
D. I and III only
E. I, II and III

3. Leptons can interact by which of the following forces?

A. strong nuclear force, and weak nuclear force
B. weak nuclear force, electromagnetic force, and gravitation
C. strong nuclear force, weak nuclear force, and electromagnetic force
D. strong nuclear force, weak nuclear force, electromagnetic force, and gravitation
E. strong nuclear force, electromagnetic force, and gravitation

4. Which of the following particles do NOT contain quarks?

I. alpha particle II. positron III. neutron

A. I only
B. II only
C. I and III only
D. II and III only
E. I, II and III

5. What are the possible charges of a quark (not an antiquark)?

A. $-2/3\ e$, $+1/3\ e$
B. $-1/3\ e$, $+2/3\ e$
C. $-e$, 0, e
D. $-2/3\ e$, $-1/3\ e$, $+1/3\ e$, $+2/3\ e$
E. $-1/3\ e$, $+1/3\ e$

6. How many quarks are in a deuteron, ${}_{1}^{2}H$?

A. 1
B. 3
C. 4
D. 6
E. 9

7. How many quarks are in a tritium isotope, ${}_{1}^{3}H$?

A. 1
B. 4
C. 5
D. 6
E. 9

8. How does the range of an exchange force depend on the mass of the exchange particle?

I. The range does not depend on the mass of the exchange particle
II. The range is shorter for a massive exchange particle than for a light exchange particle
III. The range is longer for a massive exchange particle than for a light exchange particle

A. I only
B. II only
C. III only
D. I and II only
E. I, II and III

9. An electron and a positron, both essentially at rest, annihilate each other, emitting two identical photons in the process. What is the wavelength of these photons? The mass of an electron or positron is 9.11×10^{-31} kg. (Use $h = 6.626 \times 10^{-34}$ J·s, $c = 3.00 \times 10^{8}$ m/s and $hc = 1240$ eV· nm)

A. 2.42 pm
B. 3.72 pm
C. 1.21×10^{-12} pm
D. 1.57 pm
E. 4.28 pm

10. A proton is a specific combination of three quarks. What are the electric charges on these quarks, expressed in terms of e?

A. $+2/3\ e$, $+2/3\ e$, and $-1/3\ e$
B. $-2/3\ e$, $+2/3\ e$, and $-1/3\ e$
C. $+2/3\ e$, $+2/3\ e$, and $+1/3\ e$
D. $-2/3\ e$, $-2/3\ e$, and $-1/3\ e$
E. $+2/3\ e$, $+2/3\ e$, and $+1/3\ e$

11. A positron (or antielectron) is made to stop in a sample of matter. Soon after, how many gamma rays are observed, and with what energy each?

A. none
B. one, 0.511 MeV each
C. two, 0.511 MeV each
D. one, 13.6 eV each
E. two, 13.6 eV each

12. What is the minimum energy required to produce a proton-antiproton pair? (Use the rest mass of a proton = 938 MeV/c^2)

A. 0 MeV
B. 938 MeV
C. 1876 MeV
D. 2814 MeV
E. 3752 MeV

13. An η meson with a rest energy of 548 MeV decays into two gamma rays. What is the wavelength of the gamma rays? (Use $h = 6.63 \times 10^{-34}$ J·s, and $c = 3.0 \times 10^8$ m/s and $hc = 1240$ eV nm).

A. 2.3×10^{-15} m
B. 2.3×10^{-16} m
C. 4.5×10^{-15} m
D. 4.5×10^{-16} m
E. 2.3×10^{-14} m

14. What is the speed of a proton with kinetic energy equal to 2 GeV?

A. 0.78 *c*
B. 0.84 c
C. 0.88 *c*
D. 0.93 c
E. 0.95 *c*

15. What is the speed of a 2 MeV electron?

A. 0.75 c
B. 0.84 c
C. 0.90 *c*
D. 0.94 *c*
E. 0.97 *c*

16. What is the temperature corresponding to the thermal energy of 50 GeV, the particle energy in the accelerator at SLAC?

A. 3.9×10^{11} K
B. 5.8×10^{14} K
C. 7.8×10^{11} K
D. 7.8×10^{14} K
E. 1.85×10^{15} K

17. How far is the galaxy from Earth if it recedes from Earth at a speed $v = 0.5\ c$? (Use $H_0 = 0.022$ m/s/lightyear for Hubble's constant).

A. 22.7 lightyears
B. 1.8×10^4 lightyears
C. 2.3×10^5 lightyears
D. 5.6×10^8 lightyears
E. 6.8×10^9 lightyears

18. A certain galaxy is 2.7×10^9 lightyears from Earth. How fast is it receding from Earth? (Use $H_0 = 0.022$ m/(s·lightyear) for Hubble's constant).

A. 0.2 c
B. 0.3 c
C. 0.4 c
D. 0.5 c
E. 0.7 c

19. What is the minimum energy required to produce an electron-positron pair? (Use the rest mass of an electron = 0.511 MeV/c^2).

A. 0.511 MeV
B. 1.022 MeV
C. 1.533 MeV
D. 2.044 MeV
E. 3.066 MeV

20. The rest energy of π meson is about 140 MeV. Using the uncertainty principle, estimate the range of strong force. (Use $h = 6.63 \times 10^{-34}$ J·s, $hc = 1.24$ eV μm).

A. 0.35×10^{-15} m
B. 0.7×10^{-15} m
C. 1.4×10^{-15} m
D. 2.8×10^{-15} m
E. 4.2×10^{-15} m

21. The rest energy of a Z^0 boson is 91 GeV. Using the uncertainty principle, what is the approximate interaction time when it mediates a weak interaction? (Use $h = 6.63 \times 10^{-34}$ J·s, $hc = 1.24$ eV μm).

A. 3.6×10^{-26} s
B. 2.6×10^{-27} s
C. 7.2×10^{-27} s
D. 5.2×10^{-28} s
E. 1.8×10^{-29} s

22. A μ^+ and μ^-at rest annihilate to produce two gamma rays that fly in opposite directions. What is the wavelength of each gamma-ray?
(Use the rest energy of a muon = 106 MeV, $h = 6.63 \times 10^{-34}$ J·s and $hc = 1.24$ eV μm).

A. 1.17×10^{-14} m
B. 5.63×10^{-15} m
C. 2.48×10^{-15} m
D. 1.24×10^{-15} m
E. 3.72×10^{-16} m

23. The rest energy of W^+ boson is about 80 GeV. Using the uncertainty principle, estimate the range of weak interaction. (Use $h = 6.63 \times 10^{-34}$ J·s and $hc = 1.24$ eV μm).

A. 1.3×10^{-15} m
B. 2.5×10^{-16} m
C. 2.5×10^{-17} m
D. 2.5×10^{-18} m
E. 4.0×10^{-19} m

Notes for active learning

Notes for active learning

Quantum Mechanics

1. The work function of a particular metal is 1.90 eV. What is the longest wavelength of light that can cause photoelectron emission from this metal?

A. 64 nm
B. 98 nm
C. 247 nm
D. 449 nm
E. 653 nm

2. Which of the following statements is correct if the frequency of the light in a laser beam is doubled while the number of photons per second in the beam is fixed?

I. The power in the beam does not change
II. The intensity of the beam doubles
III. The energy of individual photons does not change

A. I only
B. II only
C. III only
D. I and II only
E. I, II and III

3. Upon being struck by 240 nm photons, a material ejects electrons with a maximum kinetic energy of 2.58 eV. What is the work function of this material?

A. 1.17 eV
B. 2.04 eV
C. 2.60 eV
D. 3.26 eV
E. 4.34 eV

4. A high-energy photon collides with matter and creates an electron-positron pair. What is the minimum frequency of the photon? (Use the $m_{electron} = 9.11 \times 10^{-31}$ kg, $c = 3.00 \times 10^8$ m/s, and $h = 6.626 \times 10^{-34}$ J·s)

A. greater than 1.24×10^{12} Hz
B. greater than 2.47×10^{16} Hz
C. greater than 2.47×10^{20} Hz
D. greater than 2.47×10^{22} Hz
E. greater than 1.24×10^{24} Hz

5. What is the longest wavelength of light that can cause photoelectron emission from metal with a work function of 2.20 eV?

A. 216 nm
B. 372 nm
C. 484 nm
D. 564 nm
E. 642 nm

6. In 1932, C. D. Anderson:

A. set the limits on the probability of measurement accuracy
B. predicted the positron from relativistic quantum mechanics
C. discovered the positron using a cloud chamber
D. was the first to produce diffraction patterns of electrons in crystals
E. was the first to produce X-ray-like diffraction patterns of electrons passing through the thin metal foil

7. What is the energy of an optical photon of frequency 6.43×10^{14} Hz (Use $h = 6.626 \times 10^{-34}$ J·s and 1 eV = 1.60×10^{-19} J)

A. 1.04 eV
B. 1.86 eV
C. 2.66 eV
D. 3.43 eV
E. 5.37 eV

8. A photocathode has a work function of 2.4 eV. The photocathode is illuminated with monochromatic radiation, whose photon energy is 3.4 eV. What is the maximum kinetic energy of the photoelectrons produced? (Use 1 eV = 1.60×10^{-19} J)

A. 3.4×10^{-20} J
B. 1.6×10^{-19} J
C. 4.6×10^{-19} J
D. 5.8×10^{-19} J
E. 7.2×10^{-18} J

9. A photocathode whose work function is 2.9 eV is illuminated with white light with a continuous wavelength band from 400 nm to 700 nm. What is the range of the wavelength band in this white light illumination for which photoelectrons are NOT produced?

A. 360 to 440 nm
B. 400 to 480 nm
C. 430 to 500 nm
D. 430 to 700 nm
E. 460 to 760 nm

10. Photon A has twice the momentum of photon B as both travels in a vacuum. Which of the following statements about these photons is correct?

A. Both photons have the same speed
B. Both photons have the same wavelength
C. Photon A is traveling twice as fast as photon B
D. The energy of photon A is half as high as the energy of photon B
E. The wavelength of photon B is half as high as the wavelength of photon A

11. What is the energy of the photon emitted when an electron drops from the n = 20 state to the n = 7 state in a hydrogen atom?

A. 0.244 eV
B. 0.288 eV
C. 0.336 eV
D. 0.404 eV
E. 0.492 eV

12. Protons are accelerated in a particle accelerator. What is the de Broglie wavelength when the energy of the protons is doubled if the protons are non-relativistic (i.e., their kinetic energy is much less than mc^2)?

A. increases by a factor of 2
B. increases by a factor of 3
C. decreases by a factor of 2
D. decreases by a factor of $\sqrt{2}$
E. increases by a factor of $\sqrt{2}$

13. A certain photon, after being scattered from a free electron at rest, moves at an angle of 120° with respect to the incident direction. If the wavelength of the incident photon is 0.591 nm, what is the wavelength of the scattered photon? (Use the $m_{\text{electron}} = 9.11 \times 10^{-31}$ kg, $c = 3.00 \times 10^8$ m/s and $h = 6.626 \times 10^{-34}$ J·s)

A. 0.0 nm
B. 0.180 nm
C. 0.252 nm
D. 0.366 nm
E. 0.595 nm

14. Increasing the *brightness* of a beam of light without changing its color increases the:

A. speed of the photons
B. frequency of the light
C. number of photons per second traveling in the beam
D. energy of each photon
E. wavelength of the photons

15. What is the frequency of the light emitted by atomic Hydrogen according to the Balmer formula with $m = 4$ and $n = 9$?

A. 820 Hz
B. 1,640 Hz
C. 1,820 Hz
D. 2.24×10^{-6} m
E. 1.65×10^{14} Hz

16. One of the emission lines described by the original version of the Balmer formula has a wavelength of 377 nm. What is the value of n in the Balmer formula that gives this emission line?

A. 5
B. 7
C. 9
D. 11
E. 13

17. What is the wavelength of the most intense light emitted by a giant star of surface temperature 5000 K? (Use the constant in Wien's law = 0.00290 m·K)

A. 366 nm
B. 448 nm
C. 490 nm
D. 540 nm
E. 580 nm

18. In 1928, Paul Dirac:

A. set the limits on the probability of measurement accuracy
B. developed a wave equation for matter waves
C. suggested the existence of matter waves
D. discovered the positron using a cloud chamber
E. predicted the positron from relativistic quantum mechanics

19. A photocathode has a work function of 2.4 eV. The photocathode is illuminated with monochromatic radiation, whose photon energy is 3.4 eV. What is the maximum kinetic energy of the photoelectrons produced?

A. 0.9×10^{-19} J
B. 1.6×10^{-19} J
C. 2.8×10^{-19} J
D. 4.2×10^{-19} J
E. 5.8×10^{-19} J

20. In a Compton scattering experiment, which scattering angle produces the greatest change in wavelength?

A. 0°
B. 45°
C. 90°
D. 180°
E. 270°

21. A photocathode has a work function of 2.4 eV. The photocathode is illuminated with monochromatic radiation, whose photon energy is 3.5 eV. What is the wavelength of the illuminating radiation?

A. 280 nm
B. 325 nm
C. 350 nm
D. 395 nm
E. 420 nm

22. If the wavelength of a photon is doubled, what happens to its energy?

A. It is reduced to one-half of its original value
B. It stays the same
C. It is doubled
D. It is increased to four times its original value
E. It is reduced to one-fourth of its original value

23. A photocathode has a work function of 2.8 eV. The photocathode is illuminated with monochromatic radiation. What is the threshold frequency for the monochromatic radiation required to produce photoelectrons?

A. 1.2×10^{14} Hz
B. 2.8×10^{14} Hz
C. 4.6×10^{14} Hz
D. 6.8×10^{14} Hz
E. 8.4×10^{14} Hz

24. If the de Broglie wavelength of an electron is 380 nm, what is the speed of this electron? (Use $m_{electron} = 9.11 \times 10^{-31}$ kg and $h = 6.626 \times 10^{-34}$ J·s)

A. 0.6 km/s
B. 1.9 km/s
C. 3.4 km/s
D. 4.6 km/s
E. 5.8 km/s

25. What is the wavelength of the scattered photon if a photon of wavelength 1.50×10^{-10} m is scattered at an angle of 90° in the Compton effect?

A. 1.29×10^{-10} m
B. 1.52×10^{-10} m
C. 1.84×10^{-10} m
D. 2.42×10^{-10} m
E. 3.50×10^{-10} m

26. When the surface of a metal is exposed to blue light, electrons are emitted. Which of the following increases if the intensity of the blue light increases?

I. the maximum kinetic energy of the ejected electrons
II. the number of electrons ejected per second
III. the time lag between the onset of the absorption of light and the ejection of electrons

A. I only
B. II only
C. III only
D. I and II only
E. I, II and III

27. A proton has a speed of 7.2×10^4 m/s. What is the energy of a photon with the same wavelength as the de Broglie wavelength of this proton? (Use the $m_{proton} = 1.67 \times 10^{-27}$ kg and $c = 3.00 \times 10^8$ m/s)

A. 80 keV
B. 120 keV
C. 160 keV
D. 230 keV
E. 360 keV

28. In the Compton effect, as the scattering angle increases monotonically from 0° to 180°, the frequency of the X-rays scattered at that angle:

A. decreases by the $\sqrt{2}$
B. decreases monotonically
C. increases by the $\sqrt{2}$
D. increases monotonically
E. remain the same

29. In the spectrum of hydrogen, the lines obtained by setting m = 1 in the Rydberg formula are referred to as the Lyman series. What is the wavelength of the spectral line of the 15th member of the Lyman series?

A. 91.6 nm
B. 126.2 nm
C. 244.6 nm
D. 368.2 nm
E. 462.8 nm

30. If the frequency of the light in a laser beam is doubled while the number of photons per second in the beam is fixed, which of the following statements is correct?

I. The energy of individual photons doubles
II. The wavelength of the individual photon doubles
III. The intensity of the beam doubles

A. I only
B. II only
C. III only
D. I and III only
E. I, II and III

31. The spacing of the surface planes of a crystal is 159.0 pm. A beam directed normally at the surface of the crystal undergoes the first-order diffraction at an angle of 58° from the normal. What is the energy of the neutrons if the diffraction is done with a beam of monenergistic neutrons? (Use 1.67×10^{-27} kg for the mass of a neutron)

A. 0.0155 eV
B. 0.0106 eV
C. 0.0909 eV
D. 0.1450 eV
E. 0.2401 eV

32. A photocathode whose work function is 2.5 eV is illuminated with white light with a continuous wavelength band from 360 nm to 700 nm. What is the stopping potential for this white light illumination?

A. 0.95 V
B. 1.45 V
C. 1.90 V
D. 2.6 V
E. 3.2 V

33. If the momentum of an electron is 1.95×10^{-27} kg·m/s, what is the de Broglie wavelength? (Use $h = 6.626 \times 10^{-34}$ J·s)

A. 86.2 nm
B. 130.6 nm
C. 240.8 nm
D. 340.0 nm
E. 640.5 nm

34. A Hydrogen atom is excited to the n = 9 level. Its decay to the n = 6 level is detected on a photographic plate. What is the frequency of the light photographed?

A. 3,810 Hz
B. 7,240 Hz
C. 5.08×10^{13} Hz
D. 3.28×10^{-9} Hz
E. 2.24×10^{6} Hz

35. How much energy is carried by a photon of light having a frequency of 110 GHz? (Use $h = 6.626 \times 10^{-34}$ J·s)

A. 7.3×10^{-23} J
B. 2.9×10^{-25} J
C. 1.7×10^{-26} J
D. 1.1×10^{-21} J
E. 4.4×10^{-22} J

36. How many of the infinite number of Balmer spectrum lines are in the visible spectrum range?

A. 0
B. 2
C. 4
D. 6
E. infinite

37. Each photon in a beam of light has an energy of 4.20 eV. What is the wavelength of this light? ($c = 3.00 \times 10^{8}$ m/s, $h = 6.626 \times 10^{-34}$ J·s and 1 eV = 1.60×10^{-19} J)

A. 118.0 nm
B. 296.0 nm
C. 365.0 nm
D. 462.0 nm
E. 520.0 nm

38. Electrons are emitted from a surface when the light of wavelength 500.0 nm is shone on the surface, but electrons are not emitted for longer wavelengths of light. What is the work function of the surface?

A. 0.5 eV
B. 1.6 eV
C. 2.5 eV
D. 3.8 eV
E. 5.6 eV

39. In the Bohr theory, the orbital radius depends upon the principal quantum number in what way?

A. n
B. 1/n
C. n^2
D. n^3
E. $1/n^2$

40. The uncertainty in the position of a proton is 0.053 nm. What is the minimum uncertainty in its speed? (Use 1.67×10^{-27} kg as the proton mass)

A. 0.6×10^3 m/s
B. 1.2×10^3 m/s
C. 2.4×10^3 m/s
D. 3.6×10^3 m/s
E. 8.2×10^3 m/s

41. What is the wavelength of the light emitted by atomic hydrogen according to the Balmer formula with m = 9 and n = 11?

A. 8,500 nm
B. 14,700 nm
C. 22,300 nm
D. 31,900 nm
E. 39,400 nm

42. A specific particle's energy is known within 10^{-18} J. What is the minimum uncertainty in its arrival time at a detector?

A. 5.08×10^{-12} s
B. 4.25×10^{-13} s
C. 3.88×10^{-14} s
D. 2.55×10^{-15} s
E. 1.05×10^{-16} s

43. Upon being struck by 240.0 nm photons, a material ejects electrons with a maximum kinetic energy of 2.58 eV. What is the work function of this material?

A. 1.20 eV
B. 2.82 eV
C. 2.60 eV
D. 3.46 eV
E. 4.60 eV

Notes for active learning

Notes for active learning

Special Relativity

1. Consider three galaxies, Alpha, Beta, and Gamma. An astronomer in Beta sees each of the other two galaxies moving away from him in opposite directions at $0.70c$. At what speed would an observer in Alpha see the galaxy Gamma moving?

A. $0.94c$
B. $0.84c$
C. $1.2c$
D. $0.65c$
E. $0.98c$

2. A super high-speed train moves in a direction parallel to its length with a speed that approaches the speed of light. The height of the train, as measured by a stationary observer on the ground:

A. decreases slightly
B. increases slightly
C. approaches zero
D. approaches infinity
E. does not change due to the motion

3. A spaceship visits Alpha Centauri and returns to Earth. Alpha Centauri is 4.367 light-years from Earth. If the spaceship travels at one-half the speed of light for substantially all its expedition, how long was the ship gone according to an observer on Earth?

A. 2.184 years
B. 2.090 years
C. 4.367 years
D. 17.468 years
E. 19.010 years

4. A muon at rest decays in 2.2 μs. Moving at 99% the speed of light, it would be seen to "live" for how long?

A. 0.31 μs
B. 0.218 μs
C. 2.178 μs
D. 4.356 μs
E. 16 μs

5. A spaceship approaching an asteroid at a speed of $0.6c$ launches a scout rocket with a speed of $0.4c$. At what speed is the scout rocket approaching the asteroid from the point of view of the asteroid?

A. $0.52c$
B. $1.15c$
C. $0.66c$
D. $0.81c$
E. $0.73c$

6. An astronaut is resting on a bed inclined at an angle above the floor of a spaceship, with his head elevated near the front of the capsule while his feet are lower and facing the rear of the capsule. From the point of view of an observer who sees the spaceship moving near the speed of light parallel to the floor, the angle the bed makes with the floor:

A. could be greater or smaller than the angle observed by the astronaut depending on whether the rocket is moving to the right or the left
B. is smaller than the angle observed by the astronaut
C. is the same as the angle observed by the astronaut
D. is greater than the angle observed by the astronaut
E. cannot be determined from the information provided

7. As measured in Earth's rest frame, a spaceship traveling at 0.9640c takes 10.5 years to travel between two planets, not moving relative to each other. How many years does the trip take as measured by someone on the spaceship?

A. 2.79
B. 5.43
C. 8.42
D. 19.25
E. 28.

8. A hydrogen atom and a neutron fuse to form deuterium. If the energy released is 3.6×10^{-13} J, what is the mass of the deuterium? (Use the H atom nucleus as a single proton.).

A. 0.85×10^{-20} kg
B. 1.6×10^{-27} kg
C. 3.3×10^{-27} kg
D. 4.1×10^{-27} kg
E. 5.3×10^{-27} kg

9. Steve is moving past Earth at 0.99c and notices his heart beating 88 times/minute. What does his doctor on Earth observe his heart rate to be?

A. 8 beats/min
B. 12 beats/min
C. 22 beats/min
D. 28 beats/min
E. 33 beats/min

10. Two spaceships approach one another, each at a speed of 0.31c relative to a stationary observer on Earth. In terms of the speed of light, what speed does an observer on one spaceship record for the other approaching spaceship?

A. 0.71c
B. 0.38c
C. 0.57c
D. 0.64c
E. 0.69c

11. The special theory of relativity predicts an upper limit to the speed of a particle. It thus follows that there is an upper limit on which of the following property of a particle:

A. the kinetic energy
B. the total energy
C. the linear momentum
D. more than one of these is correct
E. none of the above

12. Two fixed navigation beacons mark the approach lane to a star. The beacons are in line with the star and are 49 million meters apart. A spaceship approaches the star with a relative velocity of $0.50c$ and passes the beacons. The passage of the ship between the beacons is timed by an observer on the ship. What is the time interval of the passage?

A. 240 ms
B. 280 ms
C. 360 ms
D. 140 ms
E. 90 ms

13. A pion is an unstable particle with a mean lifetime of 2.55×10^{-8} s. This is the time interval between its creation in a nuclear process and its extinction into decay products, as measured in a frame of reference at rest with respect to the pion. How far does it travel in its lifetime relative to Earth? (Use the value that an average pion travels at $0.230c$ relative to Earth).

A. 1.81 m
B. 1.09 m
C. 2.46 m
D. 3.83 m
E. 4.56 m

14. Astronaut Yuri is space-traveling from planet Y to planet Z at a speed of $0.60c$. When he is halfway between the planets, 1 light-hour from each as measured in the frame of the planets, nuclear devices are detonated. The explosions are simultaneous in the frame of the planets. What is the difference in the arrival time of the flashes from the explosions as observed by Yuri?

A. 45 min
B. 90 min
C. 125 min
D. 300 min
E. 385 min

15. From Earth, humans see the Enterprise approaching from the west at $0.800c$ and the Klingons approaching from the east at $0.900c$. What speed, in terms of the speed of light, does the crew of the Enterprise measure for the Klingon ship?

A. $0.721c$
B. $0.378c$
C. $0.547c$
D. $0.988c$
E. $0.659c$

16. Sofia is a passenger on a spaceship. After the speed of the spaceship has increased, she observes:

A. her watch moving slower
B. her watch moving faster
C. her watch got bigger
D. her watch got smaller
E. nothing unusual about the behavior of her watch

17. Two fixed navigation beacons mark the approach lane to a star. The beacons are in line with the star and are 40 million meters apart. A spaceship approaches the star with a relative velocity of $0.30c$ and passes the beacons. The passage of the ship between the beacons is timed by observers on the beacons. What is the time interval of the passage?

A. 610 ms
B. 560 ms
C. 444 ms
D. 240 ms
E. 180 ms

18. Calculate the kinetic energy of an electron moving at $0.737c$. (Use the rest energy of an electron = 511 keV)

A. 184 keV
B. 245 keV
C. 292 keV
D. 338 keV
E. 446 keV

19. In a common rest frame, event A happens on planet A, and event B occurs on planet B 100 years later. If the space-time interval between these two events is 90 light-years, how far apart are the 2 planets in their common rest frame?

A. 18 light-years
B. 22 light-years
C. 32 light-years
D. 44 light-years
E. 52 light-years

20. At what speed are lengths contracted to half and times dilated by a factor of 2?

A. $0.866c$
B. $0.986c$
C. $1.007c$
D. $1.134c$
E. $1.472c$

21. Suppose one twin takes a ride in a spaceship, traveling at a very high speed to a distant star and then back again while the other twin remains on Earth. Compared to the twin who remained on Earth, the twin who rode in the spaceship is:

A. younger than the Earth-twin
B. the same age as the Earth-twin
C. older than the Earth-twin
D. The ages cannot be determined from the information provided
E. Both younger and older than the Earth-twin

22. Two satellites with equal rest masses of 100 kg travel toward each other in deep space. One is traveling at $0.650c$ and the other at $0.850c$. The satellites collide and somehow manage to stick. What is the rest mass of the combined object after the collision?

A. 100 kg
B. 208 kg
C. 246 kg
D. 366 kg
E. 312 kg

23. What is the momentum of a proton when it is moving with a speed of $0.60c$? (Use the mass of a proton $m_{proton} = 1.67 \times 10^{-27}$ kg)

A. 1.4×10^{-19} kg·m/s
B. 2.6×10^{-19} kg·m/s
C. 5.3×10^{-19} kg·m/s
D. 3.8×10^{-19} kg·m/s
E. 8.8×10^{-19} kg·m/s

24. How many joules of energy are required to accelerate a 1.0 kg rock from rest to a speed of $0.866c$? (Use speed of light $c = 3.0 \times 10^8$ m/s)

A. 2.6×10^{16} J
B. 9.0×10^{16} J
C. 6.8×10^{17} J
D. 4.3×10^{16} J
E. 5.6×10^{16} J

25. A spaceship traveling at $0.50c$ away from Earth launches a secondary rocket in the forward direction at $0.50c$ relative to the spaceship. As measured from the frame of Earth, how fast is the secondary rocket moving away from Earth?

A. $0.92c$
B. $0.56c$
C. $0.80c$
D. $0.72c$
E. c

26. Observer A sees a pendulum oscillating back and forth in a relativistic rocket and measures its period to be T_A. Observer B moves along with the rocket and measures the period of the pendulum to be T_B. What is true about these two-time measurements?

A. $T_A > T_B$
B. $T_A = T_B$
C. $T_A < T_B$
D. T_A could be greater or smaller than T_A depending on the direction of the motion
E. T_A is greater than or equal to T_B

27. A spaceship travels to Alpha Centauri. The nearby star Alpha Centauri is 4.367 light-years from Earth. If the spaceship travels at one-half the speed of light for essentially all its expedition, how long was the ship gone, according to an observer on the spaceship?

A. 4.5 years
B. 7.6 years
C. 8.7 years
D. 11.3 years
E. 12.4 years

28. A particle physicist observes cosmic rays creating a new particle high in the atmosphere, and the speed of this particle is measured at 0.997c. It is unstable and is observed to decay in an average of 37.0 μs. If this particle were at rest in the laboratory, what would be its average lifetime?

A. 8.4 μs
B. 7.3 μs
C. 16.4 μs
D. 2.9 μs
E. 4.8 μs

29. If, instead of the correct formula for kinetic energy, someone carelessly uses the classical expression for kinetic energy for a particle moving at half the speed of light, by what magnitude percent will their calculation be in error? (Use the speed of light $c = 3.0 \times 10^8$ m/s)

A. 8%
B. 12%
C. 19%
D. 27%
E. 31%

30. In their typical rest frame, two stars are 90.0 light-years apart. If they appear to be 68.7 light-years apart from a spaceship, how fast is the spaceship moving?

A. 0.323c
B. 0.646c
C. 0.246c
D. 0.810c
E. 1.40c

31. Robert is riding in a spaceship with no windows, radios, or other means for him to observe or measure what is outside. How can he determine if the ship is stopped or moving at constant velocity?

A. He determines if the ship is moving by measuring the apparent velocity of light in the spaceship
B. He can determine if the ship is moving by checking his precision timepiece. If it is running slow, the ship is moving
C. He can determine if the ship is moving by lying down and measuring his height. If he is shorter than usual, the ship is moving
D. He cannot determine if the ship is stopped or moving at a constant velocity
E. He can determine if the ship is moving by lying down facing one direction and measuring his height and then turning to face a different direction and again measuring his height. If the two measurements are different, then he is moving

32. Two satellites with equal rest masses are traveling toward each other in deep space. One is traveling at 0.550*c* and the other at 0.750*c*. What is the speed of the combined object after the collision if the satellites collide and stick?

A. 0.108*c*
B. 0.124*c*
C. 0.245*c*
D. 0.175*c*
E. 0.328*c*

33. A spaceship approaches Earth with a speed of 0.50*c*. A passenger in the spaceship measures his heartbeat as 70 beats per minute. What is his heartbeat rate according to an observer that is at rest relative to Earth?

A. 61 beats per minute
B. 55 beats per minute
C. 68 beats per minute
D. 76 beats per minute
E. 82 beats per minute

34. A particle in a 453-m long particle accelerator is moving at 0.875*c*. How long does the particle accelerator appear to the particle?

A. 148 m
B. 219 m
C. 264 m
D. 462 m
E. 517 m

35. Someone in Earth's rest frame says that a spaceship's trip between two planets took 10.0 years, while an astronaut on the spaceship says that the trip took 5.78 years. What is the speed of the spaceship, assuming the planets are essentially at rest?

A. 0.244*c*
B. 0.515*c*
C. 0.816*c*
D. 1.17*c*
E. 2.37*c*

36. A super-speed train moves in a direction parallel to its length with a speed that approaches the speed of light. The length of the train, as measured by a stationary observer on the ground:

A. approaches infinity
B. increases due to the motion
C. decreases due to the motion
D. is not affected by the motion
E. cannot be measured because the train is moving

37. A spaceship enters the solar system moving toward the sun at a constant speed relative to the sun. By its clock, the time elapsed between the time it crosses the orbit of Jupiter and the time it crosses the orbit of Mars is 50.0 minutes. As measured in the coordinate system of the sun, how many minutes did it take for the spaceship to travel that distance? (Use the radius of the Jupiter orbit = 778×10^9 m and the orbit of Mars = 228×10^9 m)

A. 28.4 minutes
B. 44.8 minutes
C. 58.6 minutes
D. 66.2 minutes
E. 88.8 minutes

38. Two fixed navigation beacons mark the approach lane to a star. The beacons are in line with the star and are 65×10^6 meters apart. A spaceship approaches the star with a relative velocity of $0.90c$ and passes the beacons. As the ship passes the first beacon, the ship emits a short radar pulse toward the second beacon, and the radar echo is received at the ship. What is the time interval between the emission of the radar pulse and the reception of the radar echo as measured on the spaceship?

A. 210 ms
B. 420 ms
C. 130 ms
D. 100 ms
E. 60 ms

39. An electron that was initially at rest reaches a speed of $0.648c$. Through what potential difference would the electron need to be accelerated to reach this speed? (Use the rest energy of an electron = 511 keV)

A. 120 keV
B. 160 keV
C. 190 keV
D. 230 keV
E. 280 keV

40. A certain unstable particle has a mean lifetime of 1.52×10^{-6} s. This is the time interval between its creation in a nuclear process and its extinction into decay products, as measured in a frame of reference at rest with respect to the particle. A scientist on Earth observes an average such particle to travel 342 m in its lifetime. What is the speed of the particle relative to Earth? (Use the speed of light $c = 3.00 \times 10^8$ m/s)

A. $0.844c$
B. $0.486c$
C. $0.600c$
D. $0.754c$
E. $0.324c$

41. The relativistic kinetic energy formula is valid:

I. for speeds near the speed of light
II. for subatomic particles, such as electrons and protons
III. at all speeds

A. I only
B. II only
C. III only
D. I and II only
E. I and III only

42. A spaceship with a constant velocity of $0.800c$ relative to Earth travels to the star 4.30 light-years from Earth. Measured by a passenger on the ship, what distance does the spaceship travel?

A. 1.32 light-years
B. 2.58 light-years
C. 3.82 light-years
D. 4.60 light-years
E. 6.44 light-years

43. A spaceship moves away from an asteroid with a speed of $0.80c$ relative to the asteroid. The spaceship then fires a missile with a speed of $0.50c$ relative to the spaceship. What is the speed of the missile measured by astronauts on the asteroid if the missile is fired away from the asteroid?

A. $0.43c$
B. $0.93c$
C. $0.18c$
D. $0.37c$
E. $0.72c$

44. What is the total energy of an electron moving with a speed of $0.950c$? (Use the speed of light $c = 3.0 \times 10^8$ m/s and $m_{el} = 9.11 \times 10^{-31}$ kg)

A. 2.6×10^{-13} J
B. 7.4×10^{-14} J
C. 1.8×10^{-13} J
D. 3.6×10^{-14} J
E. 6.6×10^{-13} J

45. A spaceship carrying a light clock moves at a speed of $0.960c$ relative to an observer on Earth. If the clock on the ship advances by 1.00 s as measured by the space travelers aboard the ship, how long did that advance take as measured by the observer on Earth?

A. 0.84 s
B. 1.9 s
C. 2.4 s
D. 3.6 s
E. 4.8 s

46. In an "atom smasher," two particles collide head-on at relativistic speeds. The velocity of the first particle is $0.741c$ to the left, and the velocity of the second particle is $0.543c$ to the right (both speeds are measured in Earth's rest frame). How fast are the particles moving with respect to each other?

A. $0.916c$
B. $1.164c$
C. $0.744c$
D. $0.836c$
E. $0.448c$

47. A particle is moving at approximately $0.86c$. When expressed as a percentage error for the difference between the erroneous and correct expressions relative to the correct one, what percentage is the Newtonian expression for momentum in error?

A. 49%
B. 38%
C. 57%
D. 64%
E. 22%

48. A spaceship, traveling at $0.100c$ away from a stationary space platform, launches a secondary rocket towards the station, with a speed of $0.560c$ relative to the spaceship. What is the speed of the secondary rocket relative to the space platform?

A. $0.364c$
B. $0.487c$
C. $0.524c$
D. $0.588c$
E. $0.682c$

49. An electron slowing from $0.998c$ to $0.500c$ loses how much momentum?

A. 2.17 MeV/c
B. 4.36 MeV/c
C. 5.28 MeV/c
D. 7.77 MeV/c
E. 12.44 MeV/c

50. A star is moving toward Earth at $0.90c$. The star emits light, which moves away from the star at the speed of light. What do Earth-based astronomers measure for the speed of this light?

A. $0.88c$
B. $1.1c$
C. $1.8c$
D. c
E. $0.90c$

Diagnostic Tests

Diagnostic Test 1

This Diagnostic Test is designed to assess your proficiency on each topic and NOT to mimic the test. Use your test results and identify areas of strength and weakness to adjust your study plan and enhance your fundamental knowledge. The length of the Diagnostic Tests is optimal for a single study session.

#	Answer:					Review	#	Answer:					Review
1:	A	B	C	D	E	___	**26:**	A	B	C	D	E	___
2:	A	B	C	D	E	___	**27:**	A	B	C	D	E	___
3:	A	B	C	D	E	___	**28:**	A	B	C	D	E	___
4:	A	B	C	D	E	___	**29:**	A	B	C	D	E	___
5:	A	B	C	D	E	___	**30:**	A	B	C	D	E	___
6:	A	B	C	D	E	___	**31:**	A	B	C	D	E	___
7:	A	B	C	D	E	___	**32:**	A	B	C	D	E	___
8:	A	B	C	D	E	___	**33:**	A	B	C	D	E	___
9:	A	B	C	D	E	___	**34:**	A	B	C	D	E	___
10:	A	B	C	D	E	___	**35:**	A	B	C	D	E	___
11:	A	B	C	D	E	___	**36:**	A	B	C	D	E	___
12:	A	B	C	D	E	___	**37:**	A	B	C	D	E	___
13:	A	B	C	D	E	___	**38:**	A	B	C	D	E	___
14:	A	B	C	D	E	___	**39:**	A	B	C	D	E	___
15:	A	B	C	D	E	___	**40:**	A	B	C	D	E	___
16:	A	B	C	D	E	___	**41:**	A	B	C	D	E	___
17:	A	B	C	D	E	___	**42:**	A	B	C	D	E	___
18:	A	B	C	D	E	___	**43:**	A	B	C	D	E	___
19:	A	B	C	D	E	___	**44:**	A	B	C	D	E	___
20:	A	B	C	D	E	___	**45:**	A	B	C	D	E	___
21:	A	B	C	D	E	___	**46:**	A	B	C	D	E	___
22:	A	B	C	D	E	___	**47:**	A	B	C	D	E	___
23:	A	B	C	D	E	___	**48:**	A	B	C	D	E	___
24:	A	B	C	D	E	___	**49:**	A	B	C	D	E	___
25:	A	B	C	D	E	___	**50:**	A	B	C	D	E	___

#	Answer:					Review	#	Answer:					Review
51:	A	B	C	D	E	___	**76:**	A	B	C	D	E	___
52:	A	B	C	D	E	___	**77:**	A	B	C	D	E	___
53:	A	B	C	D	E	___	**78:**	A	B	C	D	E	___
54:	A	B	C	D	E	___	**79:**	A	B	C	D	E	___
55:	A	B	C	D	E	___	**80:**	A	B	C	D	E	___
56:	A	B	C	D	E	___	**81:**	A	B	C	D	E	___
57:	A	B	C	D	E	___	**82:**	A	B	C	D	E	___
58:	A	B	C	D	E	___	**83:**	A	B	C	D	E	___
59:	A	B	C	D	E	___	**84:**	A	B	C	D	E	___
60:	A	B	C	D	E	___	**85:**	A	B	C	D	E	___
61:	A	B	C	D	E	___	**86:**	A	B	C	D	E	___
62:	A	B	C	D	E	___	**87:**	A	B	C	D	E	___
63:	A	B	C	D	E	___	**88:**	A	B	C	D	E	___
64:	A	B	C	D	E	___	**89:**	A	B	C	D	E	___
65:	A	B	C	D	E	___	**90:**	A	B	C	D	E	___
66:	A	B	C	D	E	___	**91:**	A	B	C	D	E	___
67:	A	B	C	D	E	___	**92:**	A	B	C	D	E	___
68:	A	B	C	D	E	___	**93:**	A	B	C	D	E	___
69:	A	B	C	D	E	___	**94:**	A	B	C	D	E	___
70:	A	B	C	D	E	___	**95:**	A	B	C	D	E	___
71:	A	B	C	D	E	___	**96:**	A	B	C	D	E	___
72:	A	B	C	D	E	___	**97:**	A	B	C	D	E	___
73:	A	B	C	D	E	___	**98:**	A	B	C	D	E	___
74:	A	B	C	D	E	___	**99:**	A	B	C	D	E	___
75:	A	B	C	D	E	___	**100:**	A	B	C	D	E	___

1. When an object moves with constant acceleration, can its velocity change direction?

A. Yes, a car that starts from rest, speeds up, slows to a stop, and then backs up is an example
B. No, because it is always slowing down
C. No, because it is always speeding up
D. Yes, a Frisbee thrown straight up is an example
E. No, because a constant acceleration means the speed, and therefore the velocity, is also constant

2. A car of mass m is driving up a road with a slight incline θ above the horizontal. The driver sees a road closure and skids to a stop. The coefficient of static friction between the tires and the road is μ_s, and the coefficient of kinetic friction is μ_k. What is the magnitude of the gravity component of the force parallel to the surface of the road?

A. $mg \sin \theta$
B. $mg \tan \theta$
C. mg
D. $mg \cos \theta$
E. $m / (g \sin \theta)$

3. A change in which of the following will affect the buoyant force experienced by an object that is completely submerged in an incompressible liquid?

I. Density of the liquid
II. Density of the object
III. Depth of the object

A. I only
B. III only
C. I and III only
D. I and II only
E. I, II and III

4. What is the power output necessary for a 54 kg person to run at a constant velocity up a 10 m hillside in 4 s if the hillside is inclined at 30° above the horizontal? (Use the acceleration due to gravity $g = 9.8$ m/s^2 and the conversion of 1 hp = 745 W)

A. 1.92 hp
B. 1.12 hp
C. 0.89 hp
D. 3.94 hp
E. 2.46 hp

5. Some of a wave's energy dissipates as heat. In time, this reduces the wave's:

A. amplitude
B. frequency
C. speed
D. wavelength
E. period

6. A piano tuned with the third harmonic frequency of the C_3 string is 783 Hz. What is the frequency of the C_3 fundamental?

A. 473 Hz
B. 261 Hz
C. 387 Hz
D. 185 Hz
E. 127 Hz

7. What is the density of an object if it weighs 7.86 N when it is in air and 6.92 N when it is immersed in water? (Use the acceleration due to gravity $g = 9.8$ m/s^2 and the density of water $\rho = 1{,}000$ kg/m^3)

A. 6,042 kg/m^3
B. 7,286 kg/m^3
C. 8,333 kg/m^3
D. 9,240 kg/m^3
E. 11,868 kg/m^3

8. A water fountain pump recirculates water from a pool and pumps it up to a trough, where it flows along the trough and passes through a hole in the bottom of it. As the water falls back into the pool, it turns a water wheel. What aspect of this water fountain is analogous to an electric potential within an electric circuit?

A. Height of water
B. Volume flow rate
C. Flow velocity
D. Mass of water
E. Trough

9. If a sheet of copper is quickly passed through a strong permanent magnet with the plane of the sheet perpendicular to the magnetic field, which statement is true?

A. There is no movement because there is no magnetic force
B. The force experienced by the sheet of copper is due mainly to lead impurities in the copper since copper is not magnetic
C. There is a magnetic force opposing the motion of the sheet
D. There is a magnetic force assisting the motion of the sheet
E. B and C are true

10. A light ray in the water passes into the air, where the angle of incidence in the water is 42°. What is the angle of refraction in the air? (Use the index of refraction of air n = 1 and index of refraction of water = 1.33)

A. 32°
B. 18°
C. 46°
D. 74°
E. 63°

11. ^{3}Hydrogen can be used as a chemical tracer. What is the half-life of the radionuclide if 3,200 μg decays to 800 μg after 24.6 years?

A. 6.1 years
B. 12.3 years
C. 24.6 years
D. 49.2 years
E. 98.4 years

12. An object with a mass of 60 kg moves across a level surface with a constant speed of 13.5 m/s. If there is a frictional force, and the coefficient of kinetic friction is 0.8, which must be true about the forces acting on the object?

A. There must be an unbalanced amount of vertical force acting on the object, allowing it to move
B. No forces are doing work on the object
C. There must be some other horizontal force acting on the object
D. The force exerted on the object by kinetic friction is negligible
E. The total force exerted on the object is not balanced

13. The total mechanical energy of a system is:

A. all kinetic energy or all potential energy, at any one instant
B. constant if there are only conservative forces acting
C. found through the product of potential energy and kinetic energy
D. equally divided between kinetic energy and potential energy in every instance
E. never a constant value; it changes depending on the conditions

14. A 340 nm thick oil film floats on the surface of the water. The surface of the oil is illuminated from above at normal incidence with white light. What are the two wavelengths of light in the 400 nm to 800 nm wavelength band, which are most strongly reflected? (Use index of refraction for oil n = 1.5 and index of refraction for water n = 1.33)

A. 420 nm and 750 nm
B. 406 nm and 706 nm
C. 410 nm and 760 nm
D. 408 nm and 680 nm
E. 484 nm and 792 nm

15. A 1 m string is fixed at both ends and plucked. What is the wavelength corresponding to the fourth harmonic if the speed of the waves on this string is 4.2×10^4 m/s?

A. 1 m
B. 0.5 m
C. 4/3 m
D. 3/2 m
E. 2 m

16. Which statement is accurate?

A. Tensile stress is measured in N·m
B. Stress has a meaning similar to work
C. Strain has a meaning similar to force
D. The elastic modulus is the stress/strain ratio
E. Tensile strain is measured in Joules

17. A distance of 15 cm separates two point charges of +18 μC and -6 μC. What is the electric field E midway between the two charges? (Coulomb's constant $k = 9 \times 10^9$ N·m^2·C^{-2})

A. 28.8×10^6 N/C toward the positive charge
B. 28.8×10^6 N/C toward the negative charge
C. 38.4×10^6 N/C toward the positive charge
D. 38.4×10^6 N/C toward the negative charge
E. 38.4×10^4 N/C toward the negative charge

18. When the current through a resistor is increased by a factor of 4, by what factor does the power dissipated by the resistor change?

A. Increases by 16
B. Increases by 4
C. Decreases by 4
D. Decreases by 16
E. Increases by 2

19. The image of an object placed outside the focal point of a concave mirror is:

A. virtual and inverted
B. virtual and upright
C. real and inverted
D. real and upright
E. real, but the object can be inverted or upright

20. A futuristic design for a car is to have a large disk-like flywheel within the car storing kinetic energy. The flywheel has a mass of 370.0 kg with a radius of 0.500 m and can rotate up to 200.0 rev/s. Assuming this stored kinetic energy could be transferred to the linear velocity of the 1,500.0-kg car, what is the maximum attainable speed of the car?

A. 29.6 m/s
B. 88.4 m/s
C. 162 m/s
D. 221 m/s
E. 318 m/s

21. A 500 g empty iron pot is put on a stove. How much heat must it absorb to raise it from 20 °C to 70 °C? (Use the specific heat c of iron = 92 cal/kg·°C and 1 cal = 4.186 J)

A. 8,110 J
B. 9,628 J
C. 20,100 J
D. 12,500 J
E. 14,400 J

22. In a binary star system, two stars revolve about their combined center of mass and are attracted to each other by the force of gravity. The force of gravity between the stars (masses M_1 and M_2) is F. If the mass of one of the stars is decreased by a factor of 2, how would this affect the force between them?

A. Remains the same
B. Increases by a factor of 2
C. Decreases by a factor of 4
D. Increases by a factor of 4
E. Decreases by a factor of 2

23. In motion pictures, when a character falls off a cliff, he screams. If portrayed correctly, from the vantage point of an observer at the top of the cliff, the pitch of the scream the observer hears is:

A. higher than the actual pitch and increasing as he falls
B. lower than the actual pitch and decreasing as he falls
C. higher than the actual pitch and constant
D. lower than the actual pitch and increasing as he falls
E. higher than the actual pitch and decreasing as he falls

24. As a cubical block of marble is lowered at a steady rate into the ocean by a crane, the top and bottom faces are kept horizontal. Which graph depicts the total pressure (P) on the bottom of the block as a function of time (t) as the block just enters the water at $t = 0$ s?

A.
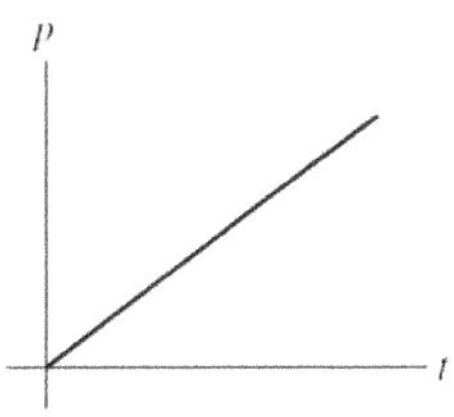

C.
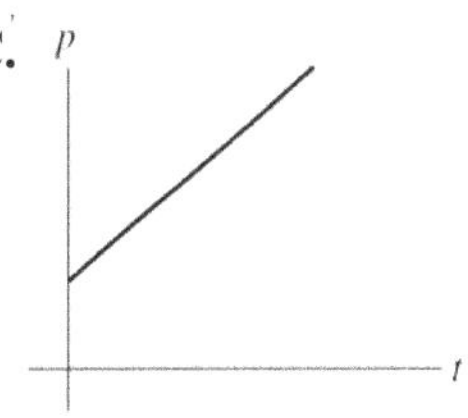

B.
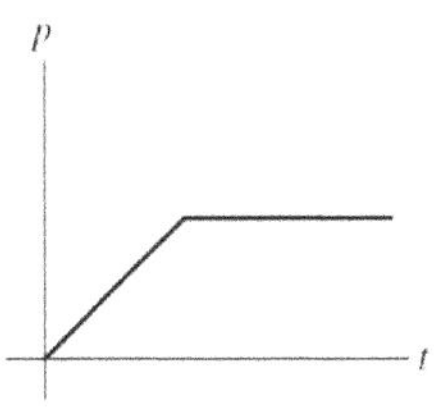

D.
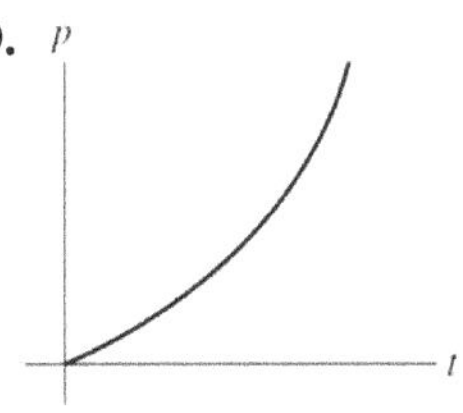

E.
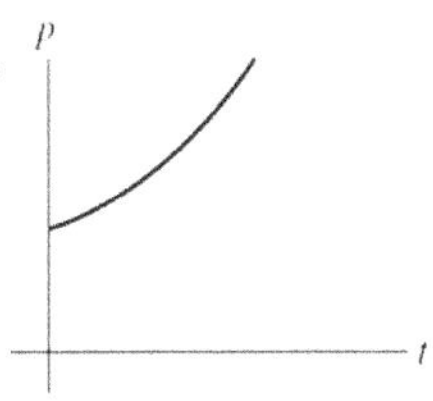

25. Two parallel metal plates separated by 0.03 m are charged to create a uniform electric field (4×10^4 N/C) between them, which points down. How does the force exerted on an α particle between the plates compare with the force exerted on a proton between the plates? (Use the acceleration due to gravity $g = 10$ m/s^2 and the charge on a proton = 1.6×10^{-19} C)

A. Twice as large and in the same direction
B. Four times as large and in the same direction
C. The same magnitude, but in the opposite direction
D. The same magnitude and in the same direction
E. Twice as large, but in the opposite direction

26. How much current flows through a 57 m length of copper wire with a radius of 5.7 mm if it is connected to a source supplying 70 V? (Use the resistivity of copper = 1.68×10^{-8} Ω·m)

A. 180 nA
B. 3,600 A
C. 7,447 A
D. 3.7×10^8 A
E. 8,200 A

27. Which statement about images is correct?

A. A real image is always upright
B. A virtual image cannot be photographed
C. A virtual image cannot be seen by the unaided eye
D. A virtual image cannot be formed on a screen
E. Mirrors always produce real images because they reflect light

28. When a β^+ particle is emitted from an unstable nucleus, the atomic number of the nucleus:

A. decreases by 2
B. increases by 2
C. decreases by 1
D. increases by 1
E. does not change

29. Which answer is correct when the first law of thermodynamics, $Q = \Delta U - \mathrm{W}$, is applied to an ideal gas taken through an isothermal process?

A. $Q = 0$
B. $\Delta \mathrm{P} = 0$
C. $\Delta U = 0$
D. $\mathrm{W} = 0$
E. None are correct

30. Consider three galaxies, Alpha, Beta, and Gamma. An observer on Gamma sees the other two galaxies, each moving away from him in opposite directions at speed $0.7c$. At what speed would an observer in Alpha see the galaxy Gamma moving?

A. $0.64c$
B. $0.70c$
C. $0.88c$
D. $0.28c$
E. $0.53c$

31. For projectile motion with no air resistance, the vertical component of a projectile's acceleration:

A. is always positive
B. remains a nonzero constant
C. continuously increases
D. continuously decreases
E. first increases and then decreases

32. A motor is connected to a power supply that supplies 5 A of current with a 25 V potential difference. The motor has a 30% efficiency rating and is used to lift a 50 kg box. How far does the motor lift the box in 60 s? (Use the acceleration due to gravity $g = 10 \text{ m/s}^2$)

A. 4.5 m
B. 7.4 m
C. 1.8 m
D. 18 m
E. 3.2 m

33. If the height that a pendulum reaches is doubled, what happens to its velocity as it passes its equilibrium position?

A. Remains the same
B. Increases by a factor of 4
C. Increases by a factor of 2
D. Increases by a factor of $\sqrt{2}$
E. Decreases by a factor of 2

34. Two metal rods are made of the same material and have the same cross-sectional area. The two rods differ only in their lengths of L and $3L$. The two rods are heated from the same initial to the same final temperature. The short rod expands its length by an amount of ΔL. What is the amount by which the length of the long rod increases?

A. $\sqrt{3}\Delta L$
B. $3\Delta L$
C. ΔL
D. $3/2\Delta L$
E. $9\Delta L$

35. If n_1 is the index of refraction for the incident medium, and n_2 is the index of the refracting medium, what conditions are necessary for the critical angle to exist?

A. $n_1 = n_2$
B. $n_1 < n_2$
C. $n_1 < 2n_2$
D. $n_1 < ½n_2$
E. $n_1 > n_2$

36. Which graph is representative of semiconductor materials? R = resistivity, T = temperature.

A. 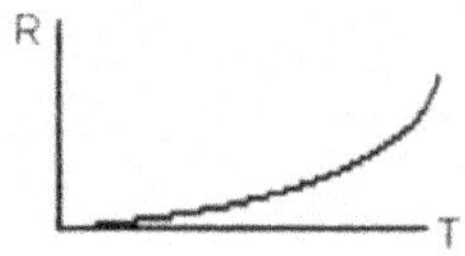

B. 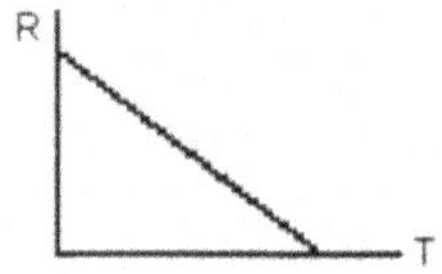

C. 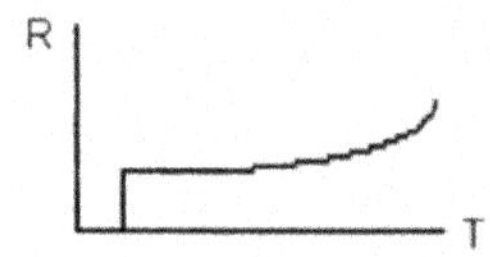

D. 

E. None of the graphs

37. Three capacitors C_1, C_2 and C_3 have equal capacitance and are connected to a battery, as shown. Which capacitor stores the most potential energy?

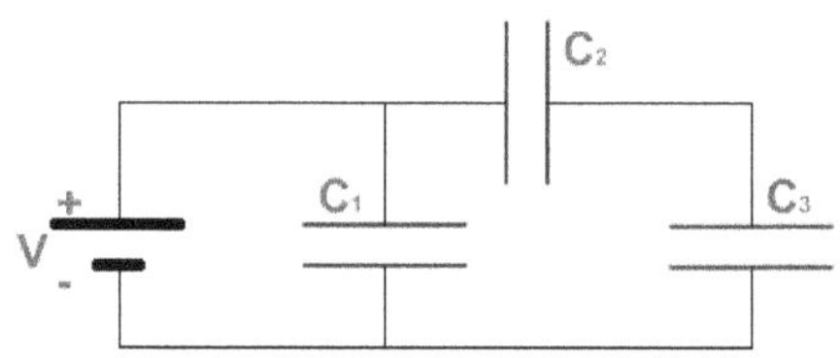

A. C_3
B. C_1
C. C_2
D. C_2 or C_3
E. All three capacitors store the same amount of energy

38. A spherical inflated balloon is submerged in a swimming pool. How is the buoyant force affected if the balloon is inflated to double its radius?

A. 6 times larger
B. 2 times larger
C. 4 times larger
D. 2 times smaller
E. 8 times larger

39. An atom of He is twice as massive as a molecule of H. How much faster is the speed of sound in H compared to the speed of sound in He?

A. 4 times faster
B. 8 times faster
C. 1.41 times faster
D. 2 times faster
E. $1/\sqrt{2}$ times faster

40. If the energy that starts a vibration increases, this increases the:

A. number of cycles/sec
B. wavelength
C. frequency
D. amplitude
E. period

41. The potential energy of a box on a shelf, relative to the floor, is a measure of the:

I. work done putting the box on the shelf from the floor
II. energy the box has because of its position above the floor
III. weight of the box × the distance above the floor

A. I only
B. II only
C. III only
D. I and III only
E. I, II and III

42. Block m has a mass of 3.3 kg. It moves on a frictionless surface with a speed (v_i) of 8.5 m/s when it makes a perfectly elastic collision with a stationary block M, which has a mass of 6.5 kg. After the collision, block m recoils with a speed (v_f) of 3 m/s. What is the speed of block M after the collision?

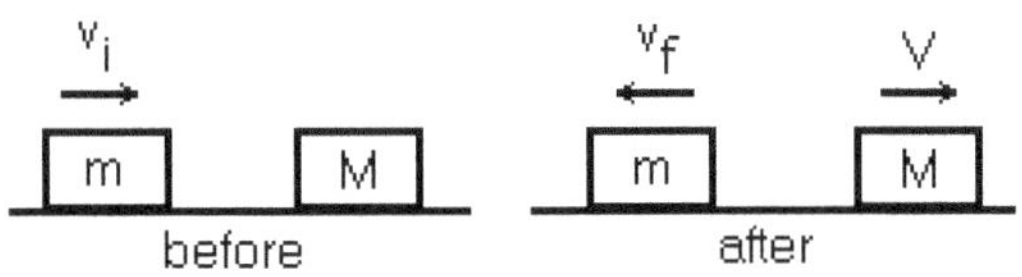

A. 5.8 m/s
B. 7.2 m/s
C. 3.6 m/s
D. 9.4 m/s
E. 12 m/s

43. A massless string connects a 10 kg block to a 70 kg mass, which hangs over the edge of the table. Ignoring friction, what is the acceleration of the 10 kg block when the other block is released? (Use the acceleration due to gravity $g = 10\ m/s^2$)

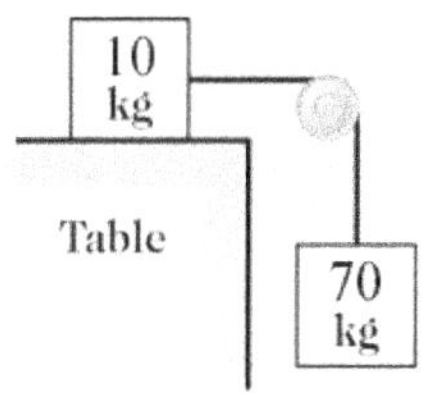

A. 5.5 m/s^2
B. 10.5 m/s^2
C. 8.8 m/s^2
D. 7.3 m/s^2
E. 16.3 m/s^2

44. An object with an initial velocity of 4 m/s moves along the x-axis with constant acceleration. How far did it travel in 2 s, if its final velocity is 16 m/s?

A. 9 m
B. 13 m
C. 32 m
D. 26 m
E. 20 m

45. When grinding a culinary knife, the 70 g of metal becomes heated to 450 °C. What is the minimum amount of 25 °C water needed if it remains liquid and does not rise above 100 °C when the hot knife is cooled in it? (Use specific heat of the knife = 0.11 cal/g·°C and the specific heat of water = 1 cal/g·°C)

A. 36 g
B. 28 g
C. 47 g
D. 18 g
E. 54 g

46. A horizontal spring-mass system oscillates on a frictionless table. If the ratio of the mass to the spring constant is 0.031 kg·m/N, and the maximum speed of the mass is 30 m/s, what is the maximum extension of the spring?

A. 0.67 m
B. 460 cm
C. 5.3 m
D. 2.6 cm
E. 4.2 m

47. What is the shortest wavelength of a photon emitted by a hydrogen atom, for which the initial state is n = 3?

A. 102.6 nm
B. 97.3 nm
C. 820.0 nm
D. 121.6 nm
E. 91.2 nm

48. The H nucleus, which has a charge of e^+, is situated to the left of a C nucleus, which has a charge of 6 e^+. Which is true regarding the direction and magnitude of the electrical force experienced by the H nucleus?

A. To the right and equal to the force exerted on the C nucleus.
B. To the left and lower than the force exerted on the C nucleus.
C. To the right and higher than the force exerted on the C nucleus.
D. To the left and equal to the force exerted on the C nucleus.
E. To the right and lower than the force exerted on the C nucleus.

49. An object is viewed at various distances using a concave mirror with a focal length of 10 m. Where is the image relative to the mirror when the object is 20 m away from the mirror?

A. 20 m behind
B. 20 m in front
C. 10 m behind
D. 10 m in front
E. 5 m behind

50. The electric field at point *P* due to a point charge *Q* a distance *R* away has magnitude *E*. What change would double the magnitude of the field at *P*?

A. Reduce the distance to $R/2$
B. Reduce the distance to $R/4$
C. Double the charge to $2Q$
D. Double the distance to $2R$
E. Double the charge to $2Q$ and at the same time reduce the distance to $R/2$

51. A 9 g bullet is shot into a stationary 4 kg block lying on a frictionless horizontal surface. The bullet remains lodged in the block and, together, they push into a spring and compress it by 3.4 cm. What is the initial velocity of the bullet? (Use the spring constant $k = 2{,}400$ N/m)

A. 664 m/s
B. 383 m/s
C. 370 m/s
D. 588 m/s
E. 274 m/s

52. What factor does the time required to complete one full cycle increase for a simple pendulum when the length is tripled?

A. 1/3
B. 3
C. 1/9
D. 9
E. $\sqrt{3}$

53. A water tank is 30 m above ground and is filled to a depth of 15 m. What is the gauge pressure at ground level in a 3 cm diameter hose? (Use the acceleration due to gravity $g = 9.8$ m/s^2 and the density of water ρ = 1,000 kg/m^3)

A. 7.2×10^3 N/m^2
B. 6.1×10^1 N/m^2
C. 1.4×10^4 N/m^2
D. 4.4×10^5 N/m^2
E. Requires the cross-sectional area of the tank

54. Consider two copper wires of equal length. How do the resistances of these two wires compare if one wire has twice the cross-sectional area?

A. The thicker wire has one-half the resistance of the thinner wire
B. The thicker wire has four times the resistance of the thinner wire
C. The thicker wire has twice the resistance of the thinner wire
D. The thicker wire has eight times the resistance of the thinner wire
E. The thicker wire has one-fourth the resistance of the thinner wire

55. For a given value of the principal quantum number n, what are the allowable orbital angular momentum quantum numbers?

A. $\ell = 1, 2, 3, \ldots, n$
B. $\ell = 0, 1, 2, \ldots, (n-1)$
C. $\ell = 0, 1, 2, \ldots, n$
D. $\ell = 1, 2, 3, \ldots, (n-1)$
E. $\ell = 1, 2, 3, \ldots, (n+1)$

56. It is necessary to determine the specific heat of a 185 g object. It is determined experimentally that it takes 14 J to raise the temperature 10° C. What is the specific heat of the object?

A. 343 J/kg·K
B. 1.6 J/kg·K
C. 25.9 J/kg·K
D. 7.6 J/kg·K
E. 13.4 J/kg·K

57. A ball is projected upward at time $t = 0$ s from a point on a roof 50 m above the ground. The ball rises, then falls and strikes the ground. The initial velocity of the ball is 24 m/s. Consider all quantities as positive in the upward direction. At time $t = 3.8$ s, the acceleration of the ball is:

A. 10 m/s^2
B. –5 m/s^2
C. zero
D. 24 m/s^2
E. –10 m/s^2

58. Two astronauts conducted an experiment where a 3,500 kg spacecraft was connected with an orbiting rocket. The rocket thrusters were fired to provide 900 N for 8 s. What was the mass of the rocket if the change in velocity of the spacecraft and rocket was 0.9 m/s?

A. 3,100 kg
B. 4,500 kg
C. 10,820 kg
D. 2,430 kg
E. 7,304 kg

59. A 0.3 kg ball hits a wall and rebounds. Initially, the ball is going 7 m/s, but after the rebound, it is going 5 m/s. What is the impulse imparted to the ball by the wall?

A. 0.6 kg·m/s
B. 1.8 kg·m/s
C. 2.1 kg·m/s
D. 3.6 kg·m/s
E. 0.9 kg·m/s

60. The Bohr model of the hydrogen atom was not able to explain:

A. the wavelengths of the emission lines in the infrared range
B. the wavelengths of the emission lines in the ultraviolet range
C. the observation that the atom does not lose energy to radiation as the electron orbits
D. the Bohr model explained the exact characteristics of a hydrogen atom
E. why some emission lines were brighter than other emission lines

61. An object is at height h above the surface of the Earth, where h is much smaller than the radius of the Earth. It takes t seconds to fall to the ground. Ignoring air resistance, at what height would this object need to be released to take $2t$ s to fall?

A. $4h$
B. $4gh$
C. $2h$
D. $2gh$
E. $2g$

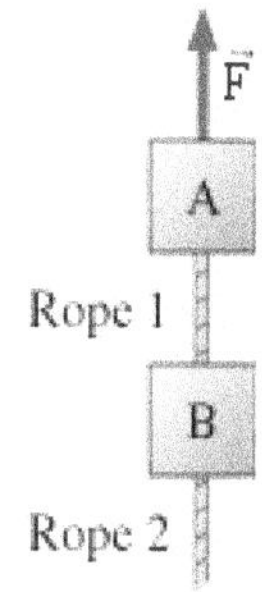

62. Two 1 kg blocks are connected by rope 1. Rope 2 hangs beneath block B. Each rope has a mass of 350 g. The entire assembly is accelerated upward at 5.5 m/s^2 by force F. What is the tension at the bottom end of rope 1? (Use the acceleration due to gravity $g = 9.8$ m/s^2)

A. 7.9 N
B. 31.5 N
C. 23 N
D. 20.6 N
E. 13 N

63. A fluid in an insulated, flexible bottle is heated by a high-resistance wire and expands. If 9 kJ of heat is applied to the system and the system does 5 kJ of work, how much does the internal energy change?

A. –4 kJ
B. 32 kJ
C. 4 kJ
D. 12 kJ
E. 52 kJ

64. One end of a spring, with a spring constant 50 N/m, is fixed at point A, while the other end is connected to a 5 kg mass. The fixed end and the mass sit on a horizontal frictionless surface so that the mass and the spring rotate about point A. The mass moves in a circle with $r = 4$ m, and the force on the mass is 20 N. How long does it take for the mass to make one complete revolution around point A?

A. 5.7 s
B. 6.3 s
C. 4.4 s
D. 3.2 s
E. 8.6 s

65. All of the following are true statements, EXCEPT:

A. Waves transport energy and matter from one region to another
B. The speed of a wave and the speed of the vibrating particles of the wave are not the same entities
C. A wave that is reflected at the same frequency as it is produced is a standing wave
D. For a transverse wave, the motion of the particles is perpendicular to the velocity vector of the wave
E. A wave in which particles move back and forth in the same direction that the wave is moving is a longitudinal wave

66. What are the wavelengths of the three lowest tones produced by an open pipe of length L?

A. $2L, L, 2L/3$
B. $4L, 4L/3, 4L/5$
C. $4L, 2L, L$
D. $2L, L, L/2$
E. $2L, L/2, L/8$

67. A 4.2 m steel wire has a diameter of 1.8 mm. The wire stretches 1.8 mm when it bears a load. What is the mass of the load? (Use Young's modulus for steel = 2×10^{11} N/m^2 and the acceleration due to gravity $g = 9.8$ m/s^2)

A. 22 kg
B. 26 kg
C. 30 kg
D. 16 kg
E. 10 kg

68. A circular loop of wire is positioned in a region of a changing magnetic field; the direction of the field remains constant, but the magnitude fluctuates. What must the loop's area vector orientation be in relation to the magnetic field direction to create the maximum induced emf?

A. An angle of 45° to the magnetic field
B. An angle of −45° to the magnetic field
C. An angle of 90° to the magnetic field
D. Parallel to the magnetic field
E. Perpendicular to the magnetic field

69. A charged, parallel-plate capacitor has an electric field E_0 between its plates. The bare nuclei of ^{1}H and ^{3}H, both at rest, are placed between the plates. Ignoring the force of gravity, how does the force, F_1, of the light ^{1}H nucleus compare with the force, F_3, of the heavy ^{3}H nucleus?

A. $F_3 = 3F_1$
B. $F_3 = \sqrt{2}F_1$
C. $F_3 = F_1$
D. $F_3 = (1/3)F_1$
E. $F_3 = ½F_1$

70. Which statement about thin, single lenses is correct?

A. A diverging lens can sometimes produce a virtual, erect image
B. A diverging lens always produces a virtual, erect image
C. A converging lens always produces a real, inverted image
D. A diverging lens always produces a real, inverted image
E. A diverging lens produces a virtual, erect image if the object is located within the focal point

71. During β^+ decay:

A. a proton is transformed to a neutron
B. an electron is released from its orbit
C. a neutron is transformed to a positron
D. a neutron is transformed to a proton
E. the number of nucleons decreases

72. Consider objects that slide friction-free along a horizontal air track. Glider A, which has a mass of 2 kg and a speed of 2 m/s, collides with glider B of 5 kg at rest. If they stick, what is their speed after the collision?

A. 0.25 m/s
B. 0.33 m/s
C. 1.2 m/s
D. 1 m/s
E. 0.6 m/s

73. While pulling a box from the ground up to his tree house with a rope, a child does 400 J of work. What is the mass of the box if the tree house is 4 m above the ground? (Use the acceleration due to gravity $g = 9.8 \text{ m/s}^2$)

A. 13.2 kg
B. 6.6 kg
C. 5.2 kg
D. 10.2 kg
E. 8.6 kg

74. An electrical motor spins at a constant of 2,695.0 rpm. If the rotor radius is 7.165 cm, what is the linear acceleration of the edge of the rotor?

A. 707.0 m/s^2
B. 1,280 m/s^2
C. 3,272 m/s^2
D. 4,028 m/s^2
E. 5,707 m/s^2

75. Some wavefronts are emitted by a source S. This diagram illustrates why:

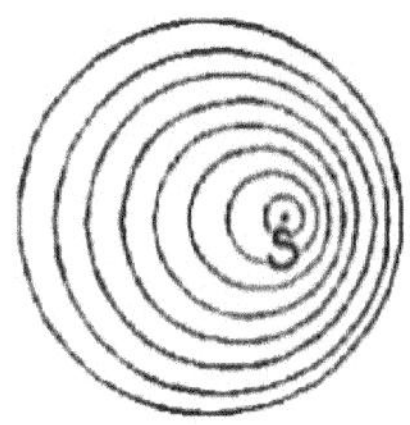

A. beats are heard
B. sonar works
C. a sound appears louder as the observer moves closer to the source
D. the siren on a fire engine truck changes its pitch as it passes the observer
E. a sound appears softer as the observer moves further from the source

76. A plastic container in the shape of a cube with 0.2 m sides is suspended in a vacuum. The container is filled to 3 atm with 10 g of N_2 gas. What is the force that the N_2 gas exerts on one face of the cube? (Use ideal gas constant $R = 0.0821$ L atm/K mol and 1 atm = 1.01×10^5 Pa)

A. 2.1×10^3 N
B. 5.6×10^4 N
C. 3.7×10^5 N
D. 4.2×10^4 N
E. 1.2×10^4 N

77. What is the maximum magnetic field for a wave if an electromagnetic wave travels in a vacuum with a maximum electric field of 1,200 V/m? (Use the speed of light c in a vacuum = 3×10^8 m/s)

A. 4×10^{-6} T
B. 2×10^{-5} T
C. 3.3×10^{-4} T
D. 8×10^{-6} T
E. 12×10^{-6} T

78. An alternating voltage oscillating at 60 Hz has a maximum value of 200 V during each cycle. What would be the reading if an RMS (root mean square) voltmeter is connected to the circuit?

A. 142 V
B. 100 V
C. 35 V
D. 284 V
E. 71 V

79. Where is the resulting image if a candle 21 cm tall is 4 m away from a diverging lens with a focal length of 3 m?

A. 12/7 m from the lens on the opposite side from the object
B. 12 m from the lens on the opposite side from the object
C. 12/7 m from the lens on the same side as the object
D. 12 m from the lens on the same side as the object
E. 4 m from the lens on the same side as the object

80. An energy level diagram of a certain atom is shown, whereby the energy difference between levels 1 and 2 is twice the energy difference between levels 2 and 3. A wavelength λ is emitted when an electron makes a transition from level 3 to 2. What possible radiation λ might be produced by other transitions between the three energy levels?

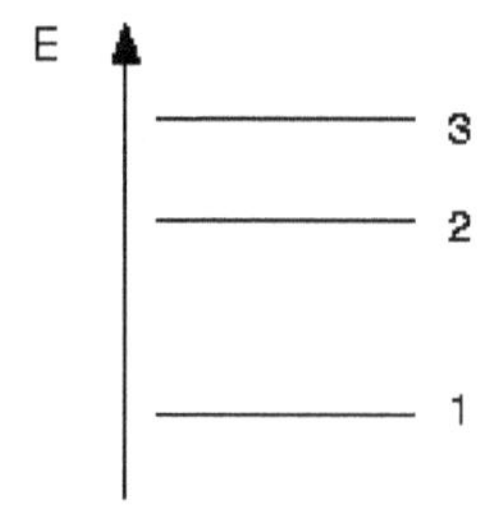

A. 2λ only
B. 2λ and 3λ
C. ½λ only
D. ½λ and λ/3
E. $\lambda/\sqrt{2}$ and $\lambda/\sqrt{3}$

81. A person running in place on an exercise machine for 10 min expends 19 kcal. Another person exercises by repeatedly lifting two 3 kg weights a distance of 50 cm. How many repetitions of this exercise are equivalent to 10 minutes of running in place? Assume that the person uses negligible energy in letting down the weights after each lift. (Use acceleration due to gravity $g = 9.8$ m/s^2 and conversion of 1 kcal = 1,000 cal and 1 cal = 4.186 J)

A. 2,300 repetitions
B. 1,800 repetitions
C. 1,360 repetitions
D. 3,940 repetitions
E. 2,705 repetitions

82. An 18 kg block is on a ramp inclined at 20° above the horizontal and is connected by a string to a 21 kg mass that hangs over the edge of the ramp. Ignoring frictional forces, what is the acceleration of the 21 kg block? (Use the acceleration due to gravity $g = 9.8$ m/s^2)

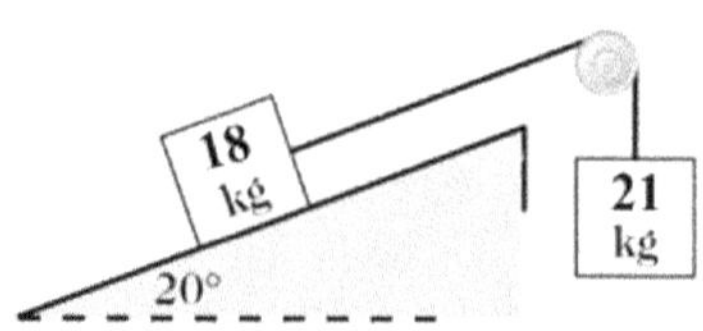

A. 3.2 m/s^2
B. 2.6 m/s^2
C. 3.7 m/s^2
D. 5.3 m/s^2
E. 7.6 m/s^2

83. A transverse wave in a string, with a wavelength of 8 m, is traveling at 4 m/s. At $t = 0$, a point on the string has a displacement of $+x$, where x is the amplitude of the wave. At what value of t is the same point on the string at a displacement of $-x$?

A. 1 s
B. 2 s
C. ¼ s
D. ½ s
E. 5 s

84. If the amount of fluid flowing through a tube remains constant, by what factor does the speed of the fluid change when the radius of the tube decreases from 16 cm to 4 cm?

A. Increases by $\sqrt{2}$
B. Increases by 16
C. Decreases by 16
D. Decreases by 4
E. Remains the same

85. Two identically-charged balls are a certain distance apart, as shown in the vector diagram.

If the charge of the ball on the left is doubled (represented by ++), which diagram represents the forces acting on the two balls?

A. 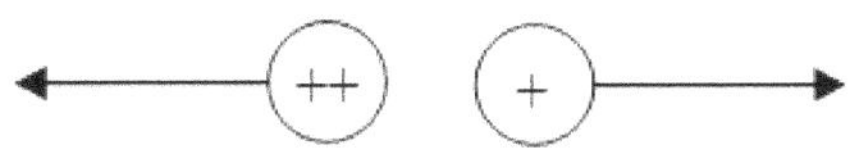

B. 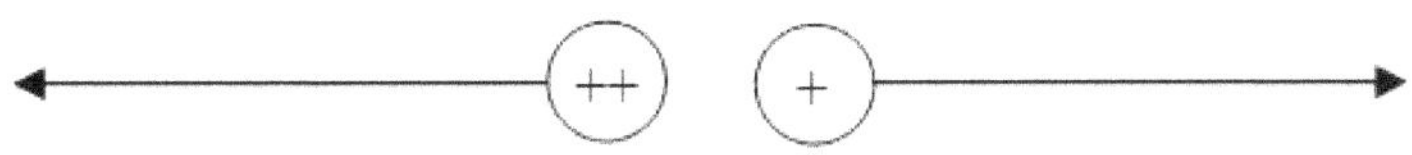

C. 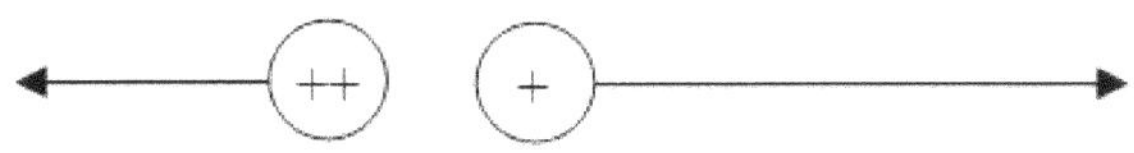

D. 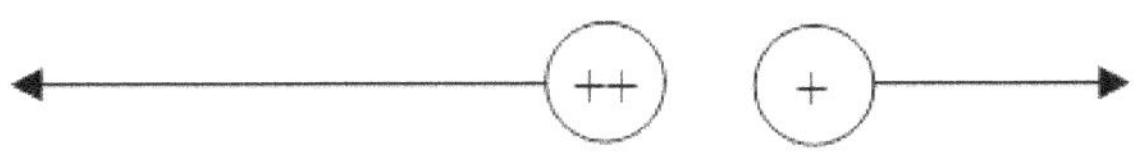

E. 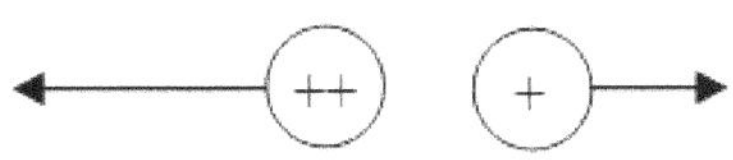

86. If Mario were in a spaceship traveling close to the speed of light with respect to Earth, he would notice that:

A. his pulse rate was higher than normal
B. his mass was higher than normal
C. his mass was lower than normal
D. some of his physical dimensions were smaller than normal
E. none of the above effects occur

87. Which statement is true if the magnification of a mirror or lens is negative?

A. Object is closer to the mirror or lens than to the image
B. Image is inverted
C. Image is erect and smaller than the object
D. Image is smaller than the object
E. Image is erect and larger than the object

88. The masses of isotopes are based on a comparison to the mass of which isotope?

A. ^{232}Uranium
B. ^{13}Carbon
C. ^{12}Carbon
D. ^{1}Hydrogen
E. ^{10}Helium

89. Ignoring air resistance, a 30 kg and 60 kg rock are thrown upward with the same initial speed (v_i). If the 30 kg rock reaches a maximum height of h, what maximum height does the 60 kg ball reach?

A. $2h$
B. h
C. $h / 2$
D. $h / 4$
E. $4h$

90. A hockey puck is set in motion across a frozen pond. Ignoring the friction of ice and air resistance, what force is required to keep the puck sliding at constant velocity?

A. mass of the puck × 9.8 m/s^2
B. weight of the puck / mass of the puck
C. weight of the puck
D. √mass of the puck
E. 0 N

91. A 0.24 kg piece of clay is thrown at a wall with an initial velocity of 16 m/s. What is the average force experienced by the clay if it stops after 91 milliseconds?

A. 84 N
B. 32 N
C. 24 N
D. 42 N
E. 64 N

92. A car drives 5 km North, then 7.3 km East, then 3.4 km Northeast, all at a constant speed. What was the magnitude of the average frictional force on the car if it performed 2.6×10^6 J of work during this trip?

A. 1.7×10^2 N
B. 4.2×10^2 N
C. 7×10^1 N
D. 5.7×10^2 N
E. 6.9×10^2 N

93. A redshift for light indicates that the light source moves:

I. at right angles to the observer
II. towards the observer
III. away from the observer

A. I only
B. II only
C. III only
D. I and II only
E. I and III only

94. An object having an emissivity of 0.867 radiates heat at a rate of 15 W when it is at a temperature T. If its temperature is doubled, what is the rate at which it radiates heat?

A. 30 W
B. 60 W
C. 80 W
D. 240 W
E. 120 W

95. A lens of focal length 50 mm is used as a magnifier to view a 6.4 mm object positioned at the lens's focal point. The user of the magnifier has a near point at 25 cm. What is the angular magnification of the magnifier?

A. 7.6
B. 6.9
C. 5.8
D. 8.1
E. 5

96. In Experiment 1, a magnet was moved toward the end of a solenoid, and a voltage was induced between the two ends of the solenoid wire. In Experiment 2, a higher voltage was observed. What might have changed between the two experiments?

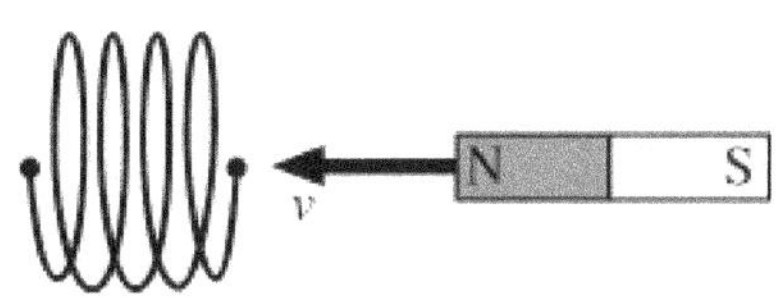

I. A stronger magnet replaced the bar magnet
II. The solenoid was replaced by one with more loops but the same length
III. The speed of the magnet increased

A. I only
B. II only
C. III only
D. I and III only
E. I, II and III

97. A car of mass m is driving up a shallow incline of angle θ with the horizontal when the driver sees a deer and quickly steps on the brakes and the tires lock up (the tires skid along the surface of the road rather than gripping it smoothly). The coefficient of static friction between the tires and the road is μ_s, and the coefficient of kinetic friction is μ_k. What is the magnitude of the net force on the car during the skid?

A. $mg\ (\mu_s \cos\theta - \sin\theta)$
B. $mg\ (\mu_s \cos\theta + \sin\theta)$
C. $mg\ \mu_s \cos\theta$
D. $mg\ (\mu_k \cos\theta + \sin\theta)$
E. $mg\ (\mu_k \cos\theta - \sin\theta)$

98. The radius of a typical nucleus is about 5.0×10^{-15} m. Assuming this to be the uncertainty in the position of a proton in the nucleus, what is the uncertainty in the proton's energy? (Use 1.67×10^{-27} kg as the proton mass)

A. 0.06 MeV
B. 0.25 MeV
C. 0.4 MeV
D. 0.8 MeV
E. 1.2 MeV

99. A 0.2 m long string vibrates in the n = 5 harmonic. What is the distance between a node and an adjacent antinode? (Use the speed of sound in air v = 340 m/s)

A. 30 mm
B. 60 mm
C. 7.5 mm
D. 20 mm
E. 120 mm

100. The graph shows position x as a function of time t for a system undergoing simple harmonic motion. Which graph represents the velocity of this system as a function of time?

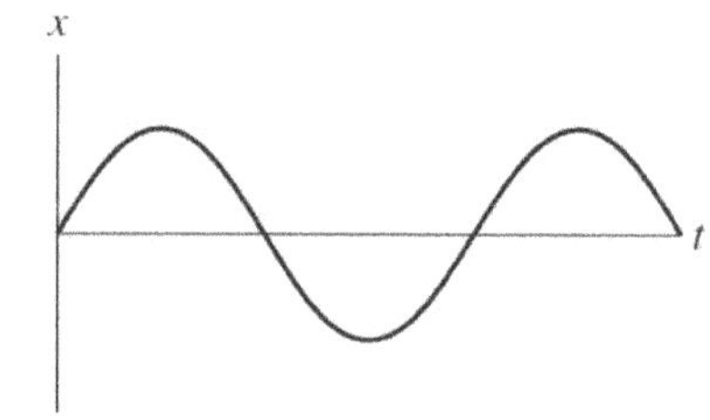

A.

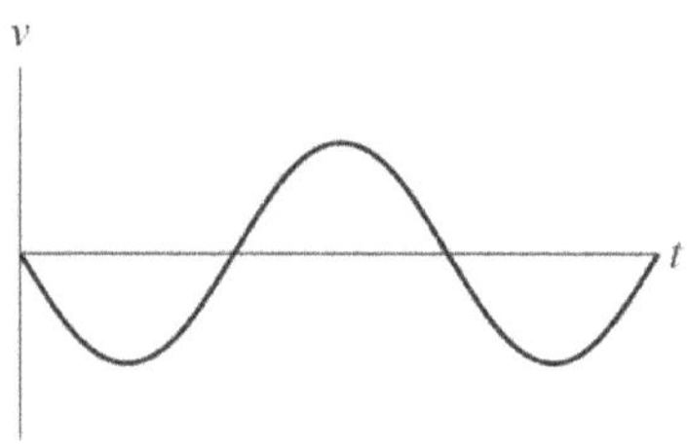

B.

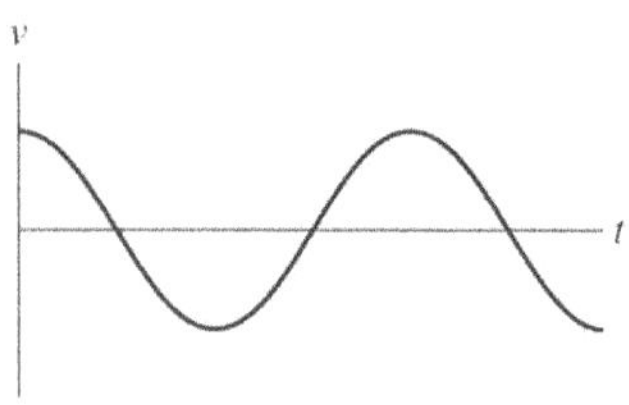

C.

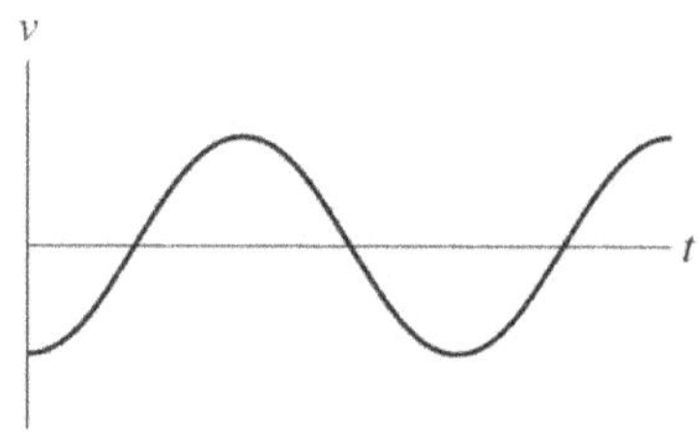

D.

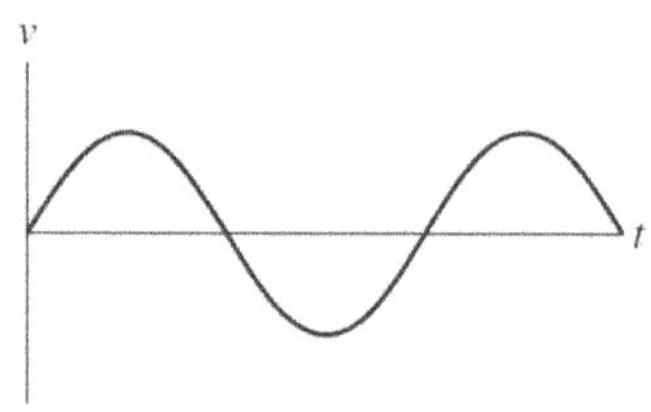

E. None of the above

Notes for active learning

Notes for active learning

Diagnostic Test 2

This Diagnostic Test is designed to assess your proficiency on each topic and NOT to mimic the test. Use your test results and identify areas of strength and weakness to adjust your study plan and enhance your fundamental knowledge. The length of the Diagnostic Tests is optimal for a single study session.

#	Answer:					Review	#	Answer:					Review
1:	A	B	C	D	E	___	**26:**	A	B	C	D	E	___
2:	A	B	C	D	E	___	**27:**	A	B	C	D	E	___
3:	A	B	C	D	E	___	**28:**	A	B	C	D	E	___
4:	A	B	C	D	E	___	**29:**	A	B	C	D	E	___
5:	A	B	C	D	E	___	**30:**	A	B	C	D	E	___
6:	A	B	C	D	E	___	**31:**	A	B	C	D	E	___
7:	A	B	C	D	E	___	**32:**	A	B	C	D	E	___
8:	A	B	C	D	E	___	**33:**	A	B	C	D	E	___
9:	A	B	C	D	E	___	**34:**	A	B	C	D	E	___
10:	A	B	C	D	E	___	**35:**	A	B	C	D	E	___
11:	A	B	C	D	E	___	**36:**	A	B	C	D	E	___
12:	A	B	C	D	E	___	**37:**	A	B	C	D	E	___
13:	A	B	C	D	E	___	**38:**	A	B	C	D	E	___
14:	A	B	C	D	E	___	**39:**	A	B	C	D	E	___
15:	A	B	C	D	E	___	**40:**	A	B	C	D	E	___
16:	A	B	C	D	E	___	**41:**	A	B	C	D	E	___
17:	A	B	C	D	E	___	**42:**	A	B	C	D	E	___
18:	A	B	C	D	E	___	**43:**	A	B	C	D	E	___
19:	A	B	C	D	E	___	**44:**	A	B	C	D	E	___
20:	A	B	C	D	E	___	**45:**	A	B	C	D	E	___
21:	A	B	C	D	E	___	**46:**	A	B	C	D	E	___
22:	A	B	C	D	E	___	**47:**	A	B	C	D	E	___
23:	A	B	C	D	E	___	**48:**	A	B	C	D	E	___
24:	A	B	C	D	E	___	**49:**	A	B	C	D	E	___
25:	A	B	C	D	E	___	**50:**	A	B	C	D	E	___

#	Answer:					Review	#	Answer:					Review
51:	A	B	C	D	E	___	**76:**	A	B	C	D	E	___
52:	A	B	C	D	E	___	**77:**	A	B	C	D	E	___
53:	A	B	C	D	E	___	**78:**	A	B	C	D	E	___
54:	A	B	C	D	E	___	**79:**	A	B	C	D	E	___
55:	A	B	C	D	E	___	**80:**	A	B	C	D	E	___
56:	A	B	C	D	E	___	**81:**	A	B	C	D	E	___
57:	A	B	C	D	E	___	**82:**	A	B	C	D	E	___
58:	A	B	C	D	E	___	**83:**	A	B	C	D	E	___
59:	A	B	C	D	E	___	**84:**	A	B	C	D	E	___
60:	A	B	C	D	E	___	**85:**	A	B	C	D	E	___
61:	A	B	C	D	E	___	**86:**	A	B	C	D	E	___
62:	A	B	C	D	E	___	**87:**	A	B	C	D	E	___
63:	A	B	C	D	E	___	**88:**	A	B	C	D	E	___
64:	A	B	C	D	E	___	**89:**	A	B	C	D	E	___
65:	A	B	C	D	E	___	**90:**	A	B	C	D	E	___
66:	A	B	C	D	E	___	**91:**	A	B	C	D	E	___
67:	A	B	C	D	E	___	**92:**	A	B	C	D	E	___
68:	A	B	C	D	E	___	**93:**	A	B	C	D	E	___
69:	A	B	C	D	E	___	**94:**	A	B	C	D	E	___
70:	A	B	C	D	E	___	**95:**	A	B	C	D	E	___
71:	A	B	C	D	E	___	**96:**	A	B	C	D	E	___
72:	A	B	C	D	E	___	**97:**	A	B	C	D	E	___
73:	A	B	C	D	E	___	**98:**	A	B	C	D	E	___
74:	A	B	C	D	E	___	**99:**	A	B	C	D	E	___
75:	A	B	C	D	E	___	**100:**	A	B	C	D	E	___

1. What is the potential energy with respect to the ground for a 2 kg mass held 4 m above the ground? (Use the acceleration due to gravity $g = 9.8$ m/s^2)

A. 20 J
B. 40 J
C. 60 J
D. 80 J
E. 120 J

2. Which point on the graph represents the maximum momentum for a diver as she leaps from a platform?

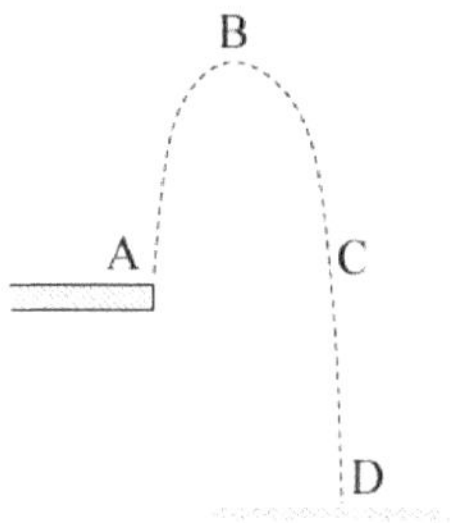

A. A
B. B
C. C
D. D
E. A and C

3. To drive a typical car at 40.0 mph on a level road for one hour requires about 3.2×10^7 J of energy. Suppose one tried to store this much energy in a spinning solid cylindrical flywheel, which was then coupled to the wheels of the car. What angular speed would be required to store 3.2×10^7 J if the flywheel has a radius of 0.60 m and mass of 400.0 kg?

A. 943.0 rad/s
B. 1,384.2 rad/s
C. 1,822.4 rad/s
D. 2,584.5 rad/s
E. 5,360.8 rad/s

4. A box is dragged to the right at a constant velocity along a level floor by a string that is horizontal with tension T. The magnitude of the frictional force is F, the gravitational force is G, and the normal force is N. Which relationship is correct?

A. $T + F = G + N$
B. $T > F$
C. $T + F = G - N$
D. $T < F$
E. $T = F$

5. A gas is confined to a rigid container that cannot expand as heat energy is added. This process is:

A. isokinetic
B. isentropic
C. isothermal
D. isometric
E. adiabatic

6. Which of the following expressions is equal to a watt?

A. kg·m^2/s^3
B. kg^2·m^2·s^2
C. kg·m/s^3
D. kg·m/s
E. kg·m^2/s^2

7. An organ pipe is a cylindrical tube that opens at both ends, whereby the air column vibrates from air flowing through the lower portion of the pipe. On an 8 °C day, the speed of sound is 3% slower than on a 20 °C day. How is the *f* affected?

A. Remains the same
B. Increases by √3%
C. Decreases by 9%
D. Decreases by 3%
E. Decreases by √3%

8. Three particles travel through a region of space where the magnetic field is pointing out of the page. What are the signs of the charges of these three particles?

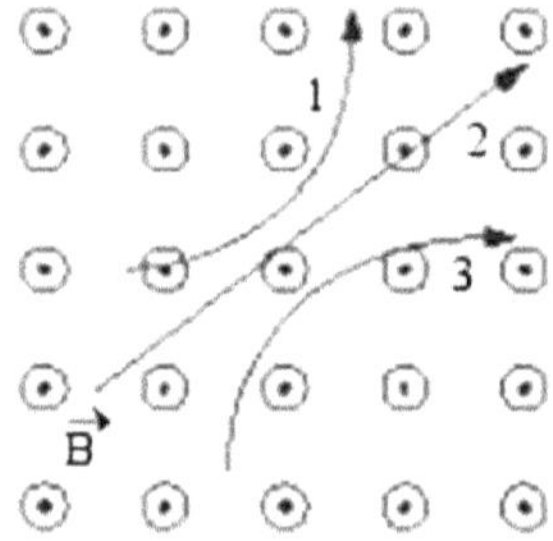

A. 1 is negative, 2 is positive, and 3 is neutral
B. 1 is positive, 2 is negative, and 3 is neutral
C. 1 is positive, 2 is neutral, and 3 is negative
D. 1 is neutral, 2 is positive, and 3 is negative
E. 1 is negative, 2 is neutral, and 3 is positive

9. Which form of electromagnetic radiation has the lowest frequency?

A. X-rays
B. γ rays
C. Radio waves
D. Microwaves
E. Infrared radiation

10. Ignoring air resistance, a purple marble is thrown upwards from a cliff with an initial speed of v_0. A grey marble is thrown downwards with the same initial speed. When the marbles reach the ground, the:

A. two marbles travel at speed proportional to their mass
B. two marbles have the same speed
C. grey marble moves faster than the purple marble
D. purple marble moves faster than the grey marble
E. two marbles travel at speed inversely proportional to their mass

11. A solid cylinder and a thin loop have the same mass and radius. Which statement is true about their moment of inertia about an axis through the exact center of the flat surfaces?

A. The solid cylinder and thin loop have the same moment of inertia
B. The solid cylinder has a greater moment of inertia
C. The thin loop has a greater moment of inertia
D. The moment of inertia cannot be determined because the volume of each is unknown
E. None of the above

12. What is the speed of transverse waves on a steel cable that lifts a 2,500 kg mass where the cable has a mass per unit length of 0.65 kg/m? (Use acceleration due to gravity $g = 9.8$ m/s^2)

A. 410 m/s
B. 920 m/s
C. 668 m/s
D. 470 m/s
E. 194 m/s

13. A viscous oil flows through a narrow pipe at a constant velocity. By what factor does the flow rate increase if the diameter of the pipe is doubled?

A. 4
B. $\sqrt{2}$
C. 6
D. 8
E. 12

14. Ignoring air resistance, two objects are thrown from a tall building. With the same initial speed, one is thrown upward and the other downward. What is the relationship between their speeds when hitting the street?

A. It is impossible to determine because the height of the building is unknown
B. They are traveling at the same speed
C. The object thrown downward is traveling faster
D. The object thrown upward is traveling faster
E. It is impossible to determine because their initial velocities are unknown

15. Which forces hold the protons in a nucleus together?

A. dipole–dipole
B. gravitational
C. nuclear
D. electrostatic attraction
E. electromagnetic

16. How much heat is required to raise the temperature of a 300 g lead ball from 20 °C to 30 °C? The specific heat of lead is 128 J/kg·K.

A. 224 J
B. 168 J
C. 576 J
D. 725 J
E. 384 J

17. Calculate the speed at which a 0.723 kg object has the same momentum as a 1.30 kg object that is moving 0.515*c*:

A. 0.548*c*
B. 0.621*c*
C. 0.734*c*
D. 0.916*c*
E. 0.416*c*

18. A photocathode has a work function of 2.4 eV. The photocathode is illuminated with monochromatic radiation, and electrons are emitted with a stopping potential of 1.1 volts. What is the wavelength of the illuminating radiation? (Use $c = 3.00 \times 10^8$ m/s, $h = 6.626 \times 10^{-34}$ J·s and 1 eV = 1.60×10^{-19} J)

A. 300 nm
B. 350 nm
C. 390 nm
D. 420 nm
E. 480 nm

19. A 0.01 kg bullet moves horizontally when it strikes a 1.5 kg block of wood suspended on a pendulum with a 2 m string. The bullet embeds into the wood, and together they swing a distance of 0.4 m upward. What was the velocity of the bullet just before it struck the wooden block? (Use acceleration due to gravity $g = 9.8$ m/s^2)

A. 423 m/s
B. 225 m/s
C. 168 m/s
D. 326 m/s
E. 526 m/s

20. Why do heavy nuclei contain more neutrons than protons?

A. Neutrons are heavier than protons
B. Neutrons and heavy nuclei are not radioactive
C. Neutrons reduce the electric repulsion of the protons
D. Neutrons are lighter than protons
E. Neutrons amplify the electric repulsion of the protons

21. How far will an object have traveled after 3 s if it starts from rest and undergoes uniform acceleration given that it reaches 5 m/s after 1 s?

A. 5 m
B. 10 m
C. 18 m
D. 22.5 m
E. 40 m

22. A block is at rest on the surface of an inclined plane as the elevation angle is gradually increased. Which statement is true about the normal force exerted by the plane on the block?

A. It is independent of the angle of elevation
B. It is inversely dependent on the coefficient of static friction between the block and plane
C. It increases as the angle of elevation increases
D. It is independent of the total amount of forces acting upon it
E. It decreases as the angle of elevation increases

23. An aluminum rod 10 cm long and a steel rod 80 cm long are joined end-to-end. Both rods are at 15 °C and have the same diameter. What is the increase in the length of the joined rod when the temperature is raised to 90 °C? (Use the coefficient of linear expansion α for aluminum = 2.4×10^{-5} K^{-1} and α for steel = 1.2×10^{-5} K^{-1})

A. 0.7 mm
B. 1.1 mm
C. 0.9 mm
D. 0.8 mm
E. 1 mm

24. Which quantity is conserved when a falling object strikes the ground?

I. Momentum of the object II. KE of the object III. Total energy

A. I only
B. III only
C. I and III only
D. II and III only
E. I, II and III

25. If the frequency of the motion of a simple harmonic oscillator is doubled, by what factor does the maximum speed of the oscillator change?

A. ½
B. √2
C. 4
D. 2
E. ¼

26. A standing wave of the third overtone is induced in a 1.2 m pipe open at one end and closed at the other. What is the number of antinodes in the standing wave? (Use the speed of sound = 340 m/s)

A. 3
B. 4
C. 5
D. 6
E. 7

27. A wire circle and solid circle of the same diameter are resting on the surface of the water. Which one can have a larger maximum mass without sinking?

A. The solid circle, by a factor of 2
B. The solid circle, by a factor of 4
C. The wire circle, by a factor of 2
D. They have the same maximum mass
E. The wire circle, by a factor of 4

28. Which statement applies to the 120 V circuit shown?

A. 120 J of energy is given to each Coulomb of charge, making up the current in the circuit
B. 120 J of energy is equally shared among Coulombs in the circuit at any instant
C. 120 C of charge flow through the lamp every second
D. 120 C of energy is converted to heat and light in the circuit every second
E. 120 J of energy is unequally shared among the Coulombs in the circuit, depending on the location

29. How many electrons pass a given point in a minute for a wire with a current of 6 mA? (Use the charge of an electron = 1.602×10^{-19} C)

A. 5.3×10^{14} electrons
B. 5.4×10^{-15} electrons
C. 2.3×10^{18} electrons
D. 3.7×10^{12} electrons
E. 720 electrons

30. How does an image appear if an object is in front of a convex mirror at a distance larger than twice the magnitude of the mirror's focal length?

A. upright and smaller
B. inverted and larger
C. inverted and smaller
D. inverted and the same size
E. upright and the same size

31. According to the de Broglie hypothesis, the idea of matter waves describes the wave-like behavior of:

I. Positively-charged stationary particles
II. Negatively-charged stationary particles
III. Particles that are moving

A. I only
B. II only
C. III only
D. I and II
E. I, II and III

32. Block I and block II slide on a frictionless level surface in one dimension and stick when they make contact. Block II is also connected to a massless, ideal spring that extends horizontally and connected to a wall. Initially, block II is at rest, and block I approaches from the right. The two blocks move before oscillating. Take the system to be the two blocks and the spring. Which statement is true?

I. Momentum is conserved
II. The sum of the spring PE and KE is conserved
III. KE is conserved

A. I only
B. II only
C. I and II only
D. II and III only
E. I, II and III

33. Ignoring air resistance, when a pebble is thrown straight upward with an initial speed v, it reaches a maximum height of h. At what speed should the pebble be thrown upward vertically for it to go twice as high?

A. $16v$
B. $8v$
C. $4v$
D. $2v$
E. $\sqrt{2}v$

34. If the mass and the length of a simple pendulum are doubled, the period:

A. increases by a factor of 4
B. increases by a factor of $\sqrt{2}$
C. remains the same
D. increases by a factor of 2
E. decreases by a factor of 2

35. Which type of organ pipe can produce only odd harmonics?

I. open II. long III. closed

A. I only
B. II only
C. III only
D. II and III only
E. None of the above

36. Determine the speed at which water exits a tank through a tiny hole in the bottom of the tank that is 20 cm in diameter and filled with water to a height of 50 cm. (Use the acceleration due to gravity $g = 9.8$ m/s^2)

A. 8.8 m/s
B. 17.8 m/s
C. 31.2 m/s
D. 21.6 m/s
E. 3.1 m/s

37. While being pulled apart, the plates of a parallel-plate capacitor are maintained with constant voltage by a battery. What happens to the strength of the electric field between the plates during this process?

A. Remains constant
B. Decreases
C. Increases by the $\sqrt{\text{voltage}}$
D. Increases by the (distance)^2
E. Increases by the $\sqrt{\text{distance}}$

38. An electron and a deuteron (^{2}H nucleus) are placed in an electric field (E_0) between the plates of a charged parallel-plate capacitor. Ignoring the force of gravity, how does the magnitude of the force on an electron F_{elec} compare with the force on the deuteron F_d? (Use the mass of proton $m_{proton} \approx 1{,}850 \times m_{electron}$)

A. $F_{elec} = \frac{1}{2}F_d$
B. $F_{elec} = 1{,}850\ F_d$
C. $F_{elec} = F_d$
D. $F_{elec} = 3{,}700\ F_d$
E. $F_{elec} = 925\ F_d$

39. A 1.8×10^{14} Hz electromagnetic wave propagates in CCl_4 with a speed of 2.43×10^8 m/s. The wave then leaves the CCl_4 and enters a vacuum. What is the wavelength of the wave in the vacuum? (Use the speed of light in a vacuum $c = 3 \times 10^8$ m/s)

A. 2,260 nm
B. 1,280 nm
C. 1,040 nm
D. 1,667 nm
E. 1,840 nm

40. How much energy does a photon of wavelength 580 nm have? (Use Planck's constant $h = 6.626 \times 10^{-34}$ J·s and the speed of light $c = 3 \times 10^8$ m/s)

A. 6.4×10^{-32} J
B. 8.8 eV
C. 2.6×10^{-38} J
D. 1.3×10^{-19} eV
E. 3.4×10^{-19} J

41. At constant pressure and temperature, Graham's Law states that the diffusion rate for a gas molecule is:

A. inversely proportional to √mass
B. inversely proportional to mass
C. proportional to the log of the mass
D. proportional to the mass
E. proportional to the √mass

42. A skier falls while skiing, and one ski, with weight w, loosens and slides down an icy slope, making an angle θ with the horizontal. Ignoring friction, what is the force that pushes the ski down the hill?

A. $w^2 \cos \theta$
B. $w \sin \theta$
C. w
D. $w \cos \theta$
E. $\sqrt{w} \tan \theta$

43. By what factor does the velocity of sound in a gas change when the absolute temperature doubles?

A. It increases by 4
B. It increases by √2
C. It remains the same
D. It increases by 2
E. It increases by 1/√2

44. A pipe with a circular cross-section has water flowing within it from point I to point II. The pipe radius at point I is 12 cm, while the radius at point II is 6 cm. If, at the end of point I, the flow rate is 0.09 m^3/s, what is the flow rate at the end of point II?

A. 0.6 m^3/s
B. 0.09 m^3/s
C. 0.15 m^3/s
D. 1.2 m^3/s
E. 2.4 m^3/s

45. What is the path for an electron moving on the page toward the right when subjected to a magnetic field pointing out of the page?

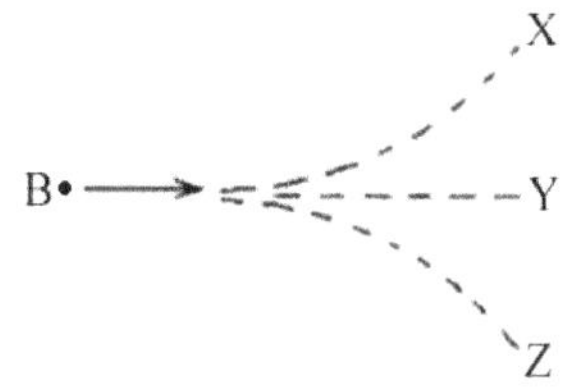

A. Curves upward (path X)
B. Continues straight ahead (path Y)
C. Curves downward (path Z)
D. Decelerates
E. Accelerates

46. A positively-charged particle travels on a path parallel to a straight wire containing a current. Which change increases the magnetic field at a point along the particle's path?

A. Increasing the speed of the particle
B. Increasing the distance of the particle from the wire
C. Decreasing the distance of the particle from the wire
D. Decreasing the current in the wire
E. Increasing the charge of the particle

47. If parallel light rays were incident on a lens of power 4 D, which statement is true for the rays on the other side of the lens?

A. The rays diverge as if from a point 4 m behind the lens
B. The rays diverge as if from a point ¼ m in front of the lens
C. The rays converge with a focal length of ¼ m
D. The rays converge with a focal length of 4 m
E. The rays diverge or converge, but more information is needed

48. The α particle has twice the electric charge of the β particle but deflects less than the β in a magnetic field because it:

A. is smaller
B. has less inertia
C. moves slower
D. is larger
E. has more inertia

49. Which statement explains why a sponge ball of the same size as a marble takes longer to reach the ground?

A. Air resistance is more significant for the sponge ball than for the marble
B. There is a stronger gravitational force between the marble ball and the ground
C. The force of gravity only acts on the marble ball
D. The force of gravity on the sponge ball is less than that on the marble ball
E. None of the above is a good explanation

50. The force of gravity attracts the Moon and the Earth. The Earth's radius is 3.7 times that of the Moon, and the Earth's mass is 80 times that of the Moon. The acceleration due to gravity on the surface of the Moon is one-sixth of the acceleration due to gravity on the Earth's surface. How does the Earth's gravitational pull on the Moon differ from the Moon's pull on the Earth?

A. 80 times greater
B. 6 times greater
C. 3.7 times greater
D. Identical
E. 6 times smaller

51. A uniform 16 kg board of length L is positioned horizontally, with its two ends supported by scales. A 4 kg package is positioned at a distance of $L / 3$ from the left end. What is the weight reading for the scale on the right? (Use the acceleration due to gravity $g = 9.8$ m/s^2 and assume mass is centered in the middle of the board)

A. 53.1 N
B. 91.5 N
C. 81.9 N
D. 43.3 N
E. 62 N

52. The brakes of a car are applied abruptly, whereby the car skids a certain distance on a straight, level road. If the car had been traveling twice as fast, what is the distance that the car would have skidded under the same conditions?

A. $\sqrt{2}$ times farther
B. Half as far
C. 4 times farther
D. Twice as far
E. Unable to determine

53. A pair of narrow slits, separated by 1.8 mm, is illuminated by a monochromatic light source. Light waves arrive at the two slits in phase. A fringe pattern is observed on a screen 4.8 m from the slits. There are 5 bright fringes per cm on the screen. What is the λ of the monochromatic light?

A. 550 nm
B. 600 nm
C. 650 nm
D. 700 nm
E. 750 nm

54. A 0.4 kg ice cube at 0 °C has sufficient heat added to it to cause total melting, and the resulting water is heated to 60 °C. How much heat is added? (Use the latent heat of fusion for water $L_f = 334{,}000$ J/kg, the latent heat of vaporization for water $L_v = 2.256 \times 10^6$ J/kg and the specific heat $c = 4.186 \times 10^3$ J/kg·°C)

A. 56 kJ
B. 84 kJ
C. 470 kJ
D. 234 kJ
E. 310 kJ

55. A concave spherical mirror has a focal length of 20 cm. Where is the image relative to the mirror if an object is 10 cm in front of the mirror?

A. 7.5 cm in front
B. 7.5 cm behind
C. 20 cm in front
D. 20 cm behind
E. 10 cm behind

56. For a DC circuit, what quantity does the slope of the graph represent?

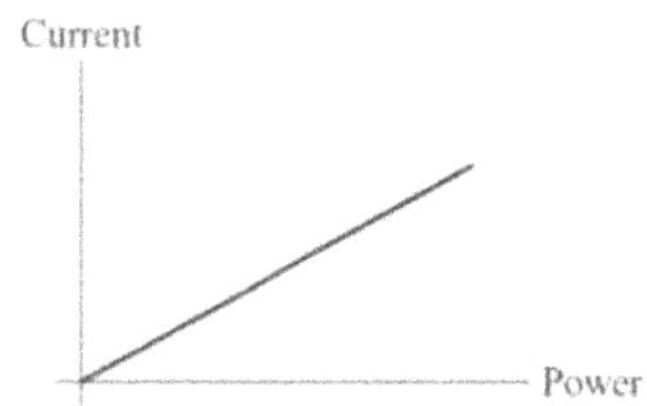

A. potential
B. (potential)^2
C. 1 / resistance
D. resistance
E. 1 / potential

57. Two identical metal balls, each with a radius of 0.15 m, are located 4 m apart, are neutral, and have an electrical potential of zero. Electrons are transferred as an isolated system from ball I to II. After the transfer, ball I has acquired a potential of 8,000 V and ball II a potential of –8,000 V. How much work is required to transfer 10^{-10} C from ball I to ball II?

A. -1.6×10^{-6} J
B. -4.7×10^{-7} J
C. 7.1×10^{-5} J
D. 3.3×10^{-6} J
E. 5.7×10^{-5} J

58. Which answer best describes what happens to a spherical lead ball with a density of 11.3 g/cm^3 when placed in a tub of mercury with a density of 13.6 g/cm^3?

A. It sinks slowly to the bottom of the mercury
B. It floats with about 17% of its volume above the surface of the mercury
C. It floats with its top exactly even with the surface of the mercury
D. It floats with about 83% of its volume above the surface of the mercury
E. It floats somewhere within the mercury, but it is impossible to determine where

59. An ambulance is moving directly towards a police station at 50 m/s. The ambulance emits a siren at 420 Hz. What is the frequency of the signal heard by police at the station? (Use the speed of sound v_s = 350 m/s)

A. 280 Hz
B. 330 Hz
C. 420 Hz
D. 490 Hz
E. 540 Hz

60. A simple pendulum has a period of 3 s on Earth. If it is taken to the Moon, where the acceleration due to gravity is 1/6 of that on Earth, what is its period on the Moon?

A. 0.5 s
B. 3.7 s
C. 7.3 s
D. 5 s
E. 12.4 s

61. An ideal, massless spring with a spring constant of 3 N/m has a resting length of 0.25 m. The spring is hanging from the ceiling when a 1.2 kg block is added to the bottom end of the spring. What is the length of the spring at static equilibrium? (Use the acceleration due to gravity $g = 10\ \text{m/s}^2$)

A. 2.35 m
B. 1.45 m
C. 4.96 m
D. 4.25 m
E. 5.5 m

62. A boy jumps with a velocity of 20 m/s at an angle of 25° above the horizontal. What is the horizontal component of the boy's velocity?

A. 8.7 m/s
B. 14.6 m/s
C. 18.1 m/s
D. 14.9 m/s
E. 23.3 m/s

63. Which of the following is true about the force due to friction when a block is sliding down the surface of an inclined plane while the elevation angle is gradually decreased?

A. Increases and the weight of the block remains constant
B. Increases and the weight of the block increases
C. Increases and the weight of the block decreases
D. Decreases and the weight of the block remains constant
E. Decreases and the weight of the block increases

64. The average velocity of an object equals the instantaneous velocity only when the velocity is:

I. increasing at a constant rate
II. constant
III. decreasing at a constant rate

A. I only
B. II only
C. III only
D. I and II only
E. II and III only

65. A sphere of surface area of 1.25 m² and emissivity 1 is at a temperature of 100 °C. What is the rate at which it radiates heat into space? (Use the Stefan-Boltzmann constant $\sigma = 5.67 \times 10^{-8}\ \text{W/m}^2\text{K}^4$)

A. 1.4 kW
B. 7.6 kW
C. 27.3 kW
D. 0.63 kW
E. 3.1 kW

66. A projectile is fired at time $t = 0$ s, from point 0 at the edge of a cliff. It has initial velocity components of $v_x = 50$ m/s and $v_y = 240$ m/s with time in flight of 60 s. The projectile lands at point P. What is the y coordinate of the projectile when its x coordinate is 1,000 m? (Use the acceleration due to gravity $g = 9.8$ m/s^2)

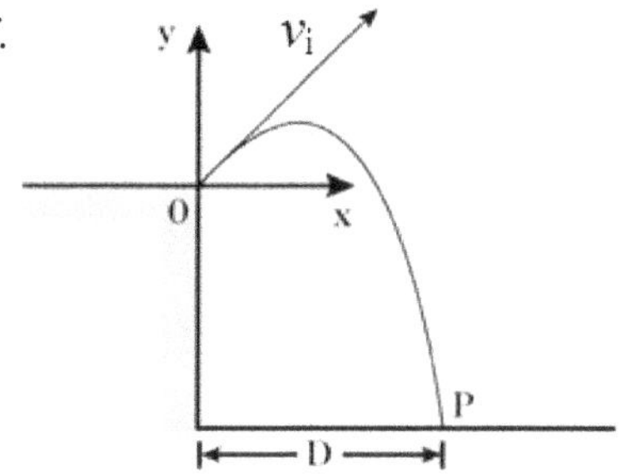

A. −160 m
B. 1,240 m
C. 2,840 m
D. 850 m
E. 1,600 m

67. A uniform, solid, cylindrical flywheel of radius 1.4 m and mass 15.0 kg rotates at 2.7 rad/s about an axis through its circular faces. What is the magnitude of the flywheel's angular momentum?

A. 22 kg·m^2/s
B. 40 kg·m^2/s
C. 64 kg·m^2/s
D. 80 kg·m^2/s
E. 140 kg·m^2/s

68. A given distance separates two particles of like charge. What is the resulting force between the particles if the charge on each particle and the distance between the two particles are doubled?

A. 4 × original
B. 2 × original
C. equal to the original
D. √2 × original
E. ¼ × original

69. If an object is placed inside the focal point of a diverging lens, the image is:

A. virtual, inverted and enlarged
B. real, upright and enlarged
C. real, inverted and enlarged
D. virtual, upright and reduced
E. real, upright and reduced

70. A heat engine takes 4 moles of an ideal gas through the reversible cycle *abca* on the pressure *vs.* volume diagram, as shown. The path *bc* is an isothermal process. The temperature at *c* is 650 K, and the volumes at *a* and *c* are 0.025 m^3 and 0.33 m^3, respectively.

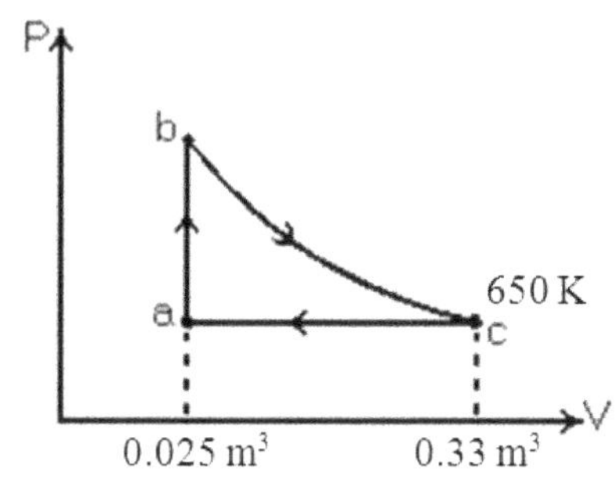

What is the work done by the gas for the path *bc*? (Use the molar heat capacity at constant volume of the gas = 18 J/mol·K and the universal gas constant R = 8.134 J/mol·K)

A. −56 kJ
B. −82 kJ
C. 56 kJ
D. 82 kJ
E. 0 kJ

71. Three objects are moving along a straight line. If the positive direction is to the right, what is the total momentum of this system?

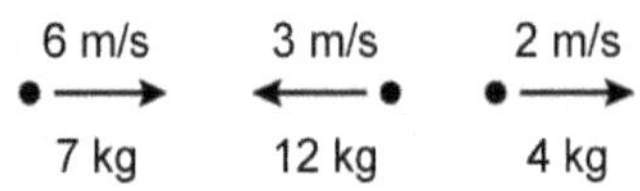

A. 86 kg·m/s
B. –86 kg·m/s
C. –14 kg·m/s
D. 14 kg·m/s
E. 0 kg·m/s

72. In which medium do sound waves travel the fastest?

A. Cool air
B. Warm air
C. A vacuum
D. Icy cold air
E. Moving air

73. The Bernoulli Equation is described by $P_1 + \frac{1}{2}\rho v_1^2 + \rho g h_1 = P_2 + \frac{1}{2}\rho v_2^2 + \rho g h_2$. What is the origin of the relationships within this expression?

A. the conservation of energy for a moving fluid
B. the continuity principle for moving particles
C. the conservation of linear momentum across an uneven surface
D. Newton's Third Law that relates equal action and reaction
E. $F = ma$, as applied to a fluid

74. How do the resistances of two copper wires compare where one has twice the length and twice the cross-sectional area?

A. The longer wire has one fourth the resistance of the shorter wire
B. The shorter wire has $\sqrt{2}$ times the resistance of the longer wire
C. The shorter wire has twice the resistance of the longer wire
D. Both wires have the same resistance
E. The shorter wire has one half the resistance of the longer wire

75. Which type of emission from the reactant nucleus causes the transformation: ${}^{15}_{8}\text{O} \rightarrow {}^{15}_{7}\text{N}$?

A. Neutron emission
B. Alpha particle
C. Electron
D. Proton emission
E. Positron emission

76. Which is an example of a reversible process?

A. Hooke's cycle
B. Carnot cycle
C. Swinn cycle
D. Boltzmann's cycle
E. Calvin cycle

77. A vehicle is driving in reverse at 6 m/s. After 10 s of uniform acceleration, the vehicle is going forward at 12 m/s. What is the acceleration?

A. 1 m/s^2
B. 2.5 m/s^2
C. 3 m/s^2
D. 4.5 m/s^2
E. 1.8 m/s^2

78. A 10 kg brick and a 1 kg book are dropped in a vacuum. What is the force of gravity on the 10 kg brick? (Use acceleration due to gravity $g = 9.8$ m/s^2)

A. Zero
B. Twice the force on the 1 kg book
C. The same force on the 1 kg book
D. 9.8 times the force on the 1 kg book
E. Ten times the force on the 1 kg book

79. A 0.06 kg golf ball is at rest on a tee. The ball has a velocity of 106 m/s immediately after being struck. If the club and ball were in contact for 0.85 ms, what is the average force exerted on the ball?

A. 4200 N
B. 5600 N
C. 3600 N
D. 7500 N
E. 6900 N

80. What happens to the atomic mass number when a β^- particle is emitted from an unstable nucleus?

A. Decreases by 2
B. Increases by 2
C. Decreases by 1
D. Increases by 1
E. Remains the same

81. A chipmunk is running along a wire with constant acceleration. If it has an initial velocity of 0.4 m/s and a final velocity of 1.8 m/s after 4 s, how far does the chipmunk run in that time?

A. 2.3 m
B. 4.4 m
C. 6.7 m
D. 9.9 m
E. 7.2 m

82. An object is moving to the right in a straight line. What happens to the object if the net force acting on the object is also directed to the right, but the magnitude of the force decreases with time?

A. It continues moving to the right with a constant speed
B. It stops and then begins moving to the left with increasing speed
C. It continues moving to the right with increasing speed
D. It continues moving to the right with decreasing speed
E. It moves to the left with increasing speed

83. What is the specific heat capacity of material if 150 kcal of heat raises the temperature of 3 kg of the material by 200 °C?

A. 0.35 kcal/kg·°C
B. 0.5 kcal/kg·°C
C. 0.75 kcal/kg·°C
D. 1.15 kcal/kg·°C
E. 0.25 kcal/kg·°C

84. It takes 40 J of work to push a large suitcase 4 m across a floor. Assuming the suitcase is pushed in the same direction as it moves, what is the magnitude of the force on the suitcase?

A. 10 N
B. 4 N
C. 16 N
D. 26 N
E. 34 N

85. A simple pendulum with a length of 58 cm has a period of 2.5 s on Mars. What is the acceleration g due to gravity on Mars?

A. 0.7 m/s^2
B. 3.7 m/s^2
C. 6.8 m/s^2
D. 9.8 m/s^2
E. 17.2 m/s^2

86. What is the source of electromagnetic waves?

A. Crystalline fluctuations
B. Vibrating nuclei
C. Accelerating electric charges
D. Changes in atomic energy levels
E. All the above

87. How far does a wavefront of an acoustical wave travel in 1 s if it has an f of 800 Hz and a λ of 0.5 m?

A. 400 m
B. 150 m
C. 75 m
D. 300 m
E. 600 m

88. Elena is a passenger on a spaceship. After the speed of the spaceship has increased, she observes:

A. no change in the length of the spaceship
B. the length of the spaceship is zero
C. the length of the spaceship got longer
D. the length of the spaceship got shorter
E. cannot be determined from the information provided

89. What is the result of connecting two identical storage batteries in parallel in a circuit?

A. Half the voltage and half the total charge compared to a circuit from a single battery
B. Half the voltage, but the same total charge compared to a circuit from a single battery
C. The same voltage and the same total charge compared to a circuit from a single battery
D. The same voltage, but half the total charge compared to a circuit from a single battery
E. The same voltage, but twice the total charge compared to a circuit from a single battery

90. What is the power of the combination for 5 D and 3 D thin lenses positioned near each other?

A. 1/8 D
B. 5/3 D
C. 3/5 D
D. 8 D
E. 8/3 D

91. Is it possible for a hydrogen nucleus to emit an α particle?

A. Yes, it usually emits α particles
B. No, because it does not contain enough nucleons
C. Yes, because α particles are the simplest form of radiation
D. No, because it would require nuclear fission of hydrogen, which is impossible
E. Yes, but hydrogen generally emits β particles instead

92. A 1,450 kg cannon fires a 90 kg cannonball at 30 m/s. Assuming that frictional forces are negligible, and the cannonball is fired horizontally, what is the recoil velocity of the cannon?

A. 5.1 m/s
B. 1.2 m/s
C. 3.4 m/s
D. 1.9 m/s
E. 2.8 m/s

93. An engineer expends 900 J to lift a block to a height of h. He then repeats the task using a simple non-motorized pulley system that reduces the input force by half. How much work must the engineer perform using the pulley system to lift the block to height h?

A. 100 J
B. 225 J
C. 900 J
D. 450 J
E. 1,400 J

94. A massless, ideal spring projects horizontally from a wall and connects to a 0.1 kg mass. The mass is oscillating in one dimension, such that it moves 0.4 m from one end of its oscillation to the other. What is the amplitude of the oscillation if it experiences 20 complete oscillations in 60 s?

A. 0.2 m
B. 0.4 m
C. 0.8 m
D. $\sqrt{0.4}$ m
E. 4 m

95. What is the wave speed of the standing wave shown if it is oscillating at 900 Hz on a string?

A. 280 m/s
B. 170 m/s
C. 360 m/s
D. 570 m/s
E. 420 m/s

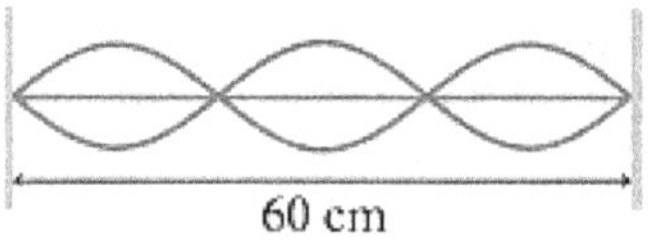

96. What is the density of water?

A. 1×10^3 kg/cm^3
B. 1×10^1 kg/cm^3
C. 1×10^2 kg/m^3
D. 1×10^{-3} g/m^3
E. 1×10^3 kg/m^3

97. Two stationary positive charges of $Q = 10^{-10}$ C are fixed and separated by a distance of $d = 4$ cm. What is the net electrostatic force on a charge $q = -10^{-9}$ C placed at a distance of $d / 2$ from each of the charges?

A. 5×10^{-10} N
B. 1.5×10^{-9} N
C. 0 N
D. 2.5×10^{-10} N
E. 1×10^{-9} N

98. A capacitor, initially having a charge of Q on the left plate and a $-Q$ on the right plate, is connected to a switch and an inductor.

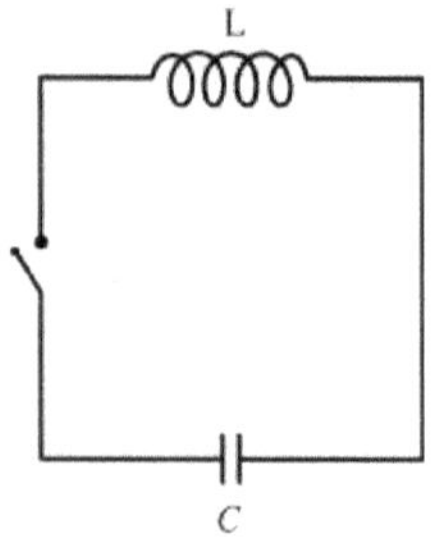

Assuming that the resistance of the circuit is zero, which statement is true when the switch is closed?

A. Current flows back and forth through the inductor, with the magnitude of the current steadily decreasing and eventually becoming zero
B. Current flows until the left plate of the capacitor has a charge $-Q$, and then current flows in the opposite direction, reversing again when the left plate has a charge of $+Q$. The cycle repeats
C. Charge flows out of the capacitor until the left plate is no longer charged, and current ceases
D. Because the switch is closed, the current flows clockwise with a constant magnitude.
E. Current flows back and forth through the inductor, with the magnitude of the current increasing at a constant rate

99. The Compton effect directly demonstrated which property of electromagnetic radiation?

I. energy content
II. particle nature
III. momenta

A. I only
B. II only
C. I and II only
D. II and III only
E. I, II and III

100. How many $2d$ electron states are possible for an atom?

A. 0
B. 8
C. 6
D. 18
E. 10

Notes for active learning

Diagnostic Test 3

This Diagnostic Test is designed to assess your proficiency on each topic and NOT to mimic the test. Use your test results and identify areas of strength and weakness to adjust your study plan and enhance your fundamental knowledge. The length of the Diagnostic Tests is optimal for a single study session.

#	Answer:					Review	#	Answer:					Review
1:	A	B	C	D	E	___	**26:**	A	B	C	D	E	___
2:	A	B	C	D	E	___	**27:**	A	B	C	D	E	___
3:	A	B	C	D	E	___	**28:**	A	B	C	D	E	___
4:	A	B	C	D	E	___	**29:**	A	B	C	D	E	___
5:	A	B	C	D	E	___	**30:**	A	B	C	D	E	___
6:	A	B	C	D	E	___	**31:**	A	B	C	D	E	___
7:	A	B	C	D	E	___	**32:**	A	B	C	D	E	___
8:	A	B	C	D	E	___	**33:**	A	B	C	D	E	___
9:	A	B	C	D	E	___	**34:**	A	B	C	D	E	___
10:	A	B	C	D	E	___	**35:**	A	B	C	D	E	___
11:	A	B	C	D	E	___	**36:**	A	B	C	D	E	___
12:	A	B	C	D	E	___	**37:**	A	B	C	D	E	___
13:	A	B	C	D	E	___	**38:**	A	B	C	D	E	___
14:	A	B	C	D	E	___	**39:**	A	B	C	D	E	___
15:	A	B	C	D	E	___	**40:**	A	B	C	D	E	___
16:	A	B	C	D	E	___	**41:**	A	B	C	D	E	___
17:	A	B	C	D	E	___	**42:**	A	B	C	D	E	___
18:	A	B	C	D	E	___	**43:**	A	B	C	D	E	___
19:	A	B	C	D	E	___	**44:**	A	B	C	D	E	___
20:	A	B	C	D	E	___	**45:**	A	B	C	D	E	___
21:	A	B	C	D	E	___	**46:**	A	B	C	D	E	___
22:	A	B	C	D	E	___	**47:**	A	B	C	D	E	___
23:	A	B	C	D	E	___	**48:**	A	B	C	D	E	___
24:	A	B	C	D	E	___	**49:**	A	B	C	D	E	___
25:	A	B	C	D	E	___	**50:**	A	B	C	D	E	___

#	Answer:					Review	#	Answer:					Review
51:	A	B	C	D	E	___	**76:**	A	B	C	D	E	___
52:	A	B	C	D	E	___	**77:**	A	B	C	D	E	___
53:	A	B	C	D	E	___	**78:**	A	B	C	D	E	___
54:	A	B	C	D	E	___	**79:**	A	B	C	D	E	___
55:	A	B	C	D	E	___	**80:**	A	B	C	D	E	___
56:	A	B	C	D	E	___	**81:**	A	B	C	D	E	___
57:	A	B	C	D	E	___	**82:**	A	B	C	D	E	___
58:	A	B	C	D	E	___	**83:**	A	B	C	D	E	___
59:	A	B	C	D	E	___	**84:**	A	B	C	D	E	___
60:	A	B	C	D	E	___	**85:**	A	B	C	D	E	___
61:	A	B	C	D	E	___	**86:**	A	B	C	D	E	___
62:	A	B	C	D	E	___	**87:**	A	B	C	D	E	___
63:	A	B	C	D	E	___	**88:**	A	B	C	D	E	___
64:	A	B	C	D	E	___	**89:**	A	B	C	D	E	___
65:	A	B	C	D	E	___	**90:**	A	B	C	D	E	___
66:	A	B	C	D	E	___	**91:**	A	B	C	D	E	___
67:	A	B	C	D	E	___	**92:**	A	B	C	D	E	___
68:	A	B	C	D	E	___	**93:**	A	B	C	D	E	___
69:	A	B	C	D	E	___	**94:**	A	B	C	D	E	___
70:	A	B	C	D	E	___	**95:**	A	B	C	D	E	___
71:	A	B	C	D	E	___	**96:**	A	B	C	D	E	___
72:	A	B	C	D	E	___	**97:**	A	B	C	D	E	___
73:	A	B	C	D	E	___	**98:**	A	B	C	D	E	___
74:	A	B	C	D	E	___	**99:**	A	B	C	D	E	___
75:	A	B	C	D	E	___	**100:**	A	B	C	D	E	___

1. A heat engine performs the reversible cycle *abca* with 9 moles of an ideal gas. Path *ca* is an adiabatic process. The temperatures at points *a* and *b* are 300 K and 500 K, respectively. The volume at point *c* is 0.2 m^3. The adiabatic constant of the gas is 1.6. What is the heat absorbed by the gas in path *ca*?

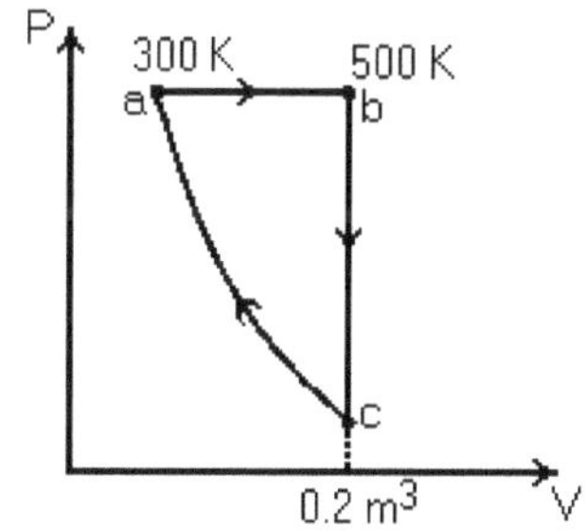

A. –10 kJ
B. 16 kJ
C. 10 kJ
D. 0 kJ
E. –16 kJ

2. Forces A, B, and C act on a body. A fourth force, *F*, keeps the body in equilibrium. What is the *x* component of the force *F*?

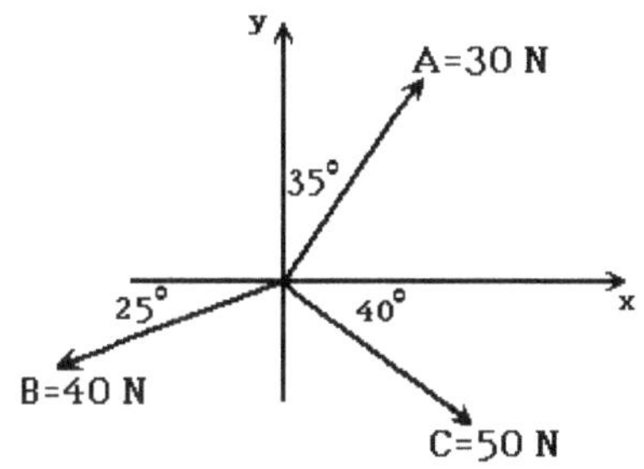

A. –20 N
B. –28 N
C. 20 N
D. 28 N
E. –32 N

3. In an organ pipe (cylindrical 0.2 m long tube 0.02 m in diameter, open at both ends), what is the λ of the 4th harmonic?

A. 0.08 m
B. 0.1 m
C. 0.05 m
D. 0.2 m
E. 0.8 m

4. What is the velocity of a fluid in a tube with a diameter of 3 mm for a pump that has a piston 15 cm in diameter moving at 2 cm/s?

A. 50 m/s
B. 21 cm/s
C. 25 m/s
D. 50 cm/s
E. 5 cm/s

5. What is the essential difference between microwaves and blue light?

A. One is a form of radiation while the other is not
B. Blue light is a beam of photons, and microwaves are positrons
C. One has a positive charge while the other does not
D. One undergoes refraction while the other does not
E. There is no essential difference other than frequency and wavelength

6. The resistance of a variable resistor can be adjusted. A battery is connected to a variable resistor, and the potential difference across the resistor and the current through it are recorded at variable resistance. Assume that the battery is an ideal potential source in series with an internal resistor. Under which circumstances is it valid to ignore the internal resistance of the battery?

I. small external resistance II. large external resistance III. low total current

A. I only
B. II only
C. III only
D. I and III only
E. II and III only

7. Which type of mirror produces an image that is always virtual, erect the same height as the object?

A. concave
B. convex
C. plane
D. spherical
E. none of above

8. Which nuclear radiation particles have the greatest mass?

A. Neutrons
B. Gamma radiation photons
C. Beta particles
D. Alpha particles
E. Protons

9. A particular motor can provide a maximum torque of 110.0 N·m. If this torque accelerates a solid, uniform, cylindrical flywheel of mass 10.0 kg and radius 3.00 m, how long will it take for the flywheel to accelerate from rest to 8.13 rad/s?

A. 2.13 s
B. 3.33 s
C. 4.65 s
D. 5.46 s
E. 6.80 s

10. A spaceship moves away from an asteroid with a speed of $0.80c$ relative to the asteroid. The spaceship then fires a missile with a speed of $0.50c$ relative to the spaceship. What is the speed of the missile measured by astronauts on the asteroid if the missile is fired toward the asteroid?

A. $0.32c$
B. $0.50c$
C. $0.92c$
D. $1.6c$
E. $0.68c$

11. Compared to falling on a wooden floor, a wine glass may not break when it falls on a carpeted floor because of the:

I. lesser impulse in stopping
II. longer time to stop
III. decreased acceleration due to gravity

A. I only
B. II only
C. III only
D. I and II only
E. I and III only

12. A 6.5 kg bag of groceries is carried 1.1 m above the ground at a constant velocity across a 3.8 m room. How much work was done on the bag in this process?

A. 183 J
B. 155 J
C. 0 J
D. 84 J
E. 22.7 J

13. A wave that has a lower frequency has a:

A. higher amplitude
B. shorter period
C. slower velocity
D. shorter wavelength
E. longer wavelength

14. A rod has a length of 2.0000 m at 20 °C. The length of the rod increases to 2.0005 m when the temperature increases to 40 °C. What is the coefficient of thermal expansion α of the material for the rod?

A. 1.25×10^{-5}/K
B. 5×10^{-3}/K
C. 0.75×10^{-5}/K
D. 2.5×10^{-5}/K
E. 0.75×10^{-3}/K

15. An object is viewed at various distances using a concave mirror with a focal length of 12 m. Relative to the mirror, where is the image when the object is 6 m in front of the mirror?

A. 12 m behind
B. 12 m in front
C. 6 m behind
D. 6 m in front
E. 24 m behind

16. What effect does the addition of resistors in parallel have on a circuit?

A. Decreases the current
B. Increases the current
C. Decreases the voltage
D. Increases the voltage
E. No change in the current

17. The electric field at point P due to a point charge of Q a distance R away has a magnitude of E. Which statement is true to double the magnitude of the field at P?

A. Reduce the charge to $Q/2$
B. Reduce the distance to R/2
C. Double the distance to 2R while reducing the charge to $Q/2$
D. Double the charge to $2Q$
E. Double the charge to $2Q$ while reducing the distance to R/2

18. If 20% of the volume of a floating buoy is above the surface of a liquid, then the density of the buoy is what percent of the density of the surrounding liquid?

A. 40%
B. 90%
C. 60%
D. 70%
E. 80%

19. A pipe of length L that is open at both ends resonates at its fundamental frequency. Which statement about the sound is correct?

A. The λ is $L/2$, and there is a displacement node at the midpoint of the pipe
B. The λ is $L/2$, and there is a displacement antinode at the midpoint of the pipe
C. The λ is $2L$, and there is a displacement node at the midpoint of the pipe
D. The λ is $2L$, and there is a displacement antinode at the midpoint of the pipe
E. The λ is $4L$, and there is a displacement node at the midpoint of the pipe

20. Wave interference occurs for:

I. sound waves II. water waves III. light waves

A. I only
B. II only
C. III only
D. I and III only
E. I, II and III

21. What is the spring constant value if a force of 30 N stretches a spring 0.75 m from equilibrium?

A. 85 N/m
B. 25 N/m
C. 35 N/m
D. 40 N/m
E. 30 N/m

22. In a perfectly inelastic collision, two moving objects of unequal mass (A and B) collide and immediately come to rest. Before colliding, object A was traveling at speed 5 times that of object B. Which is the ratio of the mass of object A to that of object B?

A. 50 : 1
B. 5 : 1
C. 1 : 5
D. 1 : 50
E. $\sqrt{5}$: 1

23. What is the wavelength of the matter-wave associated with an electron moving with a speed of 2.5×10^7 m/s? (Use $m_{electron} = 9.11 \times 10^{-31}$ kg and $h = 6.626 \times 10^{-34}$ J·s)

A. 17 pm
B. 29 pm
C. 39 pm
D. 51 pm
E. 76 pm

24. A chestnut falls straight down from a chestnut tree growing on a 20° slope. What is the component of the chestnut's impact velocity parallel to the ground if it hits the ground with a speed of 16 m/s?

A. 5.5 m/s
B. 8.6 m/s
C. 4.1 m/s
D. 11.4 m/s
E. 10.6 m/s

25. In a given reversible process, the temperature of an ideal gas is kept constant as the gas is compressed to a smaller volume. Which statement is correct?

A. The process is adiabatic
B. The pressure remains constant
C. The gas releases heat to its surroundings
D. The gas absorbs heat from its surroundings
E. None of the above

26. A heavy pile driver starting from rest smashes into a pile with a force that depends on the:

I. distance the pile driver falls
II. initial potential energy of the driver
III. initial height of the driver

A. I only
B. II only
C. III only
D. I and III only
E. I, II and III

27. Which of the following is an accurate statement?

A. The fundamental frequency is the highest frequency at which a system naturally vibrates
B. The air in an organ pipe can vibrate at an infinite number of frequencies
C. A system, like a vibrating string, has one possible frequency
D. For a singer to break a wine glass by singing, the amplitude of sound must be greater than the amplitude of vibration for a wine glass
E. None of the above

28. A satellite is in a circular orbit around a planet. What is the satellite's orbital speed if the orbital radius is 34.0 km, and the gravitational acceleration at that height is 2.3 m/s^2?

A. 26 m/s
B. 150 m/s
C. 280 m/s
D. 310 m/s
E. 390 m/s

29. A myopic girl wears eyeglasses that allow her to have clear distant vision. The power of the lenses of her eyeglasses is –3 diopters. Without eyeglasses, the far point of the girl is:

A. 0.32 m
B. 0.17 m
C. 0.75 m
D. 0.33 m
E. 0.50 m

30. A woman starts her car from rest and accelerates at a constant 2.5 m/s^2 for 8 s to get to her cruising speed. What is her velocity 8 s after beginning her trip?

A. 1.8 m/s
B. 5.5 m/s
C. 12.5 m/s
D. 7.5 m/s
E. 20 m/s

31. What is the velocity of the cars immediately after impact for a perfectly inelastic collision between a 2,500 kg car moving North at 7 m/s and a 2,000 kg car moving South at 14 m/s?

A. 2.3 m/s S
B. 9.4 m/s S
C. 5.4 m/s N
D. 9.4 m/s N
E. 11 m/s N

32. Grandfather clocks are designed to be adjusted by moving the weight at the bottom of the pendulum up or down. Suppose a grandfather clock is running slow. Which of the following adjustments of the weight would make it more accurate?

A. Move weight down
B. Move weight up
C. Remove half the mass from the weight
D. Add double the mass to the weight
E. Increase the amplitude of swing significantly

33. Why does the paper rise when the air is blown above a paper strip?

A. Air above the paper moves slower, while the pressure remains constant
B. Air above the paper moves slower, and the pressure is higher
C. Air above the paper moves faster, while the pressure remains constant
D. Air above the paper moves slower, and the pressure is lower
E. Air above the paper moves faster, and the pressure is lower

34. If a battery's internal resistance is 20 Ω, what current flows in a 4 V battery if the circuit is shorted?

A. 5 A
B. 3.2 A
C. 80 A
D. 0.2 A
E. 7.3 A

35. One Becquerel equals:

A. a mole of disintegrations/sec
B. 100 rads
C. 1 disintegration/sec
D. 3.7×10^{10} disintegrations/sec
E. 1 rem

36. A glass flask has a volume of 450 ml at a temperature of 22 °C. The flask contains 442 ml of mercury at an equilibrium temperature of 22 °C. The temperature is raised until the mercury reaches the 450 ml reference mark. What is the temperature at which this occurs? (Use the coefficients of volume expansion α of mercury = 18×10^{-5} K^{-1} and α of glass = 2×10^{-5} K^{-1})

A. 103 °C
B. 152 °C
C. 92 °C
D. 135 °C
E. 116 °C

37. Which graph represents a constant positive acceleration?

I. 

II. 

III. 

A. I only
B. II only
C. III only
D. II and III only
E. I, II and III

38. In formulating the special theory of relativity, Einstein postulated that the laws of physics are the same in reference frames that:

A. are stationary, but not in moving frames
B. oscillate
C. move at constant velocity with respect to an inertial frame
D. accelerate
E. none of the above

39. What happens to an object if a constant net torque is applied?

A. Rotates with increasing angular acceleration
B. Rotates with constant angular acceleration
C. Rotates with constant angular velocity
D. Rotates with increasing linear velocity
E. Increases its moment of inertia

40. Researchers who work around radioactivity wear dosimeter badges on their clothing to monitor their radiation exposure. These badges consist of small pieces of photographic film enclosed in a holder that protects the film against light exposure. What kind of radiation do these devices monitor?

I. alpha radiation II. beta radiation III. gamma radiation

A. I only
B. II only
C. III only
D. I and II only
E. I and III only

41. Ignoring air resistance, if it takes 16 s for a package to strike the ground, how high above the ground was the package when it was thrown upward from a stationary helicopter at 15 m/s? (Use acceleration due to gravity $g = 10\ m/s^2$)

A. 920 m
B. 740 m
C. 1,400 m
D. 1,040 m
E. 1,800 m

42. What is the mass of the car if a tow truck exerts a force of 3,000 N on a car and accelerates it at 2 m/s^2?

A. 500 kg
B. 1,000 kg
C. 1,500 kg
D. 2,000 kg
E. None of the above

43. A bimetallic strip, consisting of metal A on the top and metal B on the bottom, is rigidly attached to a wall at the left, as shown. The coefficient of linear thermal expansion of metal A is greater than that of metal B. Which statement is true if the strip is heated uniformly?

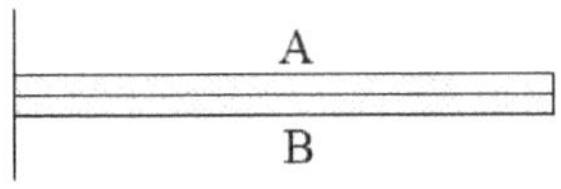

A. remains horizontal but increases in length
B. remains horizontal but decreases in length
C. curves upward
D. curves downward
E. bends in the middle

44. A massless, ideal spring has a spring constant of 4,600 N/m as it hangs from the ceiling. By how much does its potential energy increase if a 30 kg mass is attached to it? (Use the acceleration due to gravity $g = 10\ \text{m/s}^2$)

A. 9.8 J
B. 6.3 J
C. 19.4 J
D. 12.4 J
E. 4.3 J

45. What is the speed of a water wave if a floating object oscillates up and down two complete cycles in 1 s as the water wave with a wavelength of 6 m passes?

A. 1 m/s
B. 0.17 m/s
C. 12 m/s
D. 18 m/s
E. 3 m/s

46. By what value does the sound level decrease if the intensity of the sound wave decreases by a factor of 10?

A. 0.1 dB
B. 10 dB
C. 1 dB
D. 100 dB
E. 0.01 dB

47. A piece of wood is floating in a bathtub, and the second piece of wood sits on top of the first piece and does not touch the water. If the top piece is taken off and placed in the water, what happens to the water level in the tub?

A. It remains the same
B. It decreases
C. It increases
D. Requires the volumes of the two pieces of wood
E. Requires the densities of the two pieces of wood

48. A point charge Q_1 has a charge of -1 μC. What is the number of excess electrons in charge Q_1? (Use the charge of an electron $e = -1.60 \times 10^{-19}$ C)

A. 2×10^{11}
B. 2×10^{12}
C. 6.3×10^{12}
D. 6.3×10^{13}
E. 2×10^{13}

49. The capacitance of a capacitor depends on its:

I. energy stored II. charge III. potential difference across it

A. I only
B. II only
C. II and III only
D. I and III only
E. None of the above

50. What is the radius of curvature for a flat plane mirror?

A. Imaginary
B. Zero
C. Infinite
D. Negative
E. Undefined

51. During a nuclear fusion reaction, it is possible to produce neutrons with 12.5 MeV and a speed of approximately 54,300 km/s. At this speed, what is the mass of a neutron? (Use the rest mass of a neutron $m_0 = 1.675 \times 10^{-24}$ g, the speed of light $c = 3 \times 10^8$ m/s and 1 eV = 1.602×10^{-19} J)

A. 4.3×10^{-34} g
B. 6.4×10^{-21} g
C. 2.8×10^{-23} g
D. 5.1×10^{-17} g
E. 1.7×10^{-24} g

52. Consider two uniform solid spheres where both have the same diameter, but one has twice the mass of the other. What is the ratio of the larger moment of inertia to the smaller?

A. $\sqrt{2} : 1$
B. $4 : 1$
C. $10 : 1$
D. $2 : 1$
E. $16 : 1$

53. Using 1,000 J of work, an elevator is raised from the ground to the second floor in 40 s. How much power does the elevator use?

A. 100 W
B. 25 W
C. 50 W
D. 1,000 W
E. 200 W

54. A ball is attached to an ideal spring and oscillates with a period T. What is the new period if the mass of the ball is doubled?

A. $T\sqrt{2}$
B. T
C. T/2
D. 2T
E. $T/\sqrt{2}$

55. Larry drops a mailing tube to the floor, and the tube produces a musical note. The mailing tube is 1.5 m long with a cylindrical cross-sectional diameter of 3.5 cm. It is sealed at one end and open at the other. What is the wavelength of the third harmonic? (Use the speed of sound in air $v = 340$ m/s)

A. 1.5 m
B. 2 m
C. 3.5 m
D. 4 m
E. 7.5 m

56. A compressed gas with a total mass of 115 kg is stored in a spherical container with a radius of 0.6 m. What is the density of the compressed gas?

A. 147 kg/m^3
B. 183 kg/m^3
C. 151 kg/m^3
D. 127 kg/m^3
E. 267 kg/m^3

57. Electrons are made to flow in a wire when there is:

I. a potential difference across its ends
II. an imbalance between positive and negative charges in the wire
III. more final kinetic energy than initial kinetic energy in the wire

A. I only
B. II only
C. III only
D. I and II only
E. I and III only

58. A Coulomb per second is the same as a(n):

A. Newton
B. volt-second
C. ampere
D. watt
E. volt per second

59. What is observed when an object is precisely at its center of curvature in front of a concave mirror?

A. The image is not seen because it is focused at a different distance
B. The image is seen, but it appears smaller
C. The image is seen, but it appears larger
D. The image is seen as the same size, and it is upright
E. The image is seen as the same size, but it is inverted

60. Which type of radiation cannot penetrate human skin and requires only heavy paper or clothing as minimum protective shielding?

A. α
B. γ
C. β
D. Nuclide
E. X-ray

61. The process shown on the pressure *vs.* volume diagram represents which type of expansion?

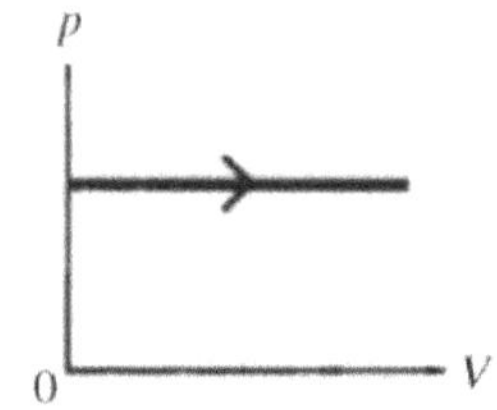

A. Isobaric
B. Isometric
C. Isothermal
D. Adiabatic
E. Isochoric

62. A block is on a frictionless table on Earth, and the block accelerates at 4 m/s^2 when a 20 N horizontal force is applied. The block and table are then set up on the Moon, where the acceleration due to gravity is 1.62 m/s^2. What is the weight of the block on the Moon?

A. 11 N
B. 9.7 N
C. 5.6 N
D. 6.2 N
E. 8.1 N

63. In the figure, the frequency is:

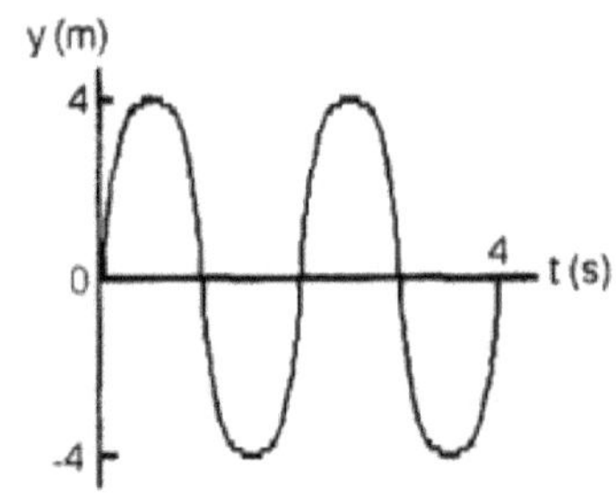

A. 2 Hz
B. 4 Hz
C. 0.5 Hz
D. 1 Hz
E. 1.5 Hz

64. Consider a tiny hole in the bottom of a tank that is 18 cm in diameter and is filled with water to a height of 80 cm. What is the speed at which the water exits the tank through the hole? (Use acceleration due to gravity $g = 9.8$ m/s^2)

A. 4 m/s
B. 17 m/s
C. 42 m/s
D. 36 m/s
E. 7 m/s

65. A coil of wire containing N turns is in an external magnetic field that is perpendicular to the plane of the coil. The magnetic field is steadily changing, and an emf (V) is induced in the coil. With other values constant, if the magnetic flux and the number of turns in the coil are doubled, by what factor does the induced emf increase or decrease?

A. V/4
B. 4V
C. V/2
D. 2V
E. V

66. A component with an 18 Ω resistor is rated for use at power levels not exceeding 16 W. How much current can safely flow through the component?

A. 0.12 A
B. 2.8 A
C. 1.31 A
D. 0.94 A
E. 12.1 A

67. How are the wavelength and frequency of light related?

A. Wavelength increases as the frequency increases
B. Wavelength increases as the frequency decreases
C. Wavelength is one-fourth of the frequency
D. Wavelength is four times the frequency
E. Wavelength is not related to the frequency

68. Which of the following has the highest ionizing power?

A. β particle
B. positron
C. α particle
D. γ particle
E. All are equal

69. A merry-go-round is 18 m in diameter, with a child sitting on the outer edge. If the merry-go-round makes 8.3 rev/min, what is the velocity of the child?

A. 1.9 m/s
B. 7.8 m/s
C. 3.9 m/s
D. 14.6 m/s
E. 18.3 m/s

70. What is the magnitude of the instantaneous force with which a ball hits a bat if a batter hits a ball pitched with a 1,700 N instantaneous force?

A. Greater than 1,700 N
B. Zero
C. Equal to √1,700 N
D. Less than 1,700 N
E. Equal to 1,700 N

71. A 2.8 kg block, moving on a frictionless surface with a speed (v_i) of 8.5 m/s, makes a perfectly elastic collision with a block of mass m at rest. After the collision, the 2.8 kg block recoils with a speed (v_f) of 1.1 m/s. What is the mass of block m?

A. 22.8 kg
B. 14.4 kg
C. 17.8 kg
D. 3.6 kg
E. 12.2 kg

72. If the height of a frictionless incline is h, determine the work done by the force due to gravity F as the crate of mass m slides down the incline.

A. $mgh\ \sin^2$
B. mgh
C. $mgh\ \cos^2$
D. $mgh\ \sin$
E. $mgh\ \tan$

73. What is the speed of a wave if it has a wavelength of 25 cm and a frequency of 1.6 kHz?

A. 400 m/s
B. 1,400 m/s
C. 0.4 m/s
D. 14 m/s
E. 28 m/s

74. A glass beaker of unknown mass contains 60 ml of water. The system absorbs 2,200 cals of heat, and the temperature rises 25 °C as a result. What is the mass of the beaker? (Use specific heat for glass = 0.18 cal/g·°C and specific heat for water = 1.0 cal/g·°C)

A. 675 g
B. 540 g
C. 460 g
D. 370 g
E. 156 g

75. What is the radius of curvature for a concave mirror with a focal length of 20 cm?

A. 20 cm
B. –40 cm
C. 40 cm
D. 10 cm
E. –20 cm

76. What is the power dissipated in the 2 Ω resistor in the circuit?

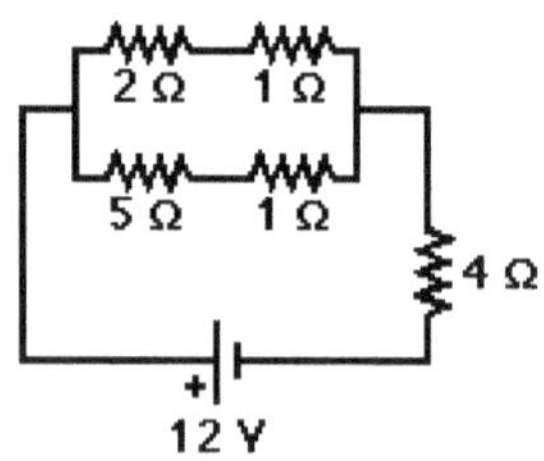

A. 6.3 W
B. 5.7 W
C. 3.5 W
D. 7.8 W
E. 8.5 W

77. When a charged particle moves with a velocity perpendicular to a magnetic field, the field acts to change the particle's:

A. v
B. KE
C. m
D. p
E. q

78. What is the gauge pressure 10 m below the surface of the ocean? (Use the acceleration due to gravity g = 9.8 m/s^2, the density of water $\rho = 10^3$ kg/m^3 and $P_{atm} = 1.01 \times 10^5$ Pa)

A. 1.4×10^5 Pa
B. 2.3×10^5 Pa
C. 9.8×10^4 Pa
D. 6.5×10^4 Pa
E. 5×10^6 Pa

79. If a 75 cm string, fixed at both ends between two adjacent nodes, supports a standing wave, what is the wavelength λ?

A. 15 cm
B. 300 cm
C. 7.5 cm
D. 200 cm
E. 150 cm

80. The speed of sound in steel is 4,900 m/s. What is the wavelength of a sound wave in steel of frequency 640 Hz?

A. 1.3 m
B. 0.83 m
C. 7.7 m
D. 4.2 m
E. 2.6 m

81. Ignoring the force of friction, what is the work done by Stacey when she carries a 15 kg mass and walks along the x-axis for 100 m with a constant velocity of 2.5 m/s?

A. 0 J
B. 5 J
C. 30 J
D. 50 J
E. 450 J

82. Hannah is in karate and delivers a swift blow to sever a cement block in two with her bare hand. The magnitude is the same for the:

I. time of impact on the block and Hannah's hand
II. force on the block and Hannah's hand
III. impulse on the block and Hannah's hand

A. I only
B. II only
C. III only
D. I and III only
E. I, II and III

83. Karen is driving a car along a road when she realizes, almost too late, that she needs to make a left-hand turn. Karen quickly turns the car's steering wheel, and the textbooks in the passenger seat go crashing against the passenger door. Which of the following statements are true?

I. The textbooks were pushed against the door by a centrifugal force
II. The textbooks were pushed against the door by a centripetal force
III. The forces acting on the textbooks when they crashed against the door were gravity, the normal force, and a force toward the right

A. I and III only
B. II and III only
C. III only
D. I, II and III
E. None of the above

84. How far does a car travel while accelerating if it accelerates from 5 m/s to 21 m/s at a rate of 4 m/s^2?

A. 112 m
B. 35 m
C. 71 m
D. 52 m
E. 153 m

85. The coefficient of linear expansion of lead is 30×10^{-6} K^{-1}. What change in temperature causes a 10 m long lead bar to change in length by 3 mm?

A. 7.5 K
B. 20 K
C. 10 K
D. 5 K
E. 40 K

86. A constant force of 10 N is applied horizontally for 15 s on a 2 kg toy cart so that the cart begins to move along the level frictionless floor. What is the kinetic energy of the cart just after the 15 s?

A. 1,540 J
B. 2,300 J
C. 3,450 J
D. 2,985 J
E. 5,625 J

87. Mary has tuned the B string of a guitar to the fundamental frequency of an E (680 Hz). The E string is left untuned. When she plucks the B string and the untuned E string together, she hears a note that changes from loud to soft to loud twice per second. What is an allowable fundamental f on the untuned E string?

A. 2 Hz
B. 340 Hz
C. 680 Hz
D. 682 Hz
E. 1,360 Hz

88. A positive reaction charge Q is held fixed at the origin. A positive charge q is on the positive x-axis and released. Assuming no friction, which statement describes the theoretical acceleration of q after its release?

A. Decreases and then increases
B. Remains constant for a while and then decreases, eventually reaching zero
C. Decreases and eventually reaches zero
D. Decreases but never reaches zero
E. Acceleration remains unchanged

89. Which statement about a single thin lens is correct?

A. A diverging lens always produces a virtual upright image
B. A converging lens sometimes produces a real upright image
C. A diverging lens sometimes produces a virtual upright image
D. A converging lens always produces a real upright image
E. A diverging lens produces a virtual upright image only if the object is between the lens and its focal point

90. What is the magnitude and direction of the combined vectors A and B if vector A has a length of 5 units and is directed to the North (N) while vector B has a length of 11 units and is directed to the South (S)?

A. 16 units, S
B. 16 units, N
C. 6 units, N
D. 6 units, S
E. 2.2 units, N

91. A laser produces a beam of 4000 nm light. A shutter allows a pulse of light for 30.0 ps to pass. What is the uncertainty in the energy of a photon in the pulse? ($h = 6.626 \times 10^{-34}$ J·s and 1 eV = 1.60×10^{-19} J)

A. 2.6×10^{-2} eV
B. 4.2×10^{-3} eV
C. 6.8×10^{-4} eV
D. 2.2×10^{-5} eV
E. 2.4×10^{-6} eV

92. At which point in a swinging pendulum's arc does it have the most energy?

I. At its lowest point, it has the most potential energy
II. The energy is the same at all points of the arc
III. At its highest point, it has the most kinetic energy

A. I only
B. II only
C. III only
D. I and II only
E. I and III only

93. If the amount of fluid flowing through a tube remains constant, how does the speed of the fluid change if the diameter of the tube increases from 3 cm to 8 cm?

A. Increases by a factor of 5
B. Increases by a factor of 9
C. Decreases by a factor of 5
D. Decreases by a factor of 7
E. Remains the same

94. Voltage can be induced in a wire by:

I. changing the current in a nearby wire
II. moving a magnet near the wire
III. moving the wire near a magnet

A. I only
B. II only
C. III only
D. II and III only
E. I, II and III

95. Ionizing radiation is dangerous to living tissue because it:

A. causes electrons to be absorbed by the cell's nucleus
B. alters the chemical structure of atoms or molecules
C. causes a large release of heat energy
D. binds protons and neutrons together
E. has a penetrating power that varies with its source

96. A container of an ideal gas at standard temperature and pressure undergoes an isothermal expansion. How much work does it do if its entropy changes by 2.6 J/K? (Use standard temperature = 273 K and pressure = 1 atm)

A. 1.4×10^3 J
B. -1×10^3 J
C. 7.1×10^2 J
D. 0 J
E. -1.4×10^3 J

97. Jack stands on a bridge and throws a rock straight down. The rock leaves Jack's hand at $t = 0$. Which of the graphs represents the velocity of the stone as a function of time?

A.
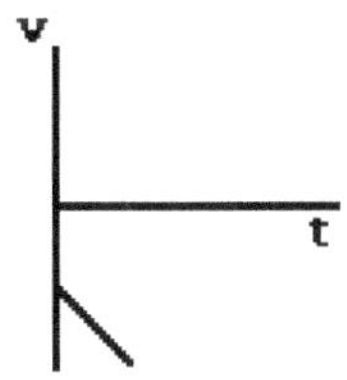

B.
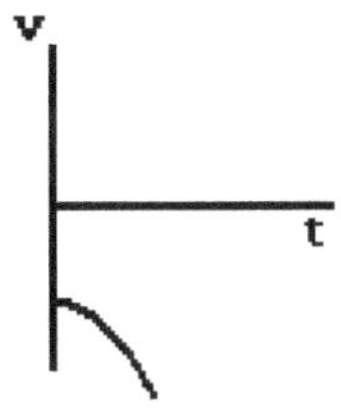

C.
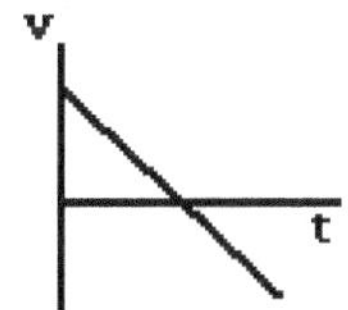

D.
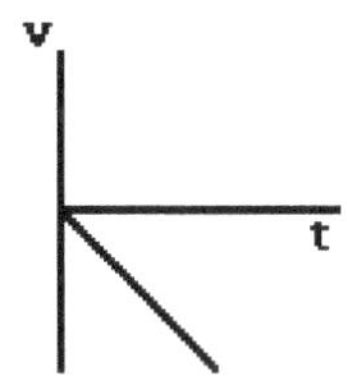

E.
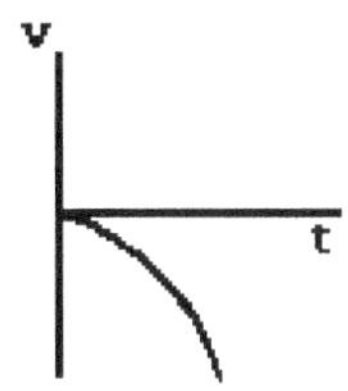

98. A 100 g ball of clay is thrown horizontally with a speed of 50 m/s toward a 900 g block resting on a frictionless surface. It hits the block and sticks. The clay exerts a constant force on the block during the 10 ms it takes for the clay to come to rest relative to the block. After 10 ms, the block and the clay are sliding along the surface as a single system. What is the force of the clay on the block during the collision?

A. 500 N
B. 2,250 N
C. 5,400 N
D. 4,500 N
E. 450 N

99. Cart A (2 kg) and Cart B (2.5 kg) run along a frictionless, level, one-dimensional track. Cart B is initially at rest, and Cart A travels 0.6 m/s toward the right when it encounters Cart B. After the collision, Cart A is at rest. What is the final velocity of Cart B?

A. 0.48 m/s
B. 0.36 m/s
C. 0.75 m/s
D. 0.64 m/s
E. 1.1 m/s

100. What is the name of the unit used to measure the amount of radiation absorbed per gram of tissue?

A. Becquerel
B. Rad
C. Curie
D. Roentgen
E. Sievert

Notes for active learning

Answer Keys
and
Detailed Explanations

Topical Practice Questions

Answer Keys

Fluids and Solids

1: C	11: A	21: E	31: C	41: A	51: A
2: D	12: A	22: C	32: A	42: D	52: B
3: A	13: A	23: D	33: B	43: B	53: D
4: A	14: E	24: E	34: D	44: E	54: D
5: E	15: C	25: D	35: C	45: A	55: C
6: C	16: D	26: B	36: D	46: C	
7: B	17: C	27: E	37: B	47: A	
8: E	18: B	28: D	38: D	48: B	
9: A	19: C	29: B	39: B	49: E	
10: C	20: D	30: B	40: E	50: D	

Heat and Thermodynamics

1: B	11: B	21: D	31: B	41: E	51: E
2: C	12: A	22: A	32: D	42: C	52: A
3: D	13: B	23: B	33: E	43: B	53: D
4: A	14: E	24: D	34: B	44: A	54: B
5: C	15: C	25: A	35: D	45: D	55: D
6: E	16: A	26: A	36: E	46: C	
7: D	17: D	27: D	37: C	47: E	
8: A	18: B	28: B	38: A	48: A	
9: C	19: E	29: E	39: B	49: D	
10: E	20: A	30: D	40: B	50: B	

Electric Circuits

1: B	11: A	21: E	31: C	41: C	51: B
2: D	12: E	22: C	32: E	42: E	52: D
3: C	13: D	23: D	33: B	43: A	53: A
4: D	14: B	24: A	34: C	44: B	54: E
5: E	15: C	25: B	35: D	45: C	55: D
6: A	16: A	26: C	36: C	46: E	
7: C	17: B	27: A	37: B	47: B	
8: D	18: D	28: D	38: E	48: D	
9: B	19: C	29: E	39: A	49: A	
10: E	20: C	30: A	40: E	50: C	

Electrostatics and Electromagnetism

1: C	11: D	21: A	31: A	41: B	51: E
2: D	12: E	22: C	32: B	42: C	52: A
3: A	13: C	23: A	33: C	43: A	53: B
4: C	14: E	24: B	34: D	44: C	54: D
5: E	15: D	25: A	35: C	45: B	55: A
6: A	16: A	26: D	36: E	46: C	
7: D	17: B	27: E	37: B	47: E	
8: B	18: D	28: C	38: A	48: A	
9: D	19: B	29: D	39: C	49: D	
10: A	20: E	30: A	40: A	50: C	

Atomic and Nuclear Structure

1: A	11: E	21: E	31: A	41: D	51: D
2: C	12: D	22: C	32: C	42: E	52: E
3: E	13: A	23: B	33: E	43: C	53: B
4: B	14: C	24: D	34: B	44: B	54: E
5: C	15: C	25: A	35: C	45: E	
6: D	16: A	26: C	36: D	46: B	
7: B	17: B	27: E	37: B	47: C	
8: E	18: D	28: D	38: E	48: C	
9: C	19: B	29: C	39: C	49: B	
10: A	20: A	30: D	40: B	50: A	

Particle Physics

1: C	11: C	21: C
2: D	12: C	22: A
3: B	13: C	23: B
4: B	14: E	
5: B	15: E	
6: D	16: B	
7: E	17: E	
8: B	18: A	
9: A	19: B	
10: A	20: C	

Quantum Mechanics

1: E	11: A	21: C	31: B	41: C
2: B	12: D	22: A	32: A	42: E
3: C	13: E	23: D	33: D	43: C
4: C	14: C	24: B	34: C	
5: D	15: E	25: B	35: A	
6: C	16: D	26: B	36: C	
7: C	17: E	27: D	37: B	
8: B	18: E	28: B	38: C	
9: D	19: B	29: A	39: C	
10: A	20: D	30: D	40: B	

Special Relativity

1: A	11: E	21: A	31: D	41: C
2: E	12: B	22: E	32: D	42: B
3: D	13: A	23: D	33: A	43: B
4: E	14: B	24: B	34: B	44: A
5: D	15: D	25: C	35: C	45: D
6: D	16: E	26: A	36: C	46: A
7: A	17: C	27: B	37: C	47: A
8: C	18: B	28: D	38: D	48: B
9: B	19: D	29: C	39: B	49: D
10: C	20: A	30: B	40: C	50: D

Notes for active learning

Fluids and Solids – Detailed Explanations

1. C is correct.

Refer to the unknown liquid as "A" and the oil as "O."

$$\rho_A h_A g = \rho_O h_O g$$

cancel g from both sides of the expression

$$\rho_A h_A = \rho_O h_O$$

$$h_A = 5 \text{ cm}$$

$$h_O = 20 \text{ cm}$$

$$h_A = \tfrac{1}{4} h_O$$

$$\rho_A (\tfrac{1}{4}) h_O = \rho_O h_O$$

$$\rho_A = 4\rho_O$$

$$\rho_A = 4(850 \text{ kg/m}^3)$$

$$\rho_A = 3{,}400 \text{ kg/m}^3$$

2. D is correct.

$$\text{P} = \rho_{oil} \times \text{V}_{oil} \times g \,/\, (\text{A}_{tube})$$

$$\text{P} = [\rho_o \pi (r_{tube})^2 \times hg] \,/\, \pi (r_{tube})^2$$

cancel $\pi(r_{tube})^2$ from the numerator and denominator

$$\text{P} = \rho_o g h$$

$$\text{P} = (850 \text{ kg/m}^3)\cdot(9.8 \text{ m/s}^2)\cdot(0.2 \text{ m})$$

$$\text{P} = 1{,}666 \text{ Pa}$$

3. A is correct.

$$m_{oil} = \rho_{oil} \text{V}_{oil}$$

$$\text{V} = \pi r^2 h$$

$$m_{oil} = \rho_{oil} \pi r^2 h$$

$$m_{oil} = \pi(850 \text{ kg/m}^3)\cdot(0.02 \text{ m})^2 \times (0.2 \text{ m})$$

$$m_{oil} = 0.21 \text{ kg}$$

$$m_{oil} = 210 \text{ g}$$

4. A is correct.

Gauge pressure is the pressure experienced by an object referenced at atmospheric pressure.

When the block is lowered, its gauge pressure increases according to:

$$P_G = \rho g h$$

At $t = 0$, the block just enters the water and $h = 0$ so $P_G = 0$.

As time passes, the height of the block below the water increases linearly, so P_G increases linearly.

5. E is correct.

Using *Bernoulli's principle* and assuming the opening of the tank is so large that the initial velocity is essentially zero:

$$\rho g h = \frac{1}{2}\rho v^2$$

cancel ρ from both sides of the expression

$$gh = \frac{1}{2}v^2$$

$$v^2 = 2gh$$

$$v^2 = 2\cdot(9.8\ \text{m/s}^2)\cdot(0.8\ \text{m})$$

$$v^2 = 15.68\ \text{m}^2/\text{s}^2$$

$$v = 3.96\ \text{m/s} \approx 4\ \text{m/s}$$

Note: the diameter is not used to solve the problem.

6. C is correct.

The *ideal gas law* is:

$$PV = nRT$$

Keeping nRT constant:

$$\text{If } P_{final} = \frac{1}{2}P_{initial}$$

$$V_{final} = \frac{1}{2}V_{initial}$$

However, in an isothermal process, there is no change in internal energy.

Therefore, because energy must be conserved:

$$\Delta U = 0$$

7. B is correct.

Hooke's Law for a spring:

$$F = kx$$

Solve for k:

$$k = F / x$$

$$k = (9.5\ \text{kg})\cdot(9.8\ \text{m/s}^2) / (0.004\ \text{m})$$

$$k = 23{,}275\ \text{N/m}$$

$$k = 2.3 \times 10^4\ \text{N/m}$$

8. E is correct.

The object sinks when the buoyant force is less than the weight of the object.

Since the buoyant force is equal to the weight of the displaced fluid, an object sinks precisely when the weight of the fluid it displaces is less than the weight of the object itself.

9. A is correct.

$$\text{P} = \rho gh$$

$$\text{P} = (10^3\ \text{kg/m}^3)\cdot(9.8\ \text{m/s}^2)\cdot(100\ \text{m})$$

$$\text{P} = 9.8 \times 10^5\ \text{N/m}^2$$

10. C is correct.

Absolute pressure = gauge pressure + atmospheric pressure

$$\text{P}_{abs} = \text{P}_G + \text{P}_{atm}$$

$$\text{P}_{abs} = \rho gh + \text{P}_{atm}$$

Atmospheric pressure is added to the total pressure at the bottom of a volume of liquid.

Therefore, if the atmospheric pressure increases, absolute pressure increases by the same amount.

11. A is correct.

Surface tension increases as temperature decreases.

Generally, the *cohesive forces* maintaining surface tension decrease as molecular thermal activity increases.

12. A is correct.

A force per unit area causes the stretch of a wire.

If force is constant, an increase in weight increases the area related to d^2.

$$E = \sigma / \varepsilon$$

where E = Young's modulus, σ = stress (F / A) and ε = strain ($\Delta L / L$)

$$E = (F / A) / (\Delta L / L)$$

$$E = FL / A\Delta L$$

Relate force and area:

$$F = (E)\cdot(\Delta L)\cdot(A) / (L)$$

$$F = (E)\cdot(\Delta L)\cdot(\pi / 4 \times d^2) / (L)$$

The force is directly proportional to the ΔL and is directly proportional to d^2.

13. A is correct.

The buoyant force upward must balance the weight downward.

Buoyant force = weight of the volume of water displaced

$$F_B = W_{object}$$

$$\rho Vg = W_{object}$$

$$W_{object} = 60\text{ N}$$

$$W_{object} = (\rho_{water})\cdot(V_{water})\cdot(g)$$

$$60\text{ N} = (1{,}000\text{ kg/m}^3)\cdot(V_{water})\cdot(10\text{ m/s}^2)$$

$$V_{water} = 60\text{ N} / (1{,}000\text{ kg/m}^3)\cdot(10\text{ m/s}^2)$$

$$V_{water} = 0.006\text{ m}^3$$

14. E is correct.

Volume flow rate:

$$Q = vA$$

$$Q = (2.5\text{ m/s})\pi r^2$$

$$Q = (2.5\text{ m/s})\cdot(0.015\text{ m})^2\pi$$

$$Q = 1.8 \times 10^{-3}\text{ m}^3\text{/s}$$

15. C is correct.

Force equation for the cork that is not accelerating:

$$F_B - mg = 0$$

Let m be the mass and V be the volume of the cork.

Replace:

$$m = \rho V$$

$$F_B = (\rho_{water})\cdot(V_{disp})\cdot(g)$$

$$(\rho_{water})\cdot(V_{disp})\cdot(g) = (\rho_{cork})\cdot(V)\cdot(g)$$

$$V_{disp} = \tfrac{3}{4}V$$

$$\rho_{water}\,(\tfrac{3}{4}Vg) = (\rho_{cork})\cdot(V)\cdot(g)$$

cancel g and V from both sides of the expression

$$\tfrac{3}{4}\rho_{water} = \rho_{cork}$$

$$\rho_{cork} / \rho_{water} = \tfrac{3}{4} = 0.75$$

16. D is correct.

volume = mass / density

$$V = (600\text{ g}) / (0.93\text{ g/cm}^3)$$

$$V = 645\text{ cm}^3$$

17. C is correct.

For monatomic gases:

$$U = 3/2k_BT$$

where U is the average KE per molecule and k_B is the Boltzmann constant

18. B is correct.

The object weighs 150 N less while immersed because the buoyant force supports 150 N of total weight.

Since the object is totally submerged, the volume of water displaced equals the volume of the object.

$$F_B = 150\text{ N}$$

$$F_B = \rho_{water} \times V_{water} \times g$$

$$V = F_B / \rho g$$

$$V = (150\text{ N}) / (1{,}000\text{ kg/m}^3)\cdot(10\text{ m/s}^2)$$

$$V = 0.015\text{ m}^3$$

19. C is correct.

$F_B / \rho_w = (m_c g) / \rho_c$

$F_B = (\rho_w m_c g) / \rho_c$

$F_B = [(1\ g/cm^3)\cdot(0.03\ kg)\cdot(9.8\ m/s^2)] / (8.9\ g/cm^3)$

$F_B = 0.033\ N$

$m_{total} = m_w + (F_B / g)$

$m_{total} = (0.14\ kg) + [(0.033\ N) / (9.8\ m/s^2)]$

$m_{total} = 0.143\ kg = 143\ g$

20. D is correct.

$v_1A_1 = v_2A_2$

$v_2 = v_1A_1 / A_2$

$v_2 = [v_1(\pi/4)d_1^2] / (\pi/4)d_2^2$

cancel $(\pi/4)$ from the numerator and denominator

$v_2 = v_1(d_1^2 / d_2^2)$

$v_2 = (1\ m/s)\cdot[(6\ cm)^2 / (3\ cm)^2]$

$v_2 = 4\ m/s$

21. E is correct.

Static fluid pressure:

$P = \rho g h$

The pressure is only dependent on gravity (g), the height (h) of the fluid above the object, and fluid density (ρ).

Pressure does depend on the depth of the object but does not depend on the surface area of the object.

Both objects are submerged to the same depth, so the fluid pressure is equal.

Note that the buoyant force on the blocks is NOT equal, but the pressure (force / area) is equal.

22. C is correct.

Surface tension force acts as the product of surface tension and the total length of contact.

$$F = \text{A}L$$

For a piece of thread, the length of contact is l as shown:

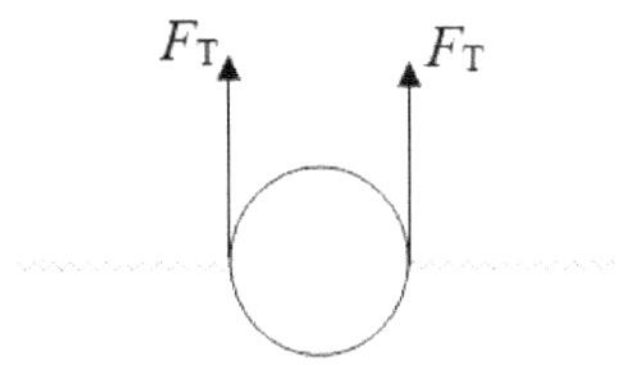

cross-sectional view

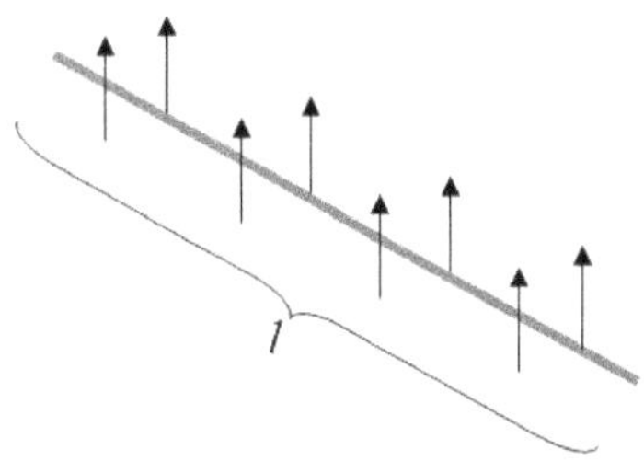

$L = 2L$ because the force acts on both sides of the thread.

Thus, for a thread rectangle, the total contact length is the total length times two.

$$L = 2(l + w + l + w)$$

$$F_{\text{max}} = 2\text{A}(l + w + l + w)$$

$$F_{\text{max}} = 2\text{A}(2l + 2w)$$

$$F_{\text{max}} = 4\text{A}(l + w)$$

23. D is correct.

Gauge pressure is the measure of pressure with respect to atmospheric pressure.

So if the pressure inside the tire is equal to the air pressure outside, the gauge reads zero.

24. E is correct.

The pressure is measured in force per unit area, which is the force divided by the area.

25. D is correct.

The shear stress is the force per unit area and has units of N/m^2.

26. B is correct.

Because the reservoir area is assumed to be virtually infinite, the velocity of the flow at the top of the tank is assumed to be zero.

Using Bernoulli's equation, find the speed through the 3 cm pipe:

$$(\tfrac{1}{2}\rho v^2 + \rho gh)_{\text{out}} = (\tfrac{1}{2}\rho v^2 + \rho gh)_{\text{in}}$$

$$\tfrac{1}{2}\rho v^2 = \rho gh$$

cancel ρ from both sides of the expression

$$\tfrac{1}{2}v^2 = gh$$

continued…

$v^2 = 2gh$

$v = \sqrt{(2gh)}$

$v = \sqrt{[2(9.8 \text{ m/s}^2)\cdot(4 \text{ m})]}$

$v_{3cm} = 8.9 \text{ m/s}$

To find the speed through the 5 cm pipe use the continuity equation:

$A_{3cm}v_{3cm} = A_{5cm}v_{5cm}$

$(\pi / 4)\cdot(3 \text{ cm})^2\cdot(8.9 \text{ m/s}) = (\pi / 4)\cdot(5 \text{ cm})^2\cdot(v_{5cm})$

$v_{5cm} = 3.2 \text{ m/s}$

27. E is correct.

The ideal gas law is:

$PV = nRT$

where n, R and T are constants

P is pressure, V is volume, n is the number of particles, R is the ideal gas law constant, and T is temperature.

If P → 3P, then V→ (1/3)V to maintain a constant temperature.

28. D is correct.

$F = PA + F_{cover}$

$P = \rho gh$

$F = \rho ghA + F_{cover}$

$F = [(1{,}000 \text{ kg/m}^3)\cdot(10 \text{ m/s}^2)\cdot(1 \text{ m})\cdot(1 \text{ m}^2)] + 1{,}500 \text{ N}$

$F = 11{,}500 \text{ N}$

29. B is correct.

The buoyant force on a totally submerged object is independent of its depth below the surface (since any increase in the water's density is ignored).

The buoyant force on the ball at a depth of 4 m is 20 N, the same as the buoyant force at 1 m.

When it sits at the bottom of the pool, the two upward forces (i.e., the buoyant force F_B and the normal force F_N) must balance the downward force of gravity.

$F_B + F_N = F_g$

$(20 \text{ N}) + F_N = 80 \text{ N}$

$F_N = 80 \text{ N} - 20 \text{ N}$

$F_N = 60 \text{ N}$

30. B is correct.

$V_{\text{Fluid Displaced}} = \frac{1}{2}V_{\text{block}}$

Buoyant force:

$F_B = \rho g V$

$\rho_F g V_F = \rho_B g V_B$

cancel g from both sides of the expression

$\rho_F V_F = \rho_B V_B$

$\rho_F(\frac{1}{2}V_B) = \rho_B V_B$

cancel V_B from both sides of the expression

$\frac{1}{2}\rho_F = \rho_B$

$\rho_F = (1.6)\rho_{\text{water}}$

$\rho_B = (1.6)\cdot(10^3\ \text{kg/m}^3)\cdot(\frac{1}{2})$

$\rho_B = 800\ \text{kg/m}^3$

31. C is correct.

Terminal velocity:

$v_t = \sqrt{(2mg / c\rho A)}$

where c is the coefficient of air drag, ρ is the density of air, and A is the projected area (πr^2)

If the masses of the three balls are equal, then the velocity of each ball is:

$r = R$:

$A = \pi r^2$

$v_1 = \sqrt{(2mg / c\rho A)}$

$r = 2R$:

$A_1 = \pi(2r)^2$

$A_1 = 4\pi r^2$

$A_1 = 4A$

$v_2 = \sqrt{(2mg / c\rho 4A)}$

$v_2 = (\frac{1}{2})\sqrt{(2mg / c\rho A)}$

$v_2 = (\frac{1}{2})v_1$

$r = 3R$:

$A_2 = \pi(3r)^2$

continued…

$A_2 = 9\pi r^2$

$A_2 = 9A$

$v_3 = \sqrt{(2mg / c\rho 9A)}$

$v_3 = (1/3)\sqrt{(2mg / c\rho A)}$

$v_3 = (1/3)v_1$

As the ball becomes larger, its terminal velocity decreases, and therefore time until impact increases.

velocity = distance / time

32. A is correct.

The pressure due to the density of a fluid surrounding an object submerged at depth d below the surface is:

$P = \rho gd$

Since the distance that the objects are below the surface is not specified, the conclusion is that object B experiences less fluid pressure than object A.

This difference is because object B is higher off the floor of the container.

Thus, its depth is less than object A.

33. B is correct.

$A_1v_1 = A_2v_2$

$A = \pi r^2$

$A_2 = \pi(2r)^2$

$A_2 = 4\pi r^2$

$A_2 = 4A_1$

If r is doubled, then area is increased by 4 times

$A_1(14\text{ m/s}) = (4A_1)v_2$

$v_2 = (A_1 \times 14\text{ m/s}) / (4 \times A_1)$

$v_2 = (14\text{ m/s}) / (4)$

$v_2 = 3.5\text{ m/s}$

Use Bernoulli's equation to find resulting pressure:

$P_1 + ½\rho v_1^2 = P_2 + ½\rho v_2^2$

$(3.5 \times 10^4\text{ Pa}) + ½(1{,}000\text{ kg/m}^3)\cdot(14\text{ m/s})^2 = P_2 + ½(1{,}000\text{ kg/m}^3)\cdot(3.5\text{ m/s})^2$

$P_2 = (13.3 \times 10^4\text{ Pa}) – (6.1 \times 10^3\text{ Pa})$

$P_2 = 12.7 \times 10^4\text{ Pa}$

34. D is correct.

The pressure due to the atmosphere is equal to its weight per unit area.

At an altitude of 2 km, there is less atmosphere pushing down than at Earth's surface.

Therefore, atmospheric pressure decreases with increasing altitude.

35. C is correct.

The buoyant force:

$F_B = \rho_{air} V_{disp} g$, and V_{disp} is the volume of the man m / ρ_{man}

$F_B = (\rho_{air} / \rho_{man}) mg$

$F_B = [(1.2 \times 10^{-3}\ g/cm^3) / (1\ g/cm^3)] \cdot (80\ kg) \cdot (9.8\ m/s^2)$

$F_B = 0.94\ N$

36. D is correct.

Graham's law: the rate at which gas diffuses is inversely proportional to the square root of its density.

37. B is correct.

If the metal rod is thought of like a spring, then Hooke's law applies:

$F = -kx$

The force is proportional to the distance that the spring is stretched.

The proportional limit is reached when the "spring" begins to deform.

The elastic limit is the point when the deformation becomes permanent.

The breaking point equals the fracture point and is when the rod snaps in two.

The elastic modulus is stress (σ) / strain (slope of the elastic region)

38. D is correct.

$A = \pi r^2$

$A_T v_T = A_P v_P$

$v_T = v_P A_P / A_T$

Ratio of the diameter^2 is equal to the ratio of area:

$v_T = v_P (d_1 / d_2)^2$

$v_T = (0.03\ m/s) \cdot [(0.12\ m) / (0.002\ m)]^2$

$v_T = 108\ m/s$

39. B is correct.

Specific gravity:

$\rho_{object} / \rho_{water}$

Archimedes' principle:

$F = \rho g\text{V}$

$\rho_{water} = F / g(0.9\text{V})$

$\rho_{object} = F / g\text{V}$

$\rho_{object} / \rho_{water} = (F / g\text{V}) / [F / g(0.9\text{V})]$

$\rho_{object} / \rho_{water} = 0.9$

V_{water} is 0.9V because 90% of the object is in the water, so 90% of the object's volume equals water displaced.

40. E is correct.

The factors considered are length, density, radius, pressure difference, and viscosity.

The *continuity equation* does not apply here because it relates velocity and radius to the flow rate.

Bernoulli's equation does not apply because it relates density and velocity.

The *Hagen-Poiseuille equation* is needed because it includes terms except for density and therefore is the most applicable to this question.

Volumetric flow rate (Q) is:

$\text{Q} = \Delta\text{P}\pi r^4 / 8\eta L$

The radius is raised to the fourth power.

A 15% change to r results in the greatest change.

41. A is correct.

A force meter provides a force, and the reading indicates what the force is.

Since the hammer is not accelerating, the force equation is:

$F_{meter} + F_B - m_h g = 0$

$m_h g = (0.68 \text{ kg})\cdot(10 \text{ m/s}^2)$

$m_h g = 6.8 \text{ N}$

The displaced volume is the volume of the hammer:

$\text{V}_{disp} = m_h / \rho_{steel}$

$\text{V}_{disp} = (680 \text{ g}) / (7.9 \text{ g/cm}^3)$

continued…

$V_{disp} = 86\ cm^3$

$F_B = \rho_{water} \times V_{disp} \times g$

$F_B = (1 \times 10^{-3}\ kg/cm^3)\cdot(86\ cm^3)\cdot(10\ m/s^2)$

$F_B = 0.86\ N$

$F_{meter} = m_h g - F_B$

$F_{meter} = 6.8\ N - 0.86\ N$

$F_{meter} = 5.9\ N$

42. D is correct.

Pascal's Principle states that pressure is transmitted undiminished in an enclosed static fluid.

43. B is correct.

The normal force (*N*) exerted by the seafloor is the net force between the weight of the submarine and the buoyant force:

$F_N = F_{net}$

$F_{net} = mg - F_B$

$F_{net} = mg - \rho g V$

$F_{net} = mg - W_{water}$

$F_N = mg - W_{water}$

44. E is correct.

$P = \rho g h$

Because the bottom of the brick is at a lower depth than the rest of the brick, it will experience the highest pressure.

45. A is correct.

Mass flow rate:

$\dot{m}$ = cross-sectional area × density × velocity

$\dot{m} = A_C \rho v$

$\dot{m} = (7\ m)\cdot(14\ m)\cdot(10^3\ kg/m^3)\cdot(3\ m/s)$

$\dot{m} = 2.9 \times 10^5\ kg/s$

46. C is correct.

Since the object is motionless:

$a = 0$

$F_{net} = 0$

The magnitude of the buoyant force upward = weight downward:

$F = W$

$F = mg$

$F = (3\text{ kg})\cdot(10\text{ m/s}^2)$

$F = 30\text{ N}$

47. A is correct.

$P = P_{atm} + \rho gh$

$P = (1.01 \times 10^5\text{ Pa}) + (10^3\text{ kg/m}^3)\cdot(10\text{ m/s}^2)\cdot(6\text{ m})$

$P = (1.01 \times 10^5\text{ Pa}) + (0.6 \times 10^5\text{ Pa})$

$P = 1.6 \times 10^5\text{ Pa}$

48. B is correct.

density = mass / volume

49. E is correct.

The bulk modulus is defined as how much material is compressed under a given external pressure:

$B = \Delta P / (\Delta V / V)$

Most solids and liquids compress slightly under external pressure.

However, gases have the greatest change in volume and, thus, the lowest value of B.

50. D is correct.

Young's Modulus is expressed as:

$E = \sigma\text{ (stress)} / \varepsilon\text{ (strain)}$

$E = (F / A) / (\Delta L / L)$

$E = (FL) / (\Delta LA)$

Solve for ΔL:

$\Delta L = FL / EA$

$\Delta L = [(8\text{ kg})\cdot(9.8\text{ m/s}^2)\cdot(2.7\text{ m})] / [(20 \times 10^{10}\text{ N/m}^2)(\pi / 4)\cdot(8 \times 10^{-4}\text{ m})^2]$

$\Delta L = 0.0021\text{ m} = 2.1\text{ mm}$

51. A is correct.

Young's Modulus is expressed as:

$E = \sigma$ (stress) / ε (strain)

$E = (F / A) / (\Delta L / L)$

$E = (FL) / (\Delta L A)$

Solve for ΔL and double the diameter:

$\Delta L = FL / EA$

$\Delta L = [(8\ \text{kg})\cdot(9.8\ \text{m/s}^2)\cdot(2.7\ \text{m})] / [(20 \times 10^{10}\ \text{N/m}^2)(\pi / 4)\cdot(2 \times 8 \times 10^{-4}\ \text{m})^2]$

$\Delta L = 0.000526$ m

$\Delta L = 0.5$ mm

52. B is correct.

Absolute pressure is measured relative to absolute zero pressure (i.e., a perfect vacuum), and gauge pressure is measured relative to atmospheric pressure.

If the atmospheric pressure increases by ΔP, the absolute pressure increases by ΔP, but the gauge pressure does not change.

Absolute pressure at an arbitrary depth h in the lake:

$P_{abs} = P_{atm} + \rho_{water}gh$

Gauge pressure at an arbitrary depth h in the lake:

$P_{gauge} = \rho_{water}gh$

53. D is correct.

Strain is defined and calculated by:

$E = \Delta L / L$

54. D is correct.

By Poiseuille's Law:

$v = \Delta P r^2 / 8\eta L$

$\eta = \Delta P r^2 / 8 L v_{effective}$

$\eta = (970\ \text{Pa})\cdot(0.0021\ \text{m})^2 / 8\cdot(1.8\ \text{m/s})\cdot(0.19\ \text{m})$

$\eta = 0.0016\ \text{N}\cdot\text{s/m}^2$

55. C is correct.

Bernoulli's equation:

$$P_1 + \frac{1}{2}\rho_1 v_1^2 + \rho_1 g h_1 = P_2 + \frac{1}{2}\rho_2 v_2^2 + \rho_2 g h_2$$

There is no height difference, so the equation reduces to:

$$P_1 + \frac{1}{2}\rho_1 v_1^2 = P_2 + \frac{1}{2}\rho_2 v_2^2$$

If the flow of air across the wing tip is v_1, then:

$$v_1 > v_2$$

Since the air that flows across the top has a higher velocity, as it travels a larger distance (curved surface of the top) over the same period, then:

$$\frac{1}{2}\rho_1 v_1^2 > \frac{1}{2}\rho_2 v_2^2$$

To keep both sides equal:

$$P_1 < P_2$$

$$\Delta P = (P_2 - P_1)$$

Thus, the lower portion of the wing experiences greater pressure and therefore lifts the wing.

Notes for active learning

Notes for active learning

Heat and Thermodynamics – Detailed Explanations

1. B is correct.

Ideal gas law:

$PV = nRT$

$P_0 = nRT / V_0$

If isothermal expansion, then n, R and T are constant

$P = nRT / (1/3\ V_0)$

$P = 3(nRT / V_0)$

$P = 3P_0$

2. C is correct.

Area expansion equation:

$\Delta A = A_0(2\alpha\Delta T)$

$\Delta A = (\pi / 4)\cdot(1.2\ \text{cm})^2\cdot(2)\cdot(19 \times 10^{-6}\ \text{K}^{-1})\cdot(200\ °\text{C})$

$\Delta A = 8.6 \times 10^{-3}\ \text{cm}^2$

$\Delta A = A_f - A_0$

$8.6 \times 10^{-3}\ \text{cm}^2 = (\pi / 4)\cdot[d_f^{\,2} - (1.2\ \text{cm})^2\,]$

$d_f = 1.2\ \text{cm}$

3. D is correct.

1 Watt = 1 J/s

Power × Time = Q

Heat needed to change the temperature of a mass:

$Q = mc\Delta T$

$P \times t = mc\Delta T$

$t = (mc\Delta T) / P$

$t = (90\ \text{g})\cdot(4.186\ \text{J/g}\cdot°\text{C})\cdot(30\ °\text{C} - 10\ °\text{C}) / (50\ \text{W})$

$t = 151\ \text{s}$

4. A is correct.

Convert 15 minutes to seconds:

$$t = (15\text{ min}/1)\cdot(60\text{ s}/1\text{ min})$$

$$t = 900\text{ s}$$

Find total energy generated:

$$Q = P \times t$$

$$Q = (1{,}260\text{ J/s})\cdot(900\text{ s})$$

$$Q = 1{,}134\text{ kJ}$$

Find mass of water needed to carry away energy:

$$Q = mL_v$$

$$m = Q / L_v$$

$$m = (1{,}134\text{ kJ}) / (22.6 \times 10^2\text{ kJ/kg})$$

$$m = 0.5\text{ kg}$$

$$m = 500\text{ g}$$

5. C is correct.

Phase changes occur at a constant temperature.

Once the phase change is complete, the temperature of the substance then increases or decreases.

For example, water remains at 0 °C until it has completely changed phase to ice before the temperature decreases further.

6. E is correct.

The amount of energy needed to melt a sample of mass m is:

$$Q = m\,L_f$$

where L_f is the latent heat of fusion

$$Q = (55\text{ kg})\cdot(334\text{ kJ/kg})$$

$$Q = 1.8 \times 10^4\text{ kJ}$$

7. D is correct.

Metals are good heat and electrical conductors because of their bonding structure.

In metallic bonding, the outer electrons are held loosely and can travel freely.

Electricity and heat require high electron mobility.

Thus, the looseness of the outer electrons in the materials allows them to be excellent conductors.

8. A is correct.

Find heat of phase change from steam to liquid:

$$Q_1 = mL_v$$

Find heat of phase change from liquid to solid:

$$Q_2 = mL_f$$

Find heat of temperature from 100 °C to 0 °C:

$$Q_3 = mc\Delta T$$

Total heat:

$$Q_{net} = Q_1 + Q_2 + Q_3$$

$$Q_{net} = mL_v + mL_f + mc\Delta T$$

To find mass:

$$Q_{net} = m(L_v + c\Delta T + L_f)$$

$$m = Q_{net} / (L_v + c\Delta T + L_f)$$

Solve:

$$Q_{net} = 200 \text{ kJ}$$

$$Q_{net} = 2 \times 10^5 \text{ J}$$

$$m = (2 \times 10^5 \text{ J}) / [(22.6 \times 10^5 \text{ J/kg}) + (4{,}186 \text{ J/kg·K})·(100 \text{ °C} - 0 \text{ °C}) + (33.5 \times 10^4 \text{ J/kg})]$$

$$m = 0.066 \text{ kg}$$

9. C is correct.

Fusion is the process whereby a substance changes from a solid to liquid (i.e., melting).

Condensation is the process whereby a substance changes from vapor to liquid.

Sublimation is when a substance changes directly from a solid to the gas phase without passing through the liquid phase.

10. E is correct.

$$Q = mc\Delta T$$

$$Q = (0.2 \text{ kg})·(14.3 \text{ J/g·K})·(1{,}000 \text{ g/kg})·(280 \text{ K} - 250 \text{ K})$$

$$Q = 86{,}000 \text{ J} = 86 \text{ kJ}$$

11. B is correct.

Heat needed to raise the temperature of aluminum:

$Q_A = m_A c_A \Delta T$

Heat needed to raise temperature of water:

$Q_W = m_W c_W \Delta T$

Total heat to raise temperature of system:

$Q_{net} = Q_A + Q_W$

$Q_{net} = m_A c_A \Delta T + m_W c_W \Delta T$

$Q_{net} = \Delta T(m_A c_A + m_W c_W)$

$Q_{net} = (98\ °C - 18\ °C)\cdot[(0.5\ kg)\cdot(900\ J/kg\cdot K) + (1\ kg)\cdot(4{,}186\ J/kg\cdot K)]$

$Q_{net} = 370{,}880\ J$

Time to produce Q_{net} with 500 W:

$Q_{net} = (500\ W)t$

$t = Q_{net} / (500\ W)$

$t = (370{,}880\ J) / 500\ W$

$t = 741.8\ s$

Convert to minutes:

$t = (741.8\ s/1)\cdot(1\ min/60\ s)$

$t = 12.4\ min \approx 12\ min$

12. A is correct.

When a substance goes through a phase change, the temperature does not change.

It can be assumed that the lower plateau is L_f and the upper plateau is L_v.

Count the columns:

$L_f = 2$, $L_v = 7$

$L_v / L_f = 7 / 2$

$L_v / L_f = 3.5$

13. B is correct.

Specific heat is the amount of heat (i.e., energy) needed to raise the temperature of the unit mass of a substance by a given amount (usually one degree).

14. E is correct.

Find ½ of KE of the BB:

$KE = ½mv^2$

$½KE = ½(½mv^2)$

$½KE_{BB} = ½(½)·(0.0045 \text{ kg})·(46 \text{ m/s})^2$

$½KE_{BB} = 2.38 \text{ J}$

The $½KE_{BB}$ is equal to energy taken to change temperature:

$Q = ½KE_{BB}$
$Q = mc\Delta T$

$mc\Delta T = ½KE_{BB}$

$\Delta T = ½KE_{BB} / mc$

Calculate to find ΔT:

$\Delta T = (2.38 \text{ J}) / (0.0045 \text{ kg})·(128 \text{ J/kg·K})$

$\Delta T = 4.1 \text{ K}$

15. C is correct.

Vaporization is the process whereby a substance changes from a liquid to a gas; boiling or evaporation.

Sublimation is the process whereby a substance changes from a solid to a gas.

16. A is correct.

Carnot efficiency:

η = work done / total energy

$\eta = W / Q_H$

$\eta = 5 \text{ J} / 18 \text{ J}$

$\eta = 0.28$

The engine's efficiency:

$\eta = (T_H - T_C) / T_H$

$0.28 = (233 \text{ K} - T_C) / 233 \text{ K}$

$(0.28)·(233 \text{ K}) = (233 \text{ K} - T_C)$

$65.2 \text{ K} = 233 \text{ K} - T_C$

$T_C = 168 \text{ K}$

17. D is correct.

Heat needed to change the temperature of a mass:

$$Q = mc\Delta T$$

Calculate to find Q:

$$Q = (0.92\ \text{kg})\cdot(113\ \text{cal/kg}\cdot{}^\circ\text{C})\cdot(96\ {}^\circ\text{C} - 18\ {}^\circ\text{C})$$

$$Q = 8{,}108.9\ \text{cal}$$

Convert to joules:

$$Q = (8{,}108.9\ \text{cal})\cdot(4.186\ \text{J/cal})$$

$$Q = 33{,}940\ \text{J}$$

18. B is correct.

During a change of state, adding heat does not change the temperature (i.e., a measure of kinetic energy).

The heat energy adds to the potential energy of the substance until the substance completely changes its state.

19. E is correct.

For a pressure *vs.* volume graph of the work done for a cyclic process carried out by gas, it is equal to the area enclosed by the cyclic process.

20. A is correct.

Specific heat of A is larger than B:

$$c_A > c_B$$

Energy to raise the temperature:

$$Q = mc\Delta T$$

If m and ΔT are equal for A and B:

$$Q_A = m_A c_A \Delta T_A$$

$$Q_B = m_B c_B \Delta T_B$$

$$Q_A > Q_B$$

This is valid because other factors are equal, and the magnitude of Q only depends on c.

21. D is correct.

Find the kinetic energy of the meteor:

$$KE = ½mv^2$$

$$KE = ½(0.0065\ kg)·(300\ m/s)^2$$

$$KE = 292.5\ J$$

Find temperature rise:

$$Q = KE$$

$$Q = mc\Delta T$$

$$mc\Delta T = KE$$

$$\Delta T = KE / mc$$

Convert KE to calories:

$$KE = (292.5\ J/1)·(1\ cal/4.186\ J)$$

$$KE = 69.9\ cal$$

Calculate ΔT:

$$\Delta T = (69.9\ cal) / [(0.0065\ kg)·(120\ cal/kg·°C)]$$

$$\Delta T = 89.6\ °C \approx 90\ °C$$

22. A is correct.

When a liquid freezes, it undergoes a phase change from liquid to solid.

For a phase change to occur, heat energy must be dissipated (removed).

During any phase change, the temperature remains constant.

23. B is correct.

For an isothermal process:

$$\Delta U = 0$$

$$\Delta U = Q - W$$

$$Q = W$$

Work to expand an ideal gas in an isothermal process:

$$W = nRT\ \ln(V_f / V_i)$$

From the ideal gas law,

$$nRT = P_f V_f$$

continued…

Giving:

$W = P_f V_f \ln(V_f / V_i)$

$W = (130 \text{ kPa})\cdot(0.2 \text{ m}^3) \ln[(0.2 \text{ m}^3) / (0.05 \text{ m}^3)]$

$W = 36 \text{ kJ}$

$Q = W$

$Q = 36 \text{ kJ}$

Since the process is isothermal, there is no change in internal energy.

Since the surroundings are doing negative work on the system:

$\Delta E = 0 = Q + W = Q - 36 \text{ kJ}$

Therefore:

$Q = 36 \text{ kJ}$

24. D is correct.

Find the potential energy of 1 kg of water:

$PE = mgh$

$PE = (1 \text{ kg})\cdot(9.8 \text{ m/s}^2)\cdot(30 \text{ m})$

$PE = 294 \text{ J}$

Assume potential energy is converted to heat for maximum temperature increase:

$PE = Q$

$Q = mc\Delta T$

$mc\Delta T = PE$

$\Delta T = PE / mc$

$\Delta T = (294 \text{ J}) / [(1 \text{ kg})\cdot(4{,}186 \text{ J/kg/K})]$

$\Delta T = 0.07 \text{ °C}$

It is not necessary to convert to Kelvin for temperature differences because a temperature change in Kelvin is equal to a temperature change in Celsius.

25. A is correct.

Find heat from phase change:

$Q = mL_f$

$Q = (0.75\text{ kg})\cdot(33{,}400\text{ J/kg})$

$Q = 25{,}050\text{ J}$

Because the water is freezing, Q should be negative due to heat being released.

$Q = -25{,}050\text{ J}$

Find the change in entropy:

$\Delta S = Q / \text{T}$

$\Delta S = -25{,}050\text{ J} / (0\text{ °C} + 273\text{ K})$

$\Delta S = -92\text{ J} / \text{K}$

A negative change in entropy indicates that the disorder of the isolated system has decreased.

When water freezes, the entropy is negative because water is more disordered than ice.

Thus, the disorder has decreased.

26. A is correct.

Copper has a larger linear expansion coefficient than iron, so it expands more than iron during a given temperature change.

The bimetallic bar bends due to the difference in expansion between the copper and iron.

27. D is correct.

Calculate heat needed to raise temperature:

$Q = mc\Delta\text{T}$

$Q = (0.110\text{ kg})\cdot(4{,}186\text{ J/kg}\cdot\text{K})\cdot(30\text{ °C} - 20\text{ °C})$

$Q = 4{,}605\text{ J}$

Calculate time needed to raise temperature with 60 W power source:

$Q = P \times t$

$Q = (60\text{ W})t$

$t = \text{Q} / (60\text{ W})$

$t = (4{,}605\text{ J}) / (60\text{ W})$

$t = 77\text{ s}$

28. B is correct.

During a change of state, the addition of heat does not change the temperature (i.e., a measure of KE).

Heat energy only adds to the potential energy of the substance until the substance completely changes its state.

29. E is correct.

Convert to Kelvin:

$T = -243\ °C + 273$

$T = 30\ K$

Double temperature:

$T_2 = (30\ K)·(2)$

$T_2 = 60\ K$

Convert back to Celsius:

$T_2 = 60\ K - 273$

$T_2 = -213\ °C$

30. D is correct.

If a researcher attempts to determine how much the temperature of a particular piece of material would rise when a known amount of heat is added to it, knowing the specific heat would be most helpful.

31. B is correct.

Convert units:

$1.7 \times 10^5\ J/kg = 170\ kJ/kg$

Change in internal energy = heat added (Q)

$Q = mL_v$

$Q = (1\ kg)·(170\ kJ/kg)$

$Q = 170\ kJ$

32. D is correct.

$Q = mc\Delta T$

If m and c are constant, the relationship is directly proportional.

To double Q, T must be doubled:

$5\ C + 273 = 278\ K$

$278\ K \times 2 = 556\ K$

$556\ K - 273 = 283\ C$

33. E is correct.

The mass of each material is required to determine the time the system takes to reach thermal equilibrium.

34. B is correct.

Body heat gives energy to the water molecules in the sweat. This energy is transferred via collisions until some molecules have enough energy to break the hydrogen bonds and escape the liquid (evaporation).

However, if a body stayed dry, the heat would not be given to the water, and the person would stay hot because the heat is not lost due to the evaporation of the water.

35. D is correct.

Calculate heat released when 0 °C water converts to 0 °C ice:

$$Q_1 = mL_f$$

$$Q_1 = (2{,}200 \text{ kg})\cdot(334 \times 10^3 \text{ J/kg})$$

$$Q_1 = 734{,}800 \text{ kJ}$$

Calculate heat released for temperature drop ΔT

$$Q_2 = mc\Delta T$$

$$Q_2 = (2{,}200 \text{ kg})\cdot(2{,}050 \text{ J/kg K})\cdot[(0 \text{ °C} - (-30 \text{ °C})]$$

ΔK = Δ°C, so units cancel:

$$Q_2 = 135{,}300 \text{ kJ}$$

Add heat released to get Q_{net}:

$$Q_{net} = Q_1 + Q_2$$

$$Q_{net} = (734{,}800 \text{ kJ}) + (135{,}300 \text{ kJ})$$

$$Q_{net} = 870{,}100 \text{ kJ}$$

36. E is correct.

Object 1 has three times the specific heat capacity and four times the mass of Object 2:

$$c_1 = 3c_2;\ m_1 = 4m_2$$

A single-phase substance obeys the *specific heat equation*:

$$Q = mc\Delta T$$

In this case, the same amount of heat is added to each substance, therefore:

$$Q_1 = Q_2$$

$$m_1c_1\Delta T_1 = m_2c_2\Delta T_2$$

continued…

$(4m_2)(3c_2)\Delta T_1 = m_2 c_2 \Delta T_2$

$12m_2c_2\Delta T_1 = m_2c_2\Delta T_2$

$12\ \Delta T_1 = \Delta T_2$

37. C is correct.

Conduction is a form of heat transfer in which the collisions of the molecules of the material transfer energy through the material. The higher temperature of the material causes the molecules to collide with more energy, which eventually is transferred throughout the material through subsequent collisions.

Radiation is a form of heat transfer in which electromagnetic waves carry energy from the emitting object and deposit the energy to the object that absorbs the radiation.

Convection is a form of heat transfer in which the mass motion of a fluid (i.e., liquids and gases) transfers energy from the source of heat.

38. A is correct.

From the ideal gas law:

$p_3V_3 = nRT_3$

$T_3 = p_3V_3 / nR$

$T_3 = 1.5p_1V_3 / nR$

$T_3 = 1.5V_3(p_1 / nR)$

Also, from the ideal gas law:

$(p_1 / nR) = T_1 / V_1$

$(p_1 / nR) = (293.2\text{ K}) / (100\text{ cm}^3)$

$(p_1 / nR) = 2.932\text{ K/cm}^3$

Calculate T_3:

$T_3 = 1.5V_3\ (2.932\text{ K/cm}^3)$

$T_3 = 1.5\ (50\text{ cm}^3)\cdot(2.932\text{ K/cm}^3)$

$T_3 = 219.9\text{ K}$

$T_3 = -53.3\ °\text{C} \approx -53\ °\text{C}$

Calculate T_4:

$T_4 = 1.5V_4\ (2.932\text{ K/cm}^3)$

$T_4 = 1.5\ (150\text{ cm}^3)\cdot(2.932\text{ K/cm}^3)$

$T_4 = 659.6\text{ K}$

$T_4 = 386.5\ °\text{C} \approx 387\ °\text{C}$

39. B is correct.

Steel is a very *conductive* material that can transfer thermal energy very well.

The steel feels colder than the plastic because its higher thermal conductivity allows it to remove more heat and thus makes touching it feel colder.

40. B is correct.

An *isobaric* process involves constant pressure.

An *isochoric* (also *isometric*) process involves a closed system at constant volume.

An *adiabatic process* occurs without the transfer of heat or matter between a system and its surroundings.

An *isothermal process* involves the change of a system in which the temperature remains constant.

An *isentropic process* is an idealized thermodynamic process that is adiabatic of a frictionless system where work is transferred such that there is no transfer of heat or matter.

41. E is correct.

Carnot coefficient of performance of a refrigeration cycle:

$C_P = T_C / (T_H - T_C)$

$C_P = Q_C / W$

$Q_C / W = T_C / (T_H - T_C)$

$W = (Q_C / T_C) \cdot (T_H - T_C)$

$W = (20 \times 10^3 \text{ J} / 293 \text{ K}) \cdot (307 \text{ K} - 293 \text{ K})$

$W = 955.6 \text{ J} = 0.956 \text{ kJ}$

Power = Work / time

$P = W / t$

$P = 0.956 \text{ kJ} / 1 \text{ s}$

$P = 0.956 \text{ kW} \approx 0.96 \text{ kW}$

42. C is correct.

Heat energy is measured in units of Joules and calories.

43. B is correct.

Convection is a form of heat transfer in which the mass motion of a fluid (i.e., liquids, gases) transfers energy from the source of heat.

44. A is correct.

Radiation is the transmission of energy as particles or waves through space or a material medium.

Examples of radiation include electromagnetic radiations such as X-rays, alpha particles, beta particles, radio waves, and visible light.

45. D is correct.

Convert P_3 to Pascals:

$$P_3 = (2 \text{ atm} / 1)\cdot(101{,}325 \text{ Pa} / 1 \text{ atm})$$

$$P_3 = 202{,}650 \text{ Pa}$$

Use the ideal gas law to find V_3:

$$PV = nRT$$

$$V = (nRT) / P$$

$$V_3 = [(0.008 \text{ mol})\cdot(8.314 \text{ J/mol}\cdot\text{K})\cdot(2{,}438 \text{ K})] / (202{,}650 \text{ Pa})$$

$$V_3 = 8 \times 10^{-4} \text{ m}^3$$

Convert to cm^3:

$$V_3 = (8 \times 10^{-4} \text{ m}^3 / 1)\cdot(100^3 \text{ cm}^3 / 1 \text{ m}^3)$$

$$V_3 = 800 \text{ cm}^3$$

46. C is correct.

An adiabatic process involves no heat added or removed from the system.

From the First Law of Thermodynamics:

$$\Delta U = Q + W$$

If $Q = 0$, then:

$$\Delta U = W$$

Because work is being done to expand the gas, it is considered negative, and then the change in internal energy is negative (decreases).

$$-\Delta U = -W$$

47. E is correct.

Standing in a breeze while wet feels colder than when dry because of water evaporation from the skin.

Water requires heat to evaporate, so this is taken from the body, making a person feel colder than if they were dry, and the evaporation did not occur.

48. A is correct.

Conduction is a form of heat transfer in which the collisions of the molecules of the material transfer energy through the material.

The higher temperature of the material causes the molecules to collide with more energy, which eventually is transferred throughout the material through subsequent collisions.

49. D is correct.

An isobaric process is a constant pressure process, so the resulting pressure is the same.

50. B is correct.

This question asks which type of surface has a higher emissivity than others and, therefore, can radiate more energy over a set period.

A *blackbody* is an idealized radiator and has the highest emissivity.

A surface most similar to a blackbody (the black surface) is the best radiator of thermal energy.

A black surface is considered an ideal blackbody and therefore has an emissivity of 1 (perfect emissivity).

The black surface will be the best radiator compared to another surface, which cannot be considered blackbodies and have an emissivity of <1.

51. E is correct.

Calculate the gap between the rods:

The gap in between the rods will be filled by both expanding, so total thermal expansion length equals 1.1 cm.

$$\Delta L = L_0 \alpha \Delta T$$

$$\Delta L_{tot} = \Delta L_B + \Delta L_A$$

$$\Delta L_{tot} = (\alpha_B L_B + \alpha_A L_A)\ \Delta T$$

Rearrange the equation for ΔT:

$$\Delta T = \Delta L_{tot} / (\alpha_B L_B + \alpha_A L_A)$$

$$\Delta T = 1.1\ \text{cm} / [(2 \times 10^{-5}\ \text{K}^{-1})\cdot(59.1\ \text{cm}) + (2.4 \times 10^{-5}\ \text{K}^{-1})\cdot(39.3\ \text{cm})]$$

$$\Delta T = 517.6\ \text{K} \approx 518\ \text{K}$$

Measuring the difference in temperature in K is the same as in °C, so it is not required to convert:

$$\Delta T = 518\ °\text{C}$$

52. A is correct.

Find seconds in a day:

t = (24 h / 1 day)·(60 min / 1 h)·(60 s / 1 min)]

t = 86,400 s

Find energy lost in a day:

E = Power × time

E = (60 W)·(86,400 s)

E = 5,184,000 J

Convert to kcal:

E = (5,184,000 J/1)·(1 cal/4.186 J)·(1 kcal/10^3 cal)

E = 1,240 kcal

53. D is correct.

Conduction is a form of heat transfer in which the collisions of the molecules of the material transfer energy through the material. The higher temperature of the material causes the molecules to collide with more energy, which eventually is transferred throughout the material through subsequent collisions.

54. B is correct.

Heat given off by warmer water is equal to that absorbed by the frozen cube.

This heat is split into heat needed to melt the cube and bring the temperature to equilibrium.

$$Q_{H2O,1} + Q_{alcohol,Temp1} + Q_{alcohol,Phase1} = 0$$

$$(mc\Delta T)_{H2O,1} + (mc\Delta T)_{alcohol,Temp1} + (mL_f)_{alcohol} = 0$$

$$Q_{H2O,2} + Q_{alcohol,Temp2} + Q_{alcohol,Phase2} = 0$$

$$(mc\Delta T)_{H2O,2} + (mc\Delta T)_{alcohol,Temp2} + (mL_f)_{alcohol} = 0$$

Set equal to cancel heat from phase change (since they are equal):

$$(mc\Delta T)_{H2O,1} + (mc\Delta T)_{alcohol,Temp1} = (mc\Delta T)_{H2O,2} + (mc\Delta T)_{alcohol,Temp2}$$

$$(m_{alcohol})(c_{alcohol})\cdot(\Delta T_{alcohol1} - \Delta T_{alcohol2}) = c_{H2O}(m\Delta T_{H2O,2} - m\Delta T_{H2O,1})$$

$$c_{alcohol} = (c_{H2O} / m_{alcohol})\cdot[(m\Delta T_{H2O,2} - m\Delta T_{H2O,1}) / (\Delta T_{alcohol1} - \Delta T_{alcohol2})]$$

Solving for $c_{alcohol}$:

$$c_{alc} = [(4{,}190 \text{ J/kg·K}) / (0.22 \text{ kg})]\cdot[(0.4 \text{ kg})\cdot(10 - 30 \text{ °C}) - (0.35 \text{ kg})\cdot(5 - 26 \text{ °C})] / [(5 \text{ °C} - (-10 \text{ °C}) - (10 \text{ °C} - (-10 \text{ °C})]$$

$$c_{alc} = (19{,}045 \text{ J/kg·K})\cdot[(-0.65 \text{ °C}) / (-5 \text{ °C})]$$

$$c_{alc} = 2{,}475 \text{ J/kg·K}$$

55. D is correct.

Using the calculated value for c_{alc} in the problem above:

$$Q_{H2O,1} + Q_{alcohol,Temp1} + Q_{alcohol,Phase1} = 0$$

$$(mc\Delta T)_{H2O,1} + (mc\Delta T)_{alcohol} + (mL_f)_{alcohol} = 0$$

$$L_{f\ alcohol} = [-(mc\Delta T)_{H2O,1} - (mc\Delta T)_{alcohol}] / m_{alcohol}$$

Solve:

$$L_{f\ alcohol} = -[(0.35\ kg)\cdot(4{,}190\ J/kg\cdot K)\cdot(5\ °C\ - 26\ °C) - (0.22\ kg)\cdot(2{,}475\ J/kg\cdot K)\cdot(\ 5\ °C\ - (-10\ °C)] / (0.22\ kg)$$

$$L_{f\ alcohol} = (30{,}796.5\ J - 8{,}167.5\ J) / (0.22\ kg)$$

$$L_{f\ alcohol} = 103 \times 10^3\ J/kg = 10.3 \times 10^4\ J/kg$$

Notes for active learning

Electric Circuits – Detailed Explanations

1. B is correct.

The resistance of a wire is given by:

$R = \rho L / A$

where ρ is the resistivity of the wire material

If the length L is doubled, the resistance R is doubled.

If the radius r is doubled, the area $A = \pi r^2$ is quadrupled, and the resistance R decreases by ¼.

If these two changes are combined:

$R_{new} = \rho(2L) / \pi(2r)^2$

$R_{new} = (2/4)\cdot(\rho L / \pi r^2)$

$R_{new} = (2/4)R = ½R$

2. D is correct.

Internal resistance of battery is in series with resistors in circuit:

$R_{eq} = R_1 + R_{battery}$

where R_{eq} is equivalent resistance and R_1 is resistor connected to battery

$V = IR_{eq}$

$V = I(R_1 + R_{battery})$

$R_{battery} = V / I - R_1$

$R_{battery} = (12\ V / 0.6\ A) - 6\ \Omega$

$R_{battery} = 14\ \Omega$

3. C is correct.

An *ohm* Ω is defined as the resistance between two points of a conductor when a constant potential difference of 1 V, applied to these points, produces in the conductor a current of 1 A.

A *series circuit* experiences the same current (i.e., flow of charge) through resistors regardless of their resistance.

However, the *voltage* across each resistor can be different.

Since the light bulbs are in series, the current through them is the same.

4. D is correct.

Electric Potential:

$$V = kQ / r$$

$$V_B = kQ / r_B$$

$$V_B = (9 \times 10^9\ N \cdot m^2/C^2) \cdot (1 \times 10^{-6}\ C) / 3.5\ m$$

$$V_B = 2{,}571\ V$$

$$V_A = kQ / r_A$$

$$V_A = (9 \times 10^9\ N \cdot m^2/C^2) \cdot (1 \times 10^{-6}\ C) / 8\ m$$

$$V_A = 1{,}125\ V$$

Potential difference:

$$\Delta V = V_B - V_A$$

$$\Delta V = 2{,}571\ V - 1{,}125\ V$$

$$\Delta V = 1{,}446\ V$$

5. E is correct.

The *capacitance of a parallel plate capacitor* demonstrates the influence of material, separation distance, and geometry in determining the overall capacitance.

$$C = k\mathcal{E}_0 A / d$$

where k = dielectric constant or permittivity of material between the plates, A = surface area of the conductor, and d = distance of plate separation

6. A is correct.

The energy transferred by the quantity of electric charge by a potential difference of V volts.

$$E = qV$$

$$E = ½m(\Delta v)^2$$

$$qV = ½m(v_f^2 - v_i^2)$$

$$v_f^2 = (2qV / m) + v_i^2$$

$$v_f^2 = [2(1.6 \times 10^{-19}\ C) \cdot (100\ V) / (1.67 \times 10^{-27}\ kg)] + (1.5 \times 10^5\ m/s)^2$$

$$v_f^2 = (1.9 \times 10^{10}\ m^2/s^2) + (2.3 \times 10^{10}\ m^2/s^2)$$

$$v_f^2 = 4.2 \times 10^{10}\ m^2/s^2$$

$$v_f = 2.04 \times 10^5\ m/s \approx 2 \times 10^5\ m/s$$

7. C is correct.

Power = current^2 × resistance

$P = I^2R$

Double current:

$P_2 = (2I)^2R$

$P_2 = 4(I^2R)$

$P_2 = 4P$

Power is quadrupled

8. D is correct.

A magnetic field is created only by electric charges in motion.

A stationary charged particle does not generate a magnetic field.

9. B is correct.

Combining the power equation with Ohm's law:

$P = (\Delta V)^2 / R$

where $\Delta V = 120$ V is a constant

To increase power, decrease the resistance.

A longer wire increases resistance, while a thicker wire decreases it:

$A = \pi r^2$

$R = \rho L / A$

A larger radius of the cross-sectional area means A is larger (denominator), which lowers R.

10. E is correct.

Electrical potential energy:

$W = kq_1q_2 / r$

$r = \Delta x$

$r = 2\text{ mm} - (-2\text{ mm})$

$r = 4\text{ mm}$

$W = [(9 \times 10^9\ \text{N·m}^2/\text{C}^2)\cdot(4 \times 10^{-6}\ \text{C})\cdot(8 \times 10^{-6}\ \text{C})] / (4 \times 10^{-3}\ \text{m})$

$W = (0.288\ \text{N·m}^2) / (4 \times 10^{-3}\ \text{m})$

$W = 72\ \text{J}$

11. A is correct.

$V = IR$

$I = V / R$

$I = (220\ V) / (400\ \Omega)$

$I = 0.55\ A$

12. E is correct.

A parallel circuit experiences the same potential difference across each resistor.

However, the current through each resistor can be different.

13. D is correct.

The energy transferred by the quantity of electric charge by a potential difference of V volts.

$E = qV$

$E = ½mv^2$

$qV = ½mv^2$

$v^2 = 2qV / m$

$v^2 = [2(1.6 \times 10^{-19}\ C)\cdot(990\ V)] / (9.11 \times 10^{-31}\ kg)$

$v^2 = 3.5 \times 10^{14}\ m^2/s^2$

$v = 1.9 \times 10^7\ m/s$

14. B is correct.

Calculate magnetic field perpendicular to loop:

$B_{Perp2} = (12\ T) \cos 30°$

$B_{Perp2} = 10.4\ T$

$B_{Perp1} = (1\ T) \cos 30°$

$B_{Perp1} = 0.87\ T$

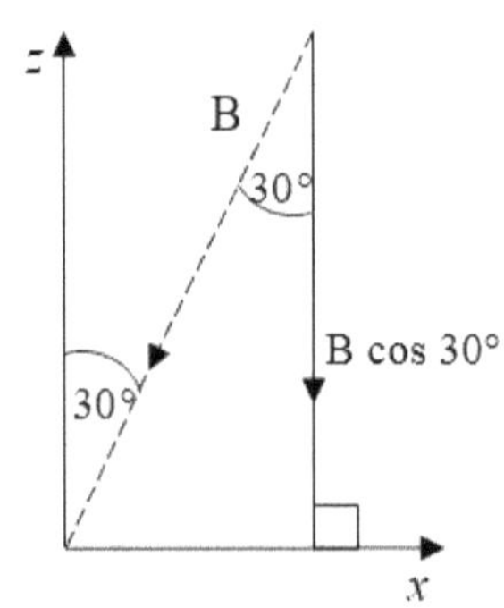

Use Faraday's Law to calculate generated voltage:

$V = N\Delta BA / \Delta t$

$V = N\Delta B(\pi r^2) / \Delta t$

$V = [(1)\cdot(10.4\ T - 0.87\ T)\cdot(\pi(0.5\ m)^2)] / (5\ s - 0\ s)$

$V = [(1)\cdot(10.4\ T - 0.87\ T)\cdot(0.785\ m^2)] / (5\ s)$

$V = 1.5\ V$

continued...

Use Ohm's Law to calculate current:

$V = IR$

$I = V / R$

$I = (1.5\ V) / (12\ \Omega)$

$I = 0.13\ A$

15. C is correct.

Ohm's Law:

$V = IR$

$V = (10\ A)\cdot(35\ \Omega)$

$V = 350\ V$

16. A is correct.

The magnitude of the acceleration is given by:

$F = ma$

$a = F / m$

$F = qE_0$

$a = qE_0 / m$

Bare nuclei = no electrons

^{1}H has 1 proton, and ^{4}He has 2 protons and 2 neutrons

Thus, ^{1}H has ½ the charge and ¼ the mass of ^{4}He.

$a_H = q_HE_0 / m_H$

$a_{He} = q_{He}E_0 / m_{He}$

$a_H = (\frac{1}{2}q_{He})E_0 / (\frac{1}{4}m_{He})$

$a_H = 2(q_{He}E_0 / m_{He})$

$a_H = 2a_{He}$

17. B is correct.

The current will change as the choice of lamp arrangement changes.

Since,

$P = V^2/R$, power increases as resistance decreases.

To rank the power in increasing order, the equivalent resistance must be ranked in decreasing order.

For arrangement B, the resistors are in series, so:

$$R_{eq} = R + R = 2R$$

For arrangement C, the resistors are in parallel, so:

$$1/R_{eq} = 1/R + 1/R$$

$$R_{eq} = R/2$$

The ranking of resistance in decreasing order is B to A to C; therefore, the ranking of power increases.

18. D is correct.

$$C = k\mathcal{E}_0\text{A} / d$$

where k = dielectric constant or permittivity of material between the plates, A = area and d = distance of plate separation

$$C = (2.1)\cdot(8.854 \times 10^{-12}\ \text{F/m})\cdot(0.01\ \text{m} \times 0.01\ \text{m}) / (0.001\ \text{m})$$

$$C = (1.9 \times 10^{-15}\ \text{F/m}) / (0.001\ \text{m})$$

$$C = 1.9 \times 10^{-12}\ \text{F}$$

$$C = 1.9\ \text{pF}$$

19. C is correct.

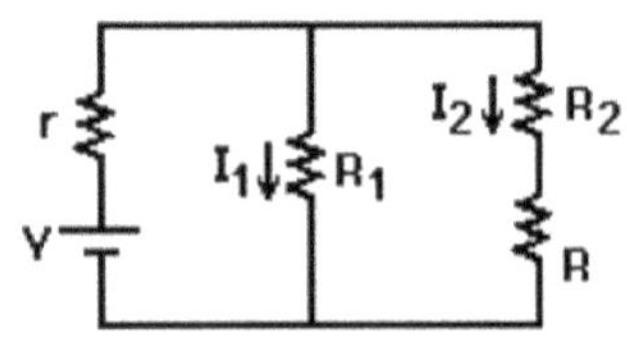

Resistor R_1 is connected directly across the battery.

Thus, the voltage across R_1 is V and is held constant at that value regardless of what happens in the circuit.

Similarly, the voltage across the series combination of R and R_2 is held constant at V.

Since the voltage across R_1 will be V, the current I_1 through R_1 will be unchanged as R changes (since R_1 did not change, $I_1 = V / R_1$ remains the same).

Since the voltage across the combination of R and R_2 will be V when R is decreased, the effective resistance of the series combination $R + R_2$ will decrease, and the current I_2 through R_2 will increase.

20. C is correct.

$V = IR$

$I = V / R$

Ohm's law: current between two points is directly proportional to the potential difference between the points.

21. E is correct.

Root mean square voltage equation:

$V_{rms} = V_{max} / \sqrt{2}$

$V_{rms} = 12 / \sqrt{2}$

$V_{rms} = (12 / \sqrt{2})\cdot(\sqrt{2} / \sqrt{2})$

$V_{rms} = (12\sqrt{2}) / 2$

$V_{rms} = 6\sqrt{2}$ V

22. C is correct.

By definition:

$V_{rms} = V_{max} / \sqrt{2}$

Therefore:

$V_{max} = V_{rms}\sqrt{2}$

$V_{max} = (150\ V)\sqrt{2}$

$V_{max} = 212$ V

23. D is correct.

Kirchhoff's junction rule states that the sum of currents coming into a junction is the sum of currents leaving a junction. A statement of conservation of charge defines that no charge is created nor destroyed in the circuit.

24. A is correct.

The capacitance of capacitors connected in parallel is the sum of the individual capacitances:

$C_{eq} = C_1 + C_2 + C_3 + C_4 = 4C$

The relationship between the total charge delivered by the battery and the voltage of the battery is:

$V = Q / C_{eq} = Q / 4C$

The charge on one capacitor is:

$Q_1 = CV$

$Q_1 = C\,(Q / 4C)$

$Q_1 = Q / 4$

25. B is correct.

If two conductors are connected by copper wire, each conductor will be at the same potential because current can flow through the wire and equalize the difference in potential.

26. C is correct.

Electromagnetic induction is the production of an electromotive force across a conductor.

When a changing magnetic field is brought near a coil, a voltage is generated in the coil, inducing a current.

The voltage generated can be calculated by Faraday's Law:

$$\text{emf} = -N\Delta\phi / \Delta t$$

where N = number of turns and $\Delta\phi$ = change in magnetic flux

27. A is correct.

Current is constant across resistors connected in series.

28. D is correct.

By convention, the direction of electric current is the direction that a positive charge migrates.

Therefore, current flows from the point of high potential to the point of lower potential.

29. E is correct.

$$PE_e = PE_1 + PE_2 + PE_3$$

$$PE_e = (kQ_1Q_2) / r_1 + (kQ_2Q_3) / r_2 + (kQ_1Q_3) / r_3$$

$$PE_e = kQ^2 [(1 / r_1) + (1 / r_2) + (1 / r_3)]$$

$r_1 = 4$ cm and $r_2 = 3$ cm are known.

Use Pythagorean Theorem to find r_3:

$$r_3^2 = r_1^2 + r_2^2$$

$$r_3^2 = (4\text{ cm})^2 + (3\text{ cm})^2$$

$$r_3^2 = 16\text{ cm}^2 + 9\text{ cm}^2$$

$$r_3^2 = 25\text{ cm}^2$$

$$r_3 = 5\text{ cm}$$

$$PE_e = (9.0 \times 10^9\text{ N·m}^2/\text{C}^2)\cdot(3.8 \times 10^{-9}\text{ C})^2 \times [(1 / 0.04\text{ m}) + (1 / 0.03\text{ m}) + (1 / 0.05\text{ m})]$$

$$PE_e = (1.2 \times 10^{-7}\text{ N·m}^2)\cdot(25\text{ m}^{-1} + 33\text{ m}^{-1} + 20\text{ m}^{-1})$$

$$PE_e = (1.2 \times 10^{-7}\text{ N·m}^2)\cdot(78\text{ m}^{-1})$$

$$PE_e = 1.0 \times 10^{-5}\text{ J}$$

30. A is correct.

The *magnetic force* acting on a charge q moving at velocity v in a magnetic field B is given by:

$$F = qv \times B$$

If q, v, and the angle between v and B are the same for both charges, then the magnitude of the force F is the same on both charges.

However, if the charges carry opposite signs, each experience oppositely-directed forces.

31. C is correct.

By convention, the direction of electric current is the direction that a positive charge migrates.

Electrons flow from regions of low potential to regions of high potential.

Electric Potential Energy:

$$U = (kQq) / r$$

Electric Potential:

$$\mathrm{V} = (kQ) / r$$

Because the charge of an electron (q) is negative, as the electron moves opposite the electric field, it must be getting closer to the positive charge Q.

An increasingly negative potential energy U is produced; thus, the potential energy decreases. Conversely, as the electron approaches Q, the electric potential V increases with less distance. This is because the product is positive, and reducing r increases V.

32. E is correct.

Magnets provide magnetic forces (i.e., attraction or repulsion that arises between electrically charged particles because of their motion).

Generators convert mechanical energy into electrical energy, turbines extract energy from fluids (e.g., air and water), and transformers transfer energy between circuits.

33. B is correct.

The potential energy of a system containing two point charges is:

$$U = kq_1q_2 / r$$

In this problem, one of the charges is positive, and the other is negative.

To account for this, write:

$$q_1 = +|q_1| \text{ and } q_2 = -|q_2|$$

The *potential energy* can be written as:

$$U = -k\,|q_1|\,|q_2| / r$$

continued...

Moreover, the *absolute value of the potential energy* is:

$$|U| = k\,|q_1|\,|q_2| / r$$

All quantities are positive.

The absolute value of the potential energy is inversely proportional to the orbital radius; therefore, the absolute value of the potential energy decreases as the orbital radius increases.

34. C is correct.

$$R = (\rho L) / (\pi r^2)$$

$$R_A = (\rho L) / (\pi r^2)$$

$$R_B = [\rho(2L)] / [\pi(2r)^2]$$

$$R_B = (2/4)\cdot[(\rho L) / (\pi r^2)]$$

$$R_B = ½[(\rho L) / (\pi r^2)]$$

$$R_B = ½R_A$$

35. D is correct.

By convention, *current flows* from high to low potential, but it represents the flow of positive charges.

Electrons flow in the opposite direction, from low potential to high potential.

36. C is correct.

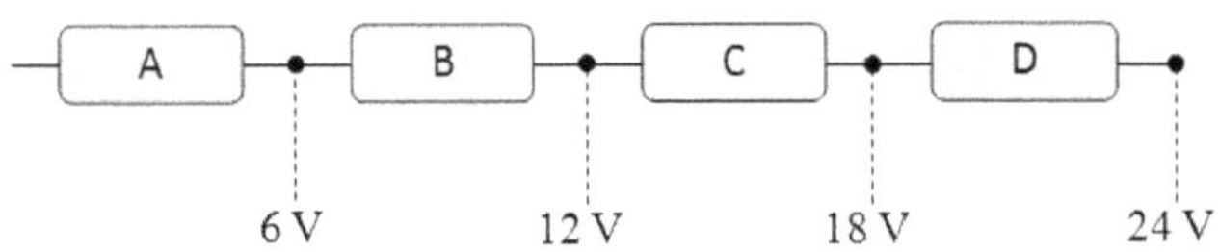

Batteries in series add voltage like resistors in series add resistance.

The resistances of the lights they power are not needed to solve the problem.

37. B is correct.

Capacitance:

$$C = (k\mathcal{E}_0 A) / d$$

$$k = (Cd) / A\mathcal{E}_0$$

If capacitance increases by a factor of 4:

$$k_2 = (4C)d / A\mathcal{E}_0$$

$$k_2 = 4(Cd / A\mathcal{E}_0)$$

$$k_2 = 4k$$

38. E is correct.

Calculate capacitance:

$C = (k\mathcal{E}_0 A) / d$

$C = [(1)\cdot(8.854 \times 10^{-12}\ \text{F/m})\cdot(0.4\ \text{m}^2)] / (0.04\ \text{m})$

$C = 8.854 \times 10^{-11}\ \text{F}$

$V = Q / C$

$V = (6.8 \times 10^{-10}\ \text{C}) / (8.854 \times 10^{-11}\ \text{F})$

$V = 7.7\ \text{V}$

39. A is correct.

Since force is the cross-product of velocity and magnetic field strength:

$F = qv \times B$

The force is at a maximum when v and B are perpendiculars:

$F = qvB \sin 90°$

$\sin 90° = 1$

$F = qvB$

40. E is correct.

Ampere is the unit of current, which is defined as the rate of flow of charge.

The current describes how much charge (in Coulombs) passes through a point every second.

So, if the current is multiplied by the number of seconds (the time interval), one can measure just how much charge passed by during that interval.

Mathematically, the units are expressed as:

$A = C / s$

$C = A\cdot s$

41. C is correct.

$\Delta V = \Delta E / q$

$\Delta V = (1 / q)\cdot(\frac{1}{2}mv_f^2 - \frac{1}{2}mv_i^2)$

$\Delta V = (m / 2q)\cdot(v_f^2 - v_i^2)$

$\Delta V = [(1.67 \times 10^{-27}\ \text{kg}) / (2)\cdot(1.6 \times 10^{-19}\ \text{C})] \times [(3.2 \times 10^5\ \text{m/s})^2 - (1.7 \times 10^5\ \text{m/s})^2]$

$\Delta V = 384\ \text{V}$

42. E is correct.

$Q = VC$

Even though the capacitors have different capacitances, the voltage across each capacitor is inversely proportional to the capacitance of that capacitor.

Like current, the charge is conserved across capacitors in series.

43. A is correct.

Calculate capacitance:

$C = k\mathcal{E}_0 A / d$

$C = [(1)\cdot(8.854 \times 10^{-12}\ \text{F/m})\cdot(0.6\ \text{m}^2)] / (0.06\ \text{m})$

$C = 8.854 \times 10^{-11}\ \text{F}$

Find potential difference:

$C = Q / V$

$V = Q / C$

$V = (7.08 \times 10^{-10}\ \text{C}) / (8.854 \times 10^{-11}\ \text{F})$

$V = 8\ \text{V}$

44. B is correct.

"In a perfect conductor" and "in the absence of resistance" have the same meaning, and current can flow in conductors of varying resistances, and a semi-perfect conductor has resistance.

45. C is correct.

Faraday's Law: a changing magnetic environment causes a voltage to be induced in a conductor.

Metal detectors send quick magnetic pulses that cause a voltage (by Faraday's Law) and subsequent current to be induced in the conductor.

By Lenz's Law, an opposing magnetic field will then arise to counter the changing magnetic field.

The detector picks up the magnetic field and notifies the operator.

Thus, metal detectors use Faraday's Law and Lenz's Law to detect metal objects.

46. E is correct.

$E = qV$

$E = (7 \times 10^{-6}\ \text{C})\cdot(3.5 \times 10^{-3}\ \text{V})$

$E = 24.5 \times 10^{-9}\ \text{J}$

$E = 24.5\ \text{nJ}$

47. B is correct.

$R_1 = \rho L_1 / A_1$

$R_2 = \rho(4L_1) / A_2$

$R_1 = R_2$

$\rho L_1 / A_1 = \rho(4L_1) / A_2$

$A_2 = 4A_1$

$(\pi / 4)d_2^{\,2} = (\pi / 4)\cdot(4)d_1^{\,2}$

$d_2^2 = 4d_1^{\,2}$

$d_2 = 2d_1$

48. D is correct.

The total resistance of a network of series resistors increases as more resistors are added.

$V = IR$

An increase in the total resistance results in a decrease in the total current through the network.

49. A is correct.

This is a circuit with two resistors in series.

Combine the two resistors into one resistor:

$R_T = R + R_{int}$

$R_T = 0.5\ \Omega + 0.1\ \Omega$

$R_T = 0.6\ \Omega$

Ohm's law:

$V = IR$

$I = V / R$

$I = 9\ V / 0.6\ \Omega$

$I = 15\ A$

50. C is correct.

Energy stored in capacitor:

$$U = ½(Q^2 / C)$$

Capacitance:

$$C = k\mathcal{E}_0 A / d$$

$$U = ½(Q^2 d) / (k\mathcal{E}_0 A)$$

$$Q = \sqrt{[(2U \times k\mathcal{E}_0 A) / d]}$$

$$Q = \sqrt{\{[(2)\cdot(10 \times 10^3\ J)\cdot(1)\cdot(8.854 \times 10^{-12}\ F/m)\cdot(2.4 \times 10^{-5}\ m^2)] / 0.0016\ m\}}$$

$$Q = 52\ \mu C$$

51. B is correct.

Potential energy:

$$U = (kQq) / r$$

Electric Potential:

$$V = (kQ) / r$$

As r increases, the potential energy U decreases, as does the electric potential V.

Movement in the direction of the electric field is a movement away from a positive charge.

52. D is correct.

Electric field energy density:

$$\eta_E = ½E^2 \times \mathcal{E}_0$$

$$\eta_E = ½(8.6 \times 10^6\ V/m)^2\cdot(8.854 \times 10^{-12}\ F/m)$$

$$\eta_E = 330\ J/m^3$$

53. A is correct.

Resistance = Ohms

$$\Omega = V / A$$

$$\Omega = [(kg\cdot m^2/s^2)/C] / [C/s]$$

$$\Omega = kg\cdot m^2/(C^2/s)$$

54. E is correct.

Ohm's Law:

$$V = IR$$

If V is constant, then I and R are inversely proportional.

An increase in R results in a decrease in I.

55. D is correct.

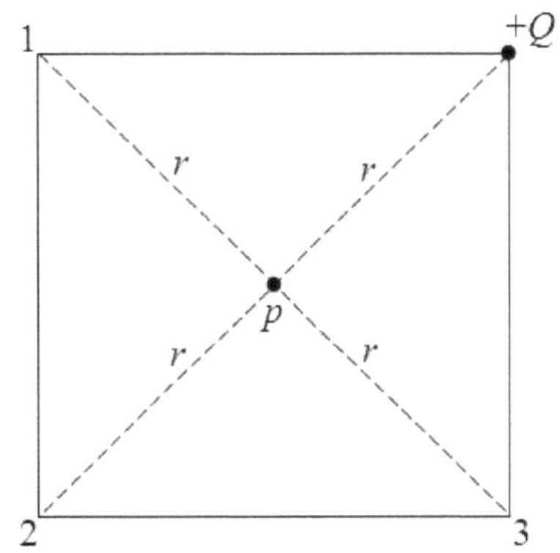

Electric Potential:

$$V = kQ / r$$

All other positions on the square (1, 2 or 3) are equidistant from point p, as is charge $+Q$.

Thus, a negative charge placed at these locations would have equal magnitude potential but an opposite sign because the new charge is negative.

Thus:

$$|V| = |-V|$$

$$V + (-V) = 0$$

Notes for active learning

Electrostatics and Electromagnetism – Detailed Explanations

1. C is correct.

Since charge is quantized, the charge Q must be a whole number (n) times the charge on a single electron:

Charge = # electrons × electron charge

$Q = n(e^-)$

$n = Q / e^-$

$n = (-1\text{ C}) / (-1.6 \times 10^{-19}\text{ C})$

$n = 6.25 \times 10^{18} \approx 6.3 \times 10^{18}$ electrons

2. D is correct.

In *Gaus's Law*, the area is a vector perpendicular to the plane.

Only the component of the electric field strength parallel is used.

Gaus's Law:

$\Phi = EA \cos \theta$

where Φ is electric flux (scalar), E is electric field strength, and A is area vector

Solve:

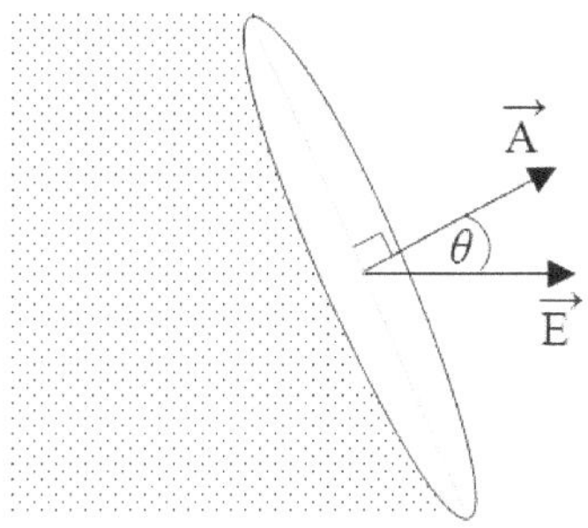

$\Phi = EA \cos (\pi / 6)$

$A = \pi r^2 = \pi D^2 / 4$

$\Phi = (740\text{ N/C})\cdot(\pi / 4)\cdot(1\text{ m})^2 \cos (\pi / 6)$

$\Phi = 160\pi\text{ N}\cdot\text{m}^2/\text{C}$

For calculation, use radians mode, not degree mode.

3. A is correct.

The magnitude of the *negative charge's electric field*:

$| E_2 | = kQ_2 / d_2^2$

$| E_2 | = (9 \times 10^9\text{ N}\cdot\text{m}^2/\text{C}^2)\cdot[(-1.3 \times 10^{-9}\text{ C}) / (+10^{-3}\text{ m})^2]$

$| E_2 | = 1.17 \times 10^7$ N/C to the left

The magnitude of the *positive charge's electric field*:

$| E_1 | = kQ_1 / d_1^2$

$| E_1 | = (9 \times 10^9\text{ N}\cdot\text{m}^2/\text{C}^2)\cdot[(1.3 \times 10^{-9}\text{ C}) / (2 \times 10^{-3}\text{ m})^2]$

$| E_1 | = 2.9 \times 10^6$ N/C to the right

continued...

$\Delta E = E_2 - E_1$

$\Delta E = (1.17 \times 10^7 \text{ N/C}) - (2.9 \times 10^6 \text{ N/C})$

$\Delta E = 8.8 \times 10^6 \text{ N/C}$, to the left

4. C is correct.

Coulomb's law: $F_1 = kQ_1Q_2 / r^2$

If r is increased by a factor of 4:

$F_e = kQ_1Q_2 / (4r)^2$

$F_e = kQ_1Q_2 / (16r^2)$

$F_e = (1/16)kQ_1Q_2 / r^2$

$F_e = (1/16)F_1$

As the distance increases by a factor of 4, the force decreases by $4^2 = 16$.

5. E is correct.

Calculate the distance between two charges using the Pythagorean Theorem:

$r^2 = (1 \text{ nm})^2 + (4 \text{ nm})^2$

$r^2 = 17 \text{ nm}^2$

$r = 4.1 \text{ nm}$

$F = kQ_1Q_2 / r^2$

$F = [(9 \times 10^9 \text{ N·m}^2/\text{C}^2)·(1.6 \times 10^{-19} \text{ C})·(1.6 \times 10^{-19} \text{ C})] / (4.1 \times 10^{-9} \text{ m})^2$

$F = 1.4 \times 10^{-11} \text{ N}$

6. A is correct.

Gamma rays have photons with the highest energy.

7. D is correct.

Use a coordinate system in which a repulsive force is in the positive direction, and an attractive force is in the negative direction.

Gravitational Force: F_g

$F_g = -Gm_1m_2 / r^2$

$F_g = -[(6.673 \times 10^{-11} \text{ N·m}^2/\text{kg}^2)·(54{,}000 \text{ kg})·(51{,}000 \text{ kg})] / (180 \text{ m})^2$

$F_g = -0.18 \text{ N·m}^2 / (32{,}400 \text{ m}^2)$

$F_g = -5.7 \times 10^{-6} \text{ N}$

continued...

Electrostatic Force: F_e

$$F_e = kQ_1Q_2 / r^2$$

$$F_e = [(9 \times 10^9\ \text{N·m}^2/\text{C}^2)·(−15 \times 10^{-6}\ \text{C})·(−11 \times 10^{-6}\ \text{C})] / (180\ \text{m})^2$$

$$F_e = (1.49\ \text{N·m}^2) / (32{,}400\ \text{m}^2)$$

$$F_e = 4.6 \times 10^{-5}\ \text{N}$$

Net Force:

$$F_{net} = F_g + F_e$$

$$F_{net} = (−5.7 \times 10^{-6}\ \text{N}) + (4.6 \times 10^{-5}\ \text{N})$$

$$F_{net} = 4 \times 10^{-5}\ \text{N}$$

F_{net} is positive, which means a net repulsive force on the asteroids; the repulsive electrostatic force between them is stronger than the attractive gravitational force.

8. B is correct.

Newton's Third Law states that for every force, there is an equal and opposite reaction force, and this applies to electrostatic forces.

Electrostatic Force:

$$F_1 = kQ_1Q_2 / r^2$$

$$F_2 = kQ_1Q_2 / r^2$$

$$F_1 = F_2$$

9. D is correct.

Forces balance to yield: $F_{electric} = F_{gravitation}$

$$F_{electric} = mg$$

The values for an electric field are provided.

$$F_{electric} = QE$$

$$F_{electric} − F_{gravitation} = 0$$

$$QE – mg = 0$$

where Q is the charge on the ball

$$QE = mg$$

$$Q = mg / E$$

$$Q = (0.008\ \text{kg})·(9.8\ \text{m/s}^2) / (3.5 \times 10^4\ \text{N/C})$$

$$Q = –2.2 \times 10^{-6}\ \text{C}$$

continued…

If the electric field points down, then a positive charge experiences a downward force.

The charge must be negative, so the electric force balances gravity.

10. A is correct.

$a = qE / m$

The electron moves against the electric field in the upward direction, so its acceleration:

$a_e = qE / m_e$

The proton moves with the electric field, which is down, so:

$a_p = qE / m_p$

However, the masses considered are small to where the gravity component is negligible.

$m_p / m_e = (1.67 \times 10^{-27}\text{ kg}) / (9.11 \times 10^{-31}\text{ kg})$

$m_p / m_e = 1{,}830$

The mass of an electron is about 1,830 times smaller than the mass of a proton.

$(1{,}830)m_e = m_p$

$a_p = qE / (1{,}830)m_e$

$a_p = a_e / (1{,}830)$

$a_e = 1{,}830a_p$

11. D is correct.

Calculate the strength of the field at point P due to one charge:

$E = kQ / r^2$

$E_1 = (9 \times 10^9\text{ N·m}^2/\text{C}^2)\cdot[(2.3 \times 10^{-11}\text{ C}) / (5 \times 10^{-3}\text{ m})^2]$

$E_1 = 8.3 \times 10^3\text{ N/C}$

Both electric field vectors point toward the negative charge, so the magnitude of each field at point P is doubled:

$E_T = 2E_1$

$E_T = 2(8.3 \times 10^3\text{ N/C})$

$E_T = 1.7 \times 10^4\text{ N/C}$

12. E is correct.

charge = # electrons × electron charge

$Q = ne^-$

$n = Q / e^-$

$n = (-10 \times 10^{-6}\text{ C}) / (-1.6 \times 10^{-19}\text{ C})$

$n = 6.3 \times 10^{13}$ electrons

13. C is correct.

Coulomb's law:

$F_e = kQ_1Q_2 / r^2$

If the separation is halved, then r decreases by ½:

$F_2 = kq_1q_2 / (½r)^2$

$F_2 = 4(kq_1q_2 / r^2)$

$F_2 = 4F_e$

14. E is correct.

Coulomb's law:

$F = kQ_1Q_2 / r^2$

Doubling the charges and distance:

$F = [k(2Q_1)\cdot(2Q_2)] / (2r)^2$

$F = [4k(Q_1)\cdot(Q_2)] / (4r^2)$

$F = (4/4)[kQ_1Q_2 / (r^2)]$

$F = kQ_1Q_2 / r^2$, remains the same

15. D is correct.

Like charges repel each other.

From Newton's Third Law, the magnitude of the force experienced by each charge is equal.

16. A is correct.

Coulomb's law:

$F = kQ_1Q_2 / r^2$

The Coulomb force between opposite charges is attractive.

Since the strength of the force is inversely proportional to the square of the separation distance (r^2), the force decreases as the charges are pulled apart.

17. B is correct.

Currents occur in a conducting circuit element when there is a potential difference between the two ends of the element, hence an established electric field.

18. D is correct.

Voltage is related to the number of coils in a wire.

More coils yield a higher voltage.

Turns ratio:

$$V_s / V_p = n_s / n_p$$

In this example:

$$n_s < n_p$$

Therefore,

$$V_s < V_p$$

The secondary voltage (V_s) is lower than the primary voltage (V_p), so it is a step-down transformer.

19. B is correct.

Coulomb's Law:

$$F_1 = kQ_1Q_2 / r^2$$

$$F_2 = kQ_1Q_2 / r^2$$

$$F_1 = F_2$$

Newton's Third Law: the force exerted by one charge on the other has the same magnitude as the force the other exerts on the first.

20. E is correct.

$$F_e = kQ_1Q_2 / r^2$$

$$F_e = (9 \times 10^9\ \text{N·m}^2/\text{C}^2)·(–1.6 \times 10^{-19}\ \text{C})·(–1.6 \times 10^{-19}\ \text{C}) / (0.03\ \text{m})^2$$

$$F_e = 2.56 \times 10^{-25}\ \text{N}$$

21. A is correct.

According to Lenz's Law, inserting a magnet into the coil causes the magnetic flux through the coil to change. This produces an emf in the coil, which drives a current through the coil:

Lenz's Law:

$$\text{emf} = –\text{N}\Delta\text{BA} / \Delta t$$

The bulb's brightness changes with a change in the current, but it cannot be known if the bulb gets brighter or dimmer without knowing the orientation of the coil with respect to the incoming magnetic pole of the magnet.

22. C is correct.

Charge = # electrons × electron charge

$Q = ne^-$

$n = Q / e^-$

$n = (8 \times 10^{-6}\ \text{C}) / (1.6 \times 10^{-19}\ \text{C})$

$n = 5 \times 10^{13}$ electrons

23. A is correct.

An object with a charge attracts another object of opposite charge or a neutral charge.

Like charges cannot attract, but the type of charge does not matter otherwise.

24. B is correct.

1 amp = 1 C / s

The *ampere* is the unit to express the flow rate of electric charge known as *current*.

One ampere of current represents one coulomb of electrical charge (6.24×10^{18} charge carriers) moving past a specific point in one second.

25. A is correct.

Initially, the current will flow clockwise, but after 180° of rotation, the current will reverse itself.

After 360° of rotation, the current will reverse itself again.

Thus, there are 2 current reverses in 1 revolution.

26. D is correct.

$W = Q\Delta V$

Electric Potential:

$V = kQ / r$

Consider the charge Q_1 fixed and move charge Q_2 from initial distance r_i to final distance r_f.

$W = Q_2(V_f - V_i)$

$W = Q_2[(kQ_1 / r_f) - (kQ_1 / r_i)]$

$W = kQ_1Q_2(1 / r_f - 1 / r_i)$

$W = (9 \times 10^9\ \text{N·m}^2/\text{C})·(2.3 \times 10^{-8}\ \text{C})·(2.5 \times 10^{-9}\ \text{C})·[(1 / 0.01\ \text{m}) - (1 / 0.1\ \text{m})]$

$W = 4.7 \times 10^{-5}\ \text{J}$

27. E is correct.

The electric field is oriented so that a positively-charged particle would be forced to move to the top because negatively-charged particles move to the bottom of the cube to be closer to the source of the electric field.

Therefore, the positively charged particles are forced upward and negatively charged downward, leaving the top surface positively charged.

28. C is correct.

The magnitude of the force between the center charge and each charge at a vertex is 5 N. The net force of these two forces is directed toward the third vertex.

To determine the magnitude of the net force, calculate the magnitude of the component of each force acting in that direction:

$$F_{net} = F_1 \sin(½\,\theta) + F_2 \sin(½\,\theta) \quad \text{(see diagram)}$$

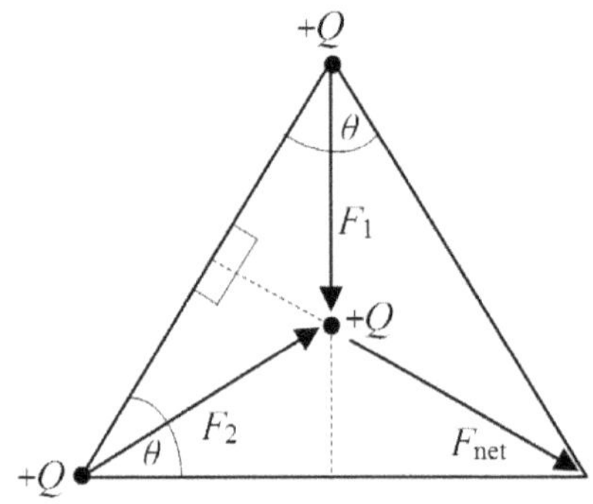

Since it is an equilateral triangle $\theta = 60°$,

$$F_{net} = (5\text{ N} \sin 30°) + (5\text{ N} \sin 30°)$$

$$F_{net} = (5\text{ N})\cdot(½) + (5\text{ N})\cdot(½)$$

$$F_{net} = 5\text{ N}$$

29. D is correct.

Equilibrium:

$$F = kq_1q_2 / r_1^2$$

$$F_{\text{attractive on } q2} = F_{\text{repulsive on } q2}$$

$$kq_1q_2 / r_1^2 = kq_2Q / r_2^2$$

$$q_1 = Qr_1^2 / r_2^2$$

$$q_1 = (7.5 \times 10^{-9}\text{ C})\cdot(0.2\text{ m})^2 / (0.1\text{ m})^2$$

$$q_1 = 30 \times 10^{-9}\text{ C}$$

30. A is correct.

Gamma rays have the highest frequency on the electromagnetic spectrum, with frequencies $> 3 \times 10^{19}$ Hz.

31. A is correct.

The strength of the electrostatic field due to a single point charge is given by:

$E = kQ / r^2$ assumes that the source charge is in a vacuum

E depends on the magnitude of the source charge Q and the distance r from Q.

The sign of the source charge affects only the direction of the electrostatic field vectors.

The sign of the source charge does not affect the strength of the field.

32. B is correct.

$F = qvB$

$F = mv^2 / r$

$mv^2 / r = qvB$

cancel v from both sides of the expression

$mv / r = qB$

$r = (mv) / (qB)$

If the velocity doubles, the radius also doubles.

33. C is correct.

Coulomb's law:

$F_e = kQ_1Q_2 / r^2$

$1 \text{ N} = kQ_1Q_2 / r^2$

Doubling charges and keeping distance constant:

$k(2Q_1)\cdot(2Q_2) / r^2 = 4kQ_1Q_2 / r^2$

$4kQ_1Q_2 / r^2 = 4F_e$

$4F_e = 4(1 \text{ N}) = 4 \text{ N}$

34. D is correct.

The Na^+ ion is positively charged and attracts the oxygen atom.

Oxygen is slightly negative because it is more electronegative than the hydrogen atoms to which it is bonded.

35. C is correct.

Charge = # of electrons × electron charge

$Q = ne^-$

$Q = (30)\cdot(-1.6 \times 10^{-19} \text{ C})$

$Q = -4.8 \times 10^{-18} \text{ C}$

36. E is correct.

Cyclotron frequency is given as:

$f = qB / 2\pi m$

This expression does not consider speed.

37. B is correct.

The repulsive force between two particles is:

$$F = kQ_1Q_2 / r^2$$

As r increases, F decreases

Using $F = ma$, a also decreases

38. A is correct.

The given unit can be written:

[kg·m²/s²] / C = J / C, which is the definition of the volt, the unit of electric potential difference.

39. C is correct.

The Coulomb is the basic unit of electrical charge in the SI unit system.

40. A is correct.

By the Law of Conservation of Charge, a charge cannot be created nor destroyed.

41. B is correct.

Coulomb's Law:

$$F = kQ_1Q_2 / r^2$$

If both charges are doubled,

$$F = k(2Q_1)\cdot(2Q_2) / r^2$$

$$F = 4kQ_1Q_2 / r^2$$

F increases by a factor of 4.

42. C is correct.

Coulomb's law:

$$F = kQ_1Q_2 / r^2$$

$$Q_1 = Q_2$$

Therefore:

$$Q_1Q_2 = Q^2$$

$$F = kQ^2 / r^2$$

Rearranging:

$$Q^2 = Fr^2 / k$$

continued...

$Q = \sqrt{(Fr^2 / k)}$

$Q = \sqrt{[(4\ \text{N})\cdot(0.01\ \text{m})^2 / (9 \times 10^9\ \text{N}\cdot\text{m}^2/\text{C}^2)]}$

$Q = 2 \times 10^{-7}$ C

43. A is correct.

Faraday's law states that electromotive force (emf) is equal to the rate of change of magnetic flux.

Magnetic flux is the product of the magnetic field and projected area:

$\Phi = BA_{\perp}$,

where $A_{\perp}$ is the area of the loop projected on a plane perpendicular to the magnetic field

In this problem, B is vertical (and constant), so the projection plane is horizontal.

Therefore, find the orientation of the axis of rotation that guarantees that as the loop rotates, the projection of its area on a horizontal plane does not change with time.

Notice that if the orientation of the axis is at an arbitrary angle to the field, the emf can be made to be zero by aligning the axis of rotation with the axis of the loop (i.e., perpendicular to the loop).

With this orientation, the projection of the area never changes, which is not true of other alignments to the loop.

Although the emf can be made to be zero, it is not *guaranteed* to be zero.

The orientation of the axis that *guarantees* that the projected area is constant is the vertical orientation.

One way to see this is to notice that because of the high symmetry of the vertical-axis orientation, rotating the loop about a vertical axis is equivalent to changing the viewer's perspective from one angle to another.

The answer cannot depend on the perspective of the viewer.

Therefore, the projected area cannot change as the loop is rotated about the vertical axis; the emf is guaranteed to be zero.

44. C is correct.

Coulomb's law:

$F = kQ_1Q_2 / r^2$

When each particle has lost ½ its charge:

$F_2 = k(\frac{1}{2}Q_1)\cdot(\frac{1}{2}Q_2) / r^2$

$F_2 = (\frac{1}{4})kQ_1Q_2 / r^2$

$F_2 = (\frac{1}{4})F$

F decreases by a factor of ¼

45. B is correct.

The time taken for one revolution around the circular path is:

$$T = 2\pi R/v$$

where R is the radius of the circle, and v is the speed of the proton

If the speed is increased, the radius also increases.

The relationship between speed and radius follows from the fact that the magnetic interaction provides the centripetal force:

$$mv^2/ R = qvB$$

Thus:

$$R = mv / qB$$

If the speed is tripled, the radius triples, other things equal.

The final period for a revolution is:

$$T_f = 2\pi R_f / v_f = 2\pi(3R) / 3v = 2\pi R / v = T$$

46. C is correct.

An *electrostatic field* shows the path that a *positively*-charged particle would take.

As this positive particle moves closer to the negatively-charged one, the force between them increases.

Coulomb's law:

$$F = kQ_1Q_2 / r^2$$

By convention, electric field vectors point towards negative source charges.

Since electrical field strength is inversely proportional to the square of the distance from the source charge, the magnitude of the electric field progressively increases as an object moves towards the source charge.

47. E is correct.

Protons are charges, so they have an electric field.

Protons have mass, so they have a gravitational field.

Protons have an intrinsic magnetic moment, so they have a magnetic field.

48. A is correct.

$W = Q\Delta V$

$V = kq / r$

$W = (kQq)\cdot(1 / r_2 - 1 / r_1)$

$W = (kQq)\cdot(1 / 2\text{ m} - 1 / 6\text{ m})$

$W = (kQq)\cdot(1 / 3\text{ m})$

$W = (9 \times 10^9\text{ N}\cdot\text{m}^2/\text{C}^2)\cdot(3.1 \times 10^{-5}\text{ C})\cdot(-10^{-6}\text{ C}) / (1 / 3\text{ m})$

$W = -0.093\text{ J} \approx -0.09\text{ J}$

The negative sign indicates that the electric field does the work on charge q.

49. D is correct.

charge = # electrons × electron charge

$Q = ne^-$

$n = Q / e^-$

$n = (-600 \times 10^{-9}\text{ C}) / (-1.6 \times 10^{-19}\text{ C})$

$n = 3.8 \times 10^{12}$ electrons

50. C is correct.

The analog to N/kg would be N/C, the unit for the electric field.

51. E is correct.

All the electromagnetic waves travel through space (vacuum) at the same speed:

$c = 3 \times 10^8$ m/s

52. A is correct.

As the proton of charge q moves in the direction of the electric field lines, it moves away from a positive charge (because field lines emanate from a positive charge).

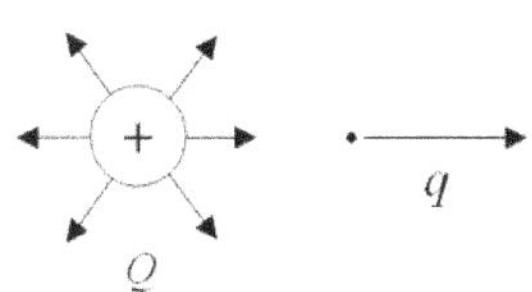

Electric Potential Energy:

$U = kQq / r$

As distance increases, the potential energy decreases because they are inversely proportional.

Electrical potential:

$V = kQ / r$

Electrical potential is inversely proportional to distance and decreases as distance increases.

53. B is correct.

1 watt = 1 J/s

54. D is correct.

An electrically polarized object has had its charge separated into opposites and thus rearranged itself within distinct regions.

55. A is correct.

Because point P is symmetric about Q_1 and Q_2 and both charges have the same positive magnitude, the electric field cancels midway between the charges.

$E = kQ / r^2$

$E_1 = -E_2$

$E_{tot} = E_1 + E_2$

$E_{tot} = (-E_2 + E_2)$

$E_{tot} = 0$ N/C

Notes for active learning

Notes for active learning

Atomic and Nuclear Structure – Detailed Explanations

1. A is correct.

Though alpha particles have low penetrating power and high ionizing power, they are not harmless.

All forms of radiation have risks and cannot be thought of as entirely harmless.

2. C is correct.

A beta particle (β) is a high-energy, high-speed electron (β^-) or positron (β^+) emitted in the radioactive decay of an atomic nucleus.

Electron emission (β^- decay) occurs in an unstable atomic nucleus with an excess of neutrons, whereby a neutron is converted into a proton, an electron, and an electron antineutrino.

Positron emission (β^+ decay) occurs in an unstable atomic nucleus with an excess of protons, whereby a proton is converted into a neutron, a positron, and an electron neutrino.

3. E is correct.

The *de Broglie wavelength* is given as:

$$\lambda = h / p$$

where h is Planck's constant and p is momentum

$$p = mv$$

$$\lambda_1 = h / mv$$

$$\lambda_2 = h / m(2v)$$

$$\lambda_2 = \tfrac{1}{2}h / mv$$

$$\lambda_2 = \tfrac{1}{2}\lambda_1$$

λ decreases by factor of 2

4. B is correct.

The *Bohr model* places electrons around the nucleus of the atom at discrete energy levels.

The *Balmer series line spectra* agreed with the Bohr model because the energy of the observed photons in each spectrum matched the transition energy of electrons within these discrete predicted states.

5. C is correct.

A *radioactive element* is an element that spontaneously emits radiation in the form of one or a combination of the following: alpha radiation, beta radiation, gamma radiation.

6. D is correct.

This is an example of an *electron capture nuclear reaction.*

When this happens, the atomic number decreases by one, but the mass number stays the same.

$$^{100}_{44}\text{Ru} + {}^{0}_{-1}\text{e}^{-} \rightarrow {}^{100}_{43}\text{Tc}$$

Ru: 100 = mass number (# protons + # neutrons)

Ru: 44 = atomic number (# protons)

From the periodic table, Tc is the element with 1 less proton than Ru.

7. B is correct.

A *nucleon* is a particle that makes up the nucleus of an atom.

The two known nucleons are protons and neutrons.

8. E is correct.

An *alpha particle* is composed of two neutrons and two protons and is identical to the nucleus of a ^{4}He atom.

Total mass of two alpha particles:

$2 \times (2 \text{ neutrons} + 2 \text{ protons}) = 8$

Mass of a ^{9}Be atom:

$5 \text{ neutrons} + 4 \text{ protons} = 9$

Mass of a ^{9}Be atom > total mass of two alpha particles

The mass of a ^{9}Be atom is greater than the mass of two alpha particles, so its mass is greater than twice the mass of a ^{4}He atom.

9. C is correct.

The *superscript* is the mass number (atomic weight), which is neutrons and protons.

The *subscript* is the atomic number, which is the number of protons.

Therefore, the number of neutrons is equal to the superscript minus the subscript.

$181 - 86 = 95$, which is the highest number of neutrons among the choices.

10. A is correct.

The number of neutrons and protons must be equal after the reaction.

Thus, the sum of the atomic number before and mass number before should be equal after the reaction.

Mass number (superscript):

$$(1 + 235) - (131 + 3) = 102$$

Atomic number (subscript):

$$92 - (53) = 39$$

$^{102}_{39}Y$ properly balances the reaction.

11. E is correct.

Balmer equation is given by:

$$\lambda = B[(n^2) / (n^2 - 2^2)]$$

$$\lambda = (3.6 \times 10^{-7})\cdot[(12^2) / (12^2 - 2^2)]$$

$$\lambda = 3.7 \times 10^{-7}\,m$$

here c is the speed of light:

$$c = \lambda f$$

Convert wavelength to frequency:

$$f = c / \lambda$$

$$f = (3 \times 10^8\,m/s) / (3.7 \times 10^{-7}\,m)$$

$$f = 8.1 \times 10^{14}\,s^{-1} = 8.1 \times 10^{14}\,Hz$$

12. D is correct.

Gamma rays are the most penetrating form of radiation because they are the highest energy and least ionizing.

A gamma-ray passes through a given amount of material without imparting as much of its energy into removing electrons from atoms and ionizing them as other forms of radiation do.

Gamma rays retain more of their energy passing through matter and can penetrate further.

13. A is correct.

Use the *Rydberg Formula*:

$$E = hf$$

$$f = c / \lambda$$

$$E = (hc) \cdot (1 / \lambda)$$

$$1 / \lambda = R(1 / n_1^2 - 1 / n_2^2)$$

where $n_1 = 1$ and $n_2 = 2$

$$E = hcR[(1 / n_1^2) - (1 / n_2^2)]$$

$$E = (4.14 \times 10^{-15}\ \text{eV·s}) \cdot (3 \times 10^8\ \text{m/s}) \cdot (1.097 \times 10^7\ \text{m}^{-1}) \cdot [(1 / 1^2) - (1 / 2^2)]$$

$$E = 13.6[1 - (1 / 4)]$$

$$E = 10.2\ \text{eV}$$

The positive energy indicates that a photon was absorbed and not emitted.

14. C is correct.

In β^- (beta minus) decay, the atomic number (subscript) increases by 1, but the atomic mass stays constant.

$${}^{87}_{37}\text{Rb} \rightarrow {}^{87}_{38}\text{Sr} + {}^{0}_{-1}\text{e} + {}^{0}_{0}\nu$$

Sr is the element with 1 more proton (subscript) than Rb.

${}^{0}_{0}\nu$ represents an electron antineutrino.

15. C is correct.

The nucleus of an atom is bound by the strong nuclear force from the nucleons within it.

The strong nuclear force must overcome the Coulomb repulsion of the protons (due to their like charges).

Neutrons help stabilize and bind the nucleus by contributing to the strong nuclear force so that it is greater than the Coulomb repulsion experienced by the protons.

16. A is correct.

Geiger-Muller counters operate using a Geiger-Muller tube, consisting of a high voltage shell and small rod in the center, filled with low-pressure inert gas (e.g., argon).

When exposed to radiation (specifically particle radiation), the radiation particles ionize atoms of the argon allowing for a brief charge to be conducted between the high voltage rod and outer shell.

The electric pulse is then displayed visually or via audio to indicate radioactivity.

17. B is correct.

The half-life calculation:

$$A = A_0(½)^{t\,/\,h}$$

where A_0 = original amount, t = time elapse and h = half-life

If three half-life pass,

$$t = 3h$$

$$A = (1)·(½)^{3h\,/\,h}$$

$$A = (1)·(½)^3$$

$$A = 0.125 = 12.5\%$$

18. D is correct.

In the *Lyman series*, electron transitions go from n ≥ 2 to n = 1.

19. B is correct.

Most of the volume of an atom is occupied by empty space.

20. A is correct.

Alpha particles are positively charged due to their protons, while beta minus particles are negatively charged (i.e., electrons).

When exposed to a magnetic field, alpha particles and electrons deflect in opposite directions due to their opposite charges and thus experience opposite forces due to the magnetic field.

21. E is correct.

Chemical reactions store and release energy in their chemical bonds but do not convert mass into energy.

However, nuclear reactions convert a small amount of mass into energy, which can be measured and calculated via the equation:

$$E = \Delta mc^2$$

22. C is correct.

The *Curie* is a non-SI unit of radioactivity equivalent to 3.7×10^{10} decays (disintegrations) per second.

The Curie is named after the early radioactivity researchers Marie and Pierre Curie.

23. B is correct.

The *atomic numbers*:

$$^{235}_{92}\text{U} \rightarrow {}^{141}_{56}\text{Ba} + {}^{92}_{36}\text{Kr}$$

The *subscripts* on each side of the expression sum to 92, so adding a proton ($^{1}_{1}$H) to the right side would not balance.

The superscripts sum to 235 on the left and sum to 233 on the right.

Add two neutrons ($^{1}_{0}$n + $^{1}_{0}$n) to the right side to balance the equation.

24. D is correct.

The *atomic number* is the subscript and represents the number of protons.

The *superscript* is the mass number and is the sum of protons and neutrons.

The number of *neutrons* can be found by taking the difference between *mass* and *atomic* numbers.

In this example, there are 16 protons and 18 neutrons.

25. A is correct.

The *half-life* calculation:

$$A = A_0(½)^{t / h}$$

where A_0 = original amount, t = time elapse and h = half-life

Consider: (2 days)·(24 hours / 1 day) = 48 hours

$$A = (1)·(½)^{(48 / 12)}$$

$$A = 0.5^4$$

$$A = 0.0625 = 1/16$$

26. C is correct.

An *alpha particle* consists of two protons and two neutrons and is identical to a helium nucleus so that it can be written as $^{4}_{2}$He

For a nuclear reaction to be written correctly, it must be balanced, and the sum of superscripts and subscripts must be equal on both sides of the reaction.

The superscripts add to 238, and the subscripts add to 92 on both sides.

Therefore, it is the only balanced answer.

27. E is correct.

The question is asking for the λ of the emitted photon so use the Rydberg Formula:

$$1 / \lambda = R(1 / n_1^2 - 1 / n_2^2)$$

$$\lambda = 1 / [R(1 / n_1^2 - 1 / n_2^2)]$$

Use $n_1 = 5$ and $n_2 = 20$ because we are solving for λ of an emitted (not absorbed) photon.

$$\lambda = 1 / [(1.097 \times 10^7 \text{ m}^{-1})\cdot(1 / 5^2 - 1 / 20^2)]$$

$$\lambda = 2.43 \text{ μm}$$

28. D is correct.

When writing a nuclear reaction, the *superscript* represents the mass number, while the *subscript* represents the atomic number.

A correct nuclear reaction is balanced when the sum of superscripts (mass number) and subscripts (atomic number) is equal on both sides of the reaction.

29. C is correct.

A *blackbody* is an ideal system that absorbs 100% of light incident upon it and reflects none.

A blackbody emits 100% of the radiation it generates; therefore, it has perfect absorption and emissivity.

30. D is correct.

Carbon dating relies upon a steady creation of ^{14}C and knowledge of the rate of creation at various points in time to determine the approximate age of objects.

If a future archeologist is unaware of nuclear bomb testing and the higher levels of ^{14}C created, then the dates they calculate for an object would be too young.

This is because a higher amount of ^{14}C would be present in samples and make them seem like they had not had time to decay and thus appear to be younger.

31. A is correct.

Gamma radiation is an electromagnetic wave and is not a particle.

Thus, when gamma radiation is emitted, the atomic number and mass number remain the same.

32. C is correct.

In β^- decay, a neutron is converted to a proton, and an electron and electron antineutrino are emitted.

In β^+ decay, a proton is converted to a neutron, and a positron and an electron neutrino are emitted.

$$^{14}_{6}C \rightarrow {}^{14}_{7}N + e^+ + \nu_e$$

33. E is correct.

In a balanced nuclear equation, the number of *nucleons* must be conserved.

The sum of mass numbers and atomic numbers must be equal on both sides of the equation.

The product (i.e., daughter nuclei) should be on the right side of the equation.

34. B is correct.

The *atomic number* is the number of protons within the nucleus that characterizes the element's nuclear and chemical properties.

35. C is correct.

The *Balmer series* is the emission spectrum of hydrogen when electrons transition from a higher state to n = 2.

Within the Balmer series, there are four visible spectral lines with colors ranging from red to violet (i.e., ROY G BIV)

36. D is correct.

Electrons were discovered through early experiments with electricity, specifically in high voltage vacuum tubes (cathode ray tubes).

Beams of electrons were observed traveling through these tubes when a high voltage was applied between the anode and cathode, and the electrons struck fluorescent material at the back of the tube.

37. B is correct.

Beams a and c deflect when an electric field is applied, indicating they have a net charge and must be particles.

Beam b is undisturbed by the applied electric field, indicating it has no net charge and must be a high-energy electromagnetic wave since other forms of radioactivity (alpha and beta radiation) are charged particles.

38. E is correct.

Beam a is composed of negatively charged particles, while beam c is composed of positively charged particles; therefore, both beams are deflected by the electric field.

Beam b is composed of particles; however, they are neutral because the electric field does not deflect them.

An example of this kind of radiation would be a *gamma-ray*, which consists of neutral photons.

39. C is correct.

A *helium nucleus* is positively charged, deflected away from the top plate, and attracted toward the negative plate.

40. B is correct.

5.37 eV is the amount of energy required to excite the electron from the ground state to the zero-energy state.

Calculate the wavelength of a photon with this energy:

$E = hf$

$f = c / \lambda$

$E = hc / \lambda$

$\lambda = hc / E$

$\lambda = (4.14 \times 10^{-15}\ \text{eV·s})·(3 \times 10^8\ \text{m/s}) / (5.37\ \text{eV})$

$\lambda = 2.3 \times 10^{-7}\ \text{m}$

41. D is correct.

Elements with atomic numbers 84 and higher are radioactive because the strong nuclear force binding the nucleus cannot overcome the Coulomb repulsion from the high number of protons within the atom.

Thus, these nuclei are unstable and emit alpha radiation to decrease the number of protons within the nucleus.

42. E is correct.

Beta particles, like all forms of ionizing radiation, cannot be considered harmless.

43. C is correct.

Einstein's theory of special relativity that expresses the equivalence of mass and energy.

$E = mc^2$

$m = E / c^2$

$m = (3.85 \times 10^{26}\ \text{J}) / (3 \times 10^8\ \text{m/s})^2$

$m = 4.3 \times 10^9\ \text{kg}$

44. B is correct.

Planck's constant quantizes the amount of energy that can be absorbed or emitted.

Therefore, it sets a discrete lowest amount of energy for energy transfer.

45. E is correct.

The decay rate of a radioactive isotope or element is constant and independent of temperature, pressure, or surface area.

46. B is correct.

Larger nuclei (atomic number above 83) tend to decay because the attractive force of the nucleons (strong nuclear force) has a limited range, and the nucleus is larger than this range.

Therefore, these nuclei tend to emit alpha particles to decrease the size of the nucleus.

Smaller nuclei are not large enough to encounter this problem, but some isotopes have an irregular ratio of neutrons to protons and become unstable.

^{14}Carbon has 8 neutrons and 6 protons, and its neutron to proton ratio is too large; therefore, it is unstable and radioactive.

47. C is correct.

Positron emission occurs during β^+ decay.

In β^+ decay, a proton converts to a neutron and emits a positron and electron neutrino.

The decay can be expressed as:

$$^{44}_{21}\text{Sc} \rightarrow {}^{44}_{20}\text{Ca} + e^+ + \nu_e$$

48. C is correct.

A scintillation counter operates by detecting light flashes from the scintillator material.

When radiation strikes the scintillator crystal (often NaI), a light flash is emitted and detected by a photomultiplier tube, which then passes an electronic signal to audio or visual identification equipment.

49. B is correct.

When a Geiger counter clicks, it indicates that it has detected the radiation from one nucleus decaying.

The click could be from an alpha, beta, or even gamma-ray source but cannot be determined without other information.

50. A is correct.

The *Pauli Exclusion Principle* states that no two electrons have the same set of quantum numbers in an atom.

Thus, every electron in an atom has a unique set of *quantum numbers*, and a particular set belongs to one electron.

51. D is correct.

The *atomic number* indicates the number of protons within the nucleus of an atom.

52. E is correct.

Gamma rays are high-energy electromagnetic radiation rays and have no charge or mass.

Due to their high energy and speed, they have a high penetrating power.

53. B is correct.

Calculate mass defect:

m_1 = (2 protons)·(1.0072764669 amu) + (2 neutrons)·(1.0086649156 amu)

m_1 = 4.031882765 amu

Δm = 4.031882765 amu – 4.002602 amu

Δm = 0.029280765 amu

Convert to kg:

Δm = (0.029280765 amu / 1)·(1.6606 × 10^{-27} kg / 1 amu)

Δm = 4.86236 × 10^{-29} kg

Find binding energy:

$E = \Delta mc^2$

E = (4.86236 × 10^{-29} kg)·(3 × 10^8 m/s)^2

E = 4.38 × 10^{-12} J ≈ 4.4 × 10^{-12} J

54. E is correct.

Boron has an atomic number of five and thus has five protons and five electrons.

Two electrons fill the 1s orbital; then, the 2*s* orbital is filled with two electrons.

Only one electron remains, and it is in the 2*p* orbital, leaving it partially filled.

The electron configuration of boron:

$1s^22s^22p$

Notes for active learning

Particle Physics – Detailed Explanations

1. C is correct.

Leptons do not interact via the strong force, so they cannot comprise the hadrons.

Quarks do interact via the strong force and are the building blocks of hadrons.

Electrons are leptons, while protons and neutrons are hadrons (since they are made of quarks).

2. D is correct.

One defining characteristic of hadrons is that they interact by the strong nuclear force.

Electrons are leptons, not hadrons (because electrons do not interact by the strong force since they have no quarks), although neutrons and protons are hadrons.

Quarks, which are neither leptons nor hadrons, are the building blocks of all hadrons.

3. B is correct.

Leptons interact by forces except for the strong nuclear force.

4. B is correct.

The positron (antielectron) is a lepton and is not made of quarks.

The neutron is a hadron.

The alpha particle is composed of two protons and two neutrons, which are hadrons made of quarks.

5. B is correct.

Hadrons have charge +1e, 0e, or −1e. They are combinations of two (mesons) or three (baryons) quarks.

The only possible way of combining two or three fractionally charged objects so that the net charge is +1e, 0e or −1e is if the charges of the objects are taken from sets (1/3 e, −2/3 e) or (−1/3 e, +2/3 e).

The former set represents the *quarks*, and the latter the *antiquarks*.

Note that it is not possible to ever see a quark by itself (i.e., not bound to other quarks/antiquarks).

Thus, any actual particle that can be observed will have an integer charge; the fractional charge for any particle that can exist by itself is never measured.

6. D is correct.

The *deuteron* is composed of two baryons, one proton and one neutron.

The *baryons* each have three quarks, so the total number of quarks in the deuteron is six.

7. E is correct.

The *tritium nucleus*, the triton, comprises three baryons, one proton, and two neutrons.

The *baryons* each have three quarks, so the total number of quarks in the triton is nine.

8. B is correct.

The more massive the particle, the more decay channels are available, resulting in a shorter lifetime and reducing the range of the exchange mediated force.

Think of this using the *uncertainty principle.*

The exchange particle is a "virtual particle," meaning it appears temporarily and disappears once the exchange is complete.

The uncertainty principle limits how long such a particle can exist: $\Delta E\ \Delta t > \hbar/2$.

Since the exchange particle's energy increases with greater mass, a heavier particle can only survive for a shorter time Δt.

A shorter lifetime means it cannot cover as much distance as a lighter particle, so the range for a heavier particle is less.

9. A is correct.

The total mass-energy of the system is the rest mass of the electron and positron combined.

Since two photons are emitted, each photon has the energy equal to the mass-energy of one electron.

The energy of each photon is, therefore:

$$E = mc^2$$

where m is the mass of an electron

The relationship between the energy of a photon and frequency and wavelength is:

$$E = hf$$

$$hf = hc/\lambda$$

$$E = hc/\lambda$$

Combining these formulas:

$$\lambda = (hc) / (mc^2)$$

$$(hc) / (mc^2) = h / mc$$

$$\lambda = h / mc$$

Entering the values given:

$$\lambda = 2.42 \times 10^{-12}\ \text{m}$$

$$\lambda = 2.42\ \text{pm}$$

10. A is correct.

The *proton* has a charge of $+1e$. Answer A is the answer in which the sum of the quark charges is $+1e$.

A proton is made of two up quarks and one down quark.

11. C is correct.

The *positron* is quickly annihilated by an electron in the sample, producing two gamma rays, each having energy equal to the mass-energy of an electron or positron (the mass of a positron is the same as the mass of an electron).

The mass-energy of an electron (and thus a positron) is 0.511 MeV.

Note that *two* photons must be emitted because a final state with only one photon would violate the conservation of linear momentum.

To see this, imagine a reference frame where the center-of-mass of the electron-positron pair is stationary.

Thus, before the annihilation, the total momentum is zero (since the center-of-mass is not moving in this reference frame).

The final state of just one photon cannot have zero momentum.

12. C is correct.

To produce a *proton-antiproton pair*, an amount of energy is necessary to equal at least the combined mass of the proton and antiproton.

The proton and antiproton have mass-energy of 938 MeV, so the minimum energy to produce a pair is twice this, 1876 MeV.

13. C is correct.

Each *gamma-ray* will have energy equal to half the mass of η.

The energy of each gamma is, therefore:

$$E = mc^2 / 2$$

where m is the mass of a η meson

The relationship between the energy of a photon and frequency and wavelength is

$$E = hf = hc/\lambda$$

Combining these:

$$\lambda = (2hc) / (mc^2)$$

Entering the values given for hc and rest energy mc^2 and converting units:

$$\lambda = (2 \times 1240 \text{ eV nm} \times 10^{-9} \text{ m/nm}) / (548 \times 10^6 \text{ eV})$$

$$\lambda = 4.5 \times 10^{-15} \text{ m}$$

14. E is correct.

The kinetic energy of a *relativistic particle* is the total energy less the rest energy.

$$K = E - mc^2 = \gamma mc^2 - mc^2 = (\gamma - 1)mc^2$$

where

$$\gamma = 1 / \sqrt{(1 - \beta^2)} \text{ and } \beta = v / c$$

Rearranging:

$$\gamma - 1 = K / mc^2$$

$$\gamma = 1 + K / mc^2$$

$$1 / \sqrt{(1 - \beta^2)} = 1 + K / mc^2$$

Square both sides and invert:

$$1 - \beta^2 = 1 / (1 + K / mc^2)^2$$

$$\beta^2 = 1 - [1 / (1 + K / mc^2)^2]$$

$$\beta = \sqrt{\{1 - [1 / (1 + K / mc^2)^2]\}}$$

The mass of a proton is $mc^2 = 0.938$ GeV.

Using this and the given value of kinetic energy, $K = 2$ GeV:

$$\beta = 0.948$$

so

$$v = 0.948c \approx 0.95c$$

15. E is correct.

The energy of a relativistic particle is:

$$E = \gamma mc^2$$

$$\gamma mc^2 = (mc^2) / \sqrt{(1 - \beta^2)}$$

$$E = (mc^2) / \sqrt{(1 - \beta^2)}$$

Where

$$\beta = v / c$$

Divide by mc^2 and square:

$$(E / mc^2)^2 = 1 / (1 - \beta^2)$$

Solve for β:

$$\beta = \sqrt{[1 - (mc^2 / E)^2]}$$

continued…

For an electron, $mc^2 = 0.511$ MeV.

Using this and the given value $E = 2$ MeV, gives:

$$\beta = 0.967$$

or

$$v = 0.97\ c$$

16. B is correct.

Using the definition of Boltzmann's constant, the relationship between temperature and energy is:

$$kT = E$$

$$T = E / k$$

where k is Boltzmann's constant, $k = 8.617 \times 10^{-5}$ eV/K

With an energy of 50 GeV = 50×10^9 eV:

$$T = (50 \times 10^9 \text{ eV}) / (8.617 \times 10^{-5} \text{ eV/K})$$

$$T = 5.8 \times 10^{14} \text{ K}$$

The commonly quoted conversion factor is 1eV = 11,600K.

17. E is correct.

Hubble's law relates the speed of a receding galaxy to its distance from Earth.

$$d = v / H$$

$$d = 0.5\ (3.0 \times 10^8 \text{ m/s}) / (0.022 \text{ m/s/lightyear})$$

$$d = 6.8 \times 10^9 \text{ lightyears}$$

18. A is correct.

Hubble's law relates the speed of a receding galaxy to its distance from Earth.

$$v = dH$$

$$v = (2.7 \times 10^9 \text{ lightyears})\cdot(0.022 \text{ m/s/lightyear})$$

$$v = 5.94 \times 10^7 \text{ m/s}$$

In terms of the speed of light $c = 3.00 \times 10^8$ m/s:

$$v = 0.2\ c$$

19. B is correct.

The minimum energy state of an electron and a positron is one in which neither has kinetic energy.

The total energy of the system is then the rest mass-energy of the electron and positron.

Each has a rest energy of 0.511 MeV, so the total minimum energy is 1.022 MeV.

20. C is correct.

The creation of the *meson* would violate the principle of conservation of energy unless its lifetime were short enough that its mass-energy fell within the limit set by Heisenberg's uncertainty principle.

$$\Delta E \Delta t \geq h / 2\pi$$

Take the minimum uncertainty to be the rest mass of the meson, as the uncertainty must be at least sufficient to create the meson.

$$\Delta t \geq h / 2\pi mc^2$$

The meson must decay before this amount of time has elapsed.

Since no particle can travel faster than the speed of light, the maximum range the meson can have is:

$$r = c\Delta t = hc / 2\pi mc^2$$

$$r = (1.24 \text{ eV } \mu\text{m}) / [2\pi\ (140 \text{ MeV})]$$

$$r = 1.4 \times 10^{-15} \text{ m}$$

21. C is correct.

The creation of the boson would violate the principle of conservation of energy unless its lifetime were short enough that its mass-energy fell within the limit set by Heisenberg's uncertainty principle.

$$\Delta E \Delta t \geq h / 2\pi$$

Take the minimum uncertainty to be the rest mass of the boson, as the uncertainty must be at least sufficient to create the boson.

$$\Delta t \geq h / 2\pi mc^2$$

For conservation of energy not to be violated, the boson must decay before this time interval has elapsed.

So, the interaction time is:

$$t = h / 2\pi mc^2$$

The rest energy of the meson in SI units is:

$$mc^2 = 91 \times 10^9 \text{ eV } (1.6 \times 10^{-19} \text{ J/eV}) = 1.456 \times 10^{-8} \text{ J}$$

Therefore:

$$t = (6.63 \times 10^{-34} \text{J s}) / [2\pi\ (1.456 \times 10^{-8} \text{ J})]$$

$$t = 7.2 \times 10^{-27} \text{ s}$$

22. A is correct.

Since the *muons* are at rest (no kinetic energy), the total energy of the system is the sum of the rest mass energies of the muons, that is, twice the mass of a single muon.

Annihilation produces two photons of equal energy, so the energy of each equals the mass-energy of 1 muon.

The wavelength of the photon of energy E is:

$$\lambda = hc / E$$

$$\lambda = (1.24 \text{ eV } \mu\text{m}) / (106 \text{ MeV})$$

$$\lambda = 1.17 \times 10^{-14} \text{ m}$$

23. B is correct.

The creation of the boson would violate the principle of conservation of energy unless its lifetime were short enough that its mass-energy fell within the limit set by Heisenberg's uncertainty principle.

$$\Delta E \Delta t \geq h / 2\pi$$

Take the minimum uncertainty to be the rest mass of the boson, as the uncertainty must be at least sufficient to create the boson.

$$\Delta t \geq h / 2\pi mc^2$$

The meson must decay before this amount of time has elapsed.

The maximum range the boson can have is:

$$r = c\Delta t = hc / 2\pi mc^2$$

$$r = (1.24 \text{ eV } \mu\text{m}) / [2\pi \cdot (80 \text{ GeV})]$$

$$r = 2.5 \times 10^{-18} \text{ m}$$

Notes for active learning

Quantum Mechanics – Detailed Explanations

1. E is correct.

The energy of each incident photon is transferred to an electron, which must then overcome the material's work function to be ejected.

Therefore:

hc / λ = Work function

$\lambda = (6.626 \times 10^{-34}\ \text{J·s} \cdot 3.00 \times 10^{8}\ \text{m/s}) / (1.90\ \text{eV} \cdot 1.60 \times 10^{-19}\ \text{J/eV})$

$\lambda = 6.53 \times 10^{-7}\ \text{m}$

λ = 653 nm

2. B is correct.

The power varies with the number of photons per second and the energy of the photons; therefore, I is incorrect.

The energy of the photons varies with the frequency of the light; therefore, III is incorrect.

The intensity of a laser beam depends on the number of photons and the energy of each photon.

The energy of the photons varies linearly with the frequency.

If the frequency doubles, the energy doubles. Since the number of photons is unchanged, the intensity doubles.

3. C is correct.

The energy of each incident photon is transferred to an electron, which must then overcome the material's work function to be ejected.

Therefore, the maximum kinetic energy remaining in an electron is:

E = Energy in 240 nm photon − work function

2.5 eV = hc / λ − work function

Work function = $(6.626 \times 10^{-34}\ \text{J·s} \cdot 3.00 \times 10^{8}\ \text{m/s}) / (240 \times 10^{-9}\ \text{m}) - 2.58\ \text{eV}$

Work function = $(8.28 \times 10^{-19}\ \text{J}) / (1.60 \times 10^{-19}\ \text{J/eV}) - 2.58\ \text{eV}$

Work function = 2.6 eV

4. C is correct.

The minimum energy of the electron-positron pair is its rest mass ($E = 2m_{\text{electron}}\,c^2$).

The photon that creates the pair must have more than the minimum energy.

The energy of the photon (hv) must therefore be:

$hv > 2m_{\text{electron}}\,c^2$

continued…

$v > 2m_{electron}\, c^2 / h$

$v > 2 \cdot (9.11 \times 10^{-31}\ \text{kg}) \cdot (3.00 \times 10^8\ \text{m/s})^2 / (6.626 \times 10^{-34}\ \text{J·s})$

$v > 2.47 \times 10^{20}$ Hz

5. D is correct.

The energy of each incident photon is transferred to an electron, which must then overcome the material's work function to be ejected.

Therefore:

$hc / \lambda >$ Work function

$\lambda < (6.626 \times 10^{-34}\ \text{J·s} \cdot 3.00 \times 10^8\ \text{m/s}) / (2.20\ \text{eV} \cdot 1.60 \times 10^{-19}\ \text{J/eV})$

$\lambda < 564$ nm

6. C is correct.

Anderson used a cloud chamber to observe the effects of cosmic rays.

A cloud chamber is a device filled with saturated water or alcohol vapor and held in a magnetic field.

As the cosmic rays interact with the molecules in the vapor, they form high-energy charged particles.

These high-energy charged particles move out from their point of creation in curved paths because of the magnetic field.

Their path is visualized as they condense droplets from the saturated vapor.

Anderson observed an event in which a pair of charged particles was created at the same point, moving out in opposite directions and with equal but opposite curvature.

This corresponded to the earlier prediction by Dirac of a positively charged electron (a positron) based upon an extra solution to the equations of a relativistic invariant form of Schrodinger's Equation.

The electron and positron pair (pair production) was produced in a collision of a cosmic ray photon with a heavy nucleus in the vapor.

7. C is correct.

$E = hv$

$E = (6.626 \times 10^{-34}\ \text{J·s}) \cdot (6.43 \times 10^{14}\ \text{Hz}) / (1.60 \times 10^{-19}\ \text{J/eV})$

$E = 2.66$ eV

8. B is correct.

The kinetic energy of the emitted electrons (KE) is the energy of the incident photon minus, at least, the work function of the photocathode surface.

continued…

Therefore:

$$KE = 3.4\ eV - 2.4\ eV$$

$$KE = (1.0\ eV)\cdot(1.60 \times 10^{-19}\ J / eV)$$

$$KE = 1.60 \times 10^{-19}\ J$$

9. D is correct.

The energy of each incident photon is transferred to an electron, which must then overcome the material's work function to be ejected.

Therefore, to eject an electron:

$$hv = hc / \lambda > \text{Work function}$$

$$\lambda < hc / (\text{Work function})$$

$$\lambda < (6.626 \times 10^{-34}\ J\cdot s)\cdot(3.00 \times 10^{8}\ m/s) / (2.9\ eV \cdot 1.60 \times 10^{-19}\ J/eV)$$

$$\lambda < 428 \times 10^{-9}\ m$$

The illumination range of 400 nm–700 nm that does not satisfy this requirement is:

$$\lambda > 428\ nm$$

10. A is correct.

Photons move at the speed of light because they are light.

Since the momentum of photon A is twice as high as the momentum of photon B, its energy is twice as high, and therefore its wavelength is half as high.

11. A is correct.

The Balmer formula for Hydrogen is:

$$1 / \lambda = (1 / 91.2\ nm)\cdot(1 / m^2 - 1 / n^2)$$

The energy difference of the n = 20 and n = 7 state corresponds to a photon of wavelength:

$$1 / \lambda = (1 / 91.2\ nm)\cdot(1 / 7^2 - 1 / 20^2)$$

$$\lambda = (91.2\ nm) / (1 / 7^2 - 1 / 20^2)$$

$$\lambda = 5{,}092\ nm$$

The energy of a photon with wavelength λ is:

$$E = hc / \lambda$$

$$E = (6.626 \times 10^{-34}\ J\cdot s)\cdot(3.00 \times 10^{8}\ m/s) / (5{,}092 \times 10^{-9}\ m)$$

$$E = (3.93 \times 10^{-20}\ J) / (1.60 \times 10^{-19}\ J/eV)$$

$$E = 0.244\ eV$$

12. D is correct.

The de Broglie wavelength is given by:

$$\lambda = h / p$$

When the energy ($E = p^2 / 2m$ for a non-relativistic proton) is doubled, the momentum is increased by $\sqrt{2}$; its de Broglie wavelength decreases by $\sqrt{2}$.

13. E is correct.

Trivially, it is known that the incoming photon must lose energy to the electron in the scattering, and therefore, its wavelength must increase. There is one answer with a longer wavelength.

Using the Compton equation at 120°:

$$\Delta\lambda = \lambda_{Compton} (1 - \cos \theta)$$

where $\lambda_{Compton} = 2.43 \times 10^{-12}$ m

$$\Delta\lambda = .00243 \text{ nm } (1.5)$$

$$\Delta\lambda = 0.00365 \text{ nm}$$

$$\lambda = 0.591 \text{ nm} + 0.00365 \text{ nm}$$

$$\lambda = 0.595 \text{ nm}$$

14. C is correct.

The brightness of a beam of light is linearly proportional to the energy of the photons in the beam and the number of photons in the beam.

If the color of the light beam, which is dependent on its frequency distribution or energy distribution, is unchanged, then the frequency and energy distribution of the light beam must be unchanged.

15. E is correct.

The Balmer formula for hydrogen is:

$$1 / \lambda = (1 / 91.2 \text{ nm})\cdot(1 / m^2 - 1 / n^2)$$

$$1 / \lambda = (1 / 91.2 \text{ nm})\cdot(1 / 4^2 - 1 / 9^2)$$

$$1 / \lambda = (1 / 91.2 \text{ nm})\cdot(1 / m^2 - 1 / n^2)$$

$$1 / \lambda = (1 / 1818 \text{ nm})$$

The frequency of light is given by:

$$v = c / \lambda$$

$$v = (3.00 \times 10^8 \text{ m/s}) / (1818 \times 10^{-9} \text{ m})$$

$$v = 1.65 \times 10^{14} \text{/s}$$

$$v = 1.65 \times 10^{14} \text{ Hz}$$

16. D is correct.

The *Balmer formula* for hydrogen for the emission of radiation from the nth level down to the mth (i.e., $n > m$) is:

$$1 / \lambda = (1 / 91.2 \text{ nm})\cdot(1 / m^2 - 1 / n^2)$$

Therefore:

$$(1 / m^2 - 1 / n^2) = 91.2 \text{ nm} / 377 \text{ nm}$$

$$(1 / m^2 - 1 / n^2) = 0.2419$$

$$1 / n^2 = 1 / m^2 - 0.2419$$

For n to be real, it must be m = 1 or m = 2.

If m = 1:

$$1 / n^2 = 1 - 0.2419$$

$$1 / n^2 = 0.7581$$

n is not an integer (i.e., the photon has less energy than the emission photon expected in a transition from n = 2 to m = 1).

Therefore, m must be 2 (i.e., the emission is from the n level down to the m = 2 level) and:

$$1 / n^2 = 1 / 4 - 0.2419$$

$$1 / n^2 = 0.0081$$

$$n = 11$$

17. E is correct.

Wien's displacement law describes the wavelength of maximum emission of radiation of a black body at temperature T. It is:

$$\lambda_{max} \cdot T = \text{constant} = 0.00290 \text{ m}\cdot\text{K}$$

At T = 5000K:

$$\lambda_{max} = 0.00290 \text{ m}\cdot\text{K} / 5000 \text{ K}$$

$$\lambda_{max} = 580 \text{ nm}$$

18. E is correct.

Dirac factored the Schrödinger equation into a simpler equation with two solutions; one described the electron, and the other described an identical particle with an opposite electric charge.

Four years later, such a "positively charged" electron was found in cloud chamber pictures of cosmic rays.

19. B is correct.

The electron absorbs the full energy of the photon and loses the work function energy as it escapes from the material.

Therefore, it has a maximum kinetic energy of:

Max KE = E – Work function

Max KE = 3.4 eV – 2.4 eV = 1.0 eV · (1.60 × 10^{-19} J/eV)

Max KE = 1.60 × 10^{-19} J

20. D is correct.

Consider the problem from the center of the mass frame, a frame moving in the same direction as the incident photon. In this frame, the electron starts moving with the speed of the center of the mass frame and in the opposite direction of the center of the mass frame.

In the center of the mass frame, the electron is scattered with the same speed into one direction, and the photon is scattered in the opposite direction, the scattering angle.

Transforming back into the laboratory frame, the electron has a final velocity equal to the vector sum of its scattered velocity in the center of the mass frame and the velocity of the center of mass.

This sum is highest when the scattering angle of the electron is in the same direction as the velocity of the center of mass frame, the direction of the incident photon.

The photon is scattered in the opposite direction of the center of the mass frame at a scattering angle of 180°.

If the velocity of the electron is maximal at that scattering angle, then the energy of the photon is minimal at that scattering angle.

Since wavelength is inversely proportional to energy, the maximal change in wavelength occurs when the photon is scattered at 180°.

21. C is correct.

This is a trick question that has nothing to do with the photocathode or the work function.

If the radiation has energy of 3.5 eV, then its wavelength is given by:

E = hv = hc / λ = 3.5 eV

λ = hc / (3.5 eV)

λ = (6.626 × 10^{-34} J·s)·(3.00 × 10^{8} m/s) / (3.5 eV)

λ = (6.626 × 10^{-34} J·s)·(3.00 × 10^{8} m/s) / (3.5 eV)

λ = (5.679 × 10^{-26} J·m /eV) / (1.6 × 10^{-19} J/eV)

λ = 355 nm

22. A is correct.

The *energy of a photon* is given by:

$$E = hv$$

$$hv = hc / \lambda$$

$$E = hc / \lambda$$

Therefore, if the wavelength is doubled, the energy is halved.

23. D is correct.

The energy of each incident photon is transferred to an electron, which must then overcome the material's work function to be ejected.

Therefore:

$$hv > \text{Work function}$$

$$v > (2.8 \text{ eV} \cdot 1.60 \times 10^{-19} \text{ J/eV}) / (6.626 \times 10^{-34} \text{ J·s})$$

$$v > 6.8 \times 10^{-14} \text{/s}$$

24. B is correct.

The *de Broglie wavelength* of a matter wave is:

$$\lambda = h / p = h / mv$$

$$v = h / (m\lambda)$$

$$v = (6.626 \times 10^{-34} \text{ J·s}) / [(9.11 \times 10^{-31} \text{ kg})·(380 \times 10^{-9} \text{ m})]$$

$$v = 1.91 \times 10^{3} \text{ m/s}$$

25. B is correct.

Using the Compton Equation:

$$\Delta\lambda = \lambda_{Compton} (1 - \cos \theta)$$

where $\lambda_{Compton} = 2.43 \times 10^{-12}$ m

For $\theta = 90°$:

$$\Delta\lambda = 2.43 \times 10^{-12} \text{ m}$$

and

$$\lambda_{scattered} = \lambda_{incident} + \Delta\lambda$$

$$\lambda_{scattered} = (1.50 \times 10^{-10} \text{ m}) + (2.43 \times 10^{-12} \text{ m})$$

$$\lambda_{scattered} = 1.5243 \times 10^{-10} \text{ m}$$

26. B is correct.

As the intensity of light increases, the photon flux increases but not the energy of the photons.

The kinetic energy of the ejected electrons depends solely on the energy of the incident photons and the work function of the metal.

Since the energy of the incident photons is unchanged, the kinetic energy of the ejected electrons does not change.

The probability of an electron being ejected and the number of electrons ejected per second depends on the flux of incident photons. It increases as the intensity of the light increases.

The electron is ejected at the same instance as the light is absorbed, independent of the intensity of the light.

Therefore, the time lag does not change.

Note: the time lag between the illumination of the surface (not the absorption of light) and the ejection of the first electron depends on the probability of ejection, which depends on the intensity of the incident light.

27. D is correct.

The de Broglie wavelength of a matter wave is:

$$\lambda = h / p$$

$$\lambda = h / (mv)$$

The energy of a photon with this wavelength is:

$$E = hv = hc / \lambda$$

Substituting in for λ:

$$E = hc / [h / (mv)]$$

$$E = mcv$$

$$E = (1.67 \times 10^{-27}\text{ kg})\cdot(3.00 \times 10^{8}\text{ m/s})\cdot(7.2 \times 10^{4}\text{ m/s})$$

$$E = 36.1 \times 10^{-15}\text{ J}$$

$$E = (36.1 \times 10^{-15}\text{ J}) / (1.60 \times 10^{-19}\text{ J/eV})$$

$$E = 225 \times 10^{3}\text{ eV}$$

28. B is correct.

The *Compton effect* measures the change in energy as an x-ray scatters off of an electron.

The total momentum and energy of the x-ray and the electron, initially at rest, must be conserved.

As the scattering angle of the x-ray increases monotonically, its change in momentum increases, and therefore the momentum imparted to the electron must increase.

If the momentum of the electron increases, then its energy must increase, and the energy of the X-ray decreases.

If the energy of the X-ray decreases, then the frequency, which is proportional to its energy, must decrease.

29. A is correct.

The *Rydberg formula* for hydrogen for the emission of radiation from the nth level to the mth (i.e., n > m) is:

$$1 / \lambda = (1 / 91.2 \text{ nm})\cdot(1 / m^2 - 1 / n^2)$$

If n = 16 (since the first spectral line is from n = 2) and m = 1,

then:

$$1 / \lambda = (1 / 91.2 \text{ nm})\cdot(1 / 1^2 - 1 / 16^2)$$

$$\lambda = 91.2 \text{ nm} / (1 - 0.004)$$

$$\lambda = 91.6 \text{ nm}$$

30. D is correct.

If the frequency of light in a laser beam is doubled, the energy of each photon in that laser beam doubles since each photon has energy $E = hv$.

The wavelength of each photon is divided by 2, since $\lambda = c / v$.

The intensity of the laser beam is the power per unit area.

The power is the energy delivered per unit time.

The energy delivered is proportional to the energy of each photon times the number of photons.

If the energy of each photon is doubled and the number of photons remains unchanged (assuming the area of the laser beam is unchanged), then the intensity doubles.

Therefore, I and III are correct.

31. B is correct.

Order diffraction occurs when the incident and scattered beams hit the crystal at the same angle (i.e., in this case, the crystal planes are oriented at 58° / 2 to the normal).

The neutron matter waves add coherently (i.e., constructive or destructive interference) if the extra distance that the neutrons must travel as they scatter off the next plane of the crystal is equal to an integer number of de Broglie wavelengths.

At 58°, the extra distance is:

$$2 \cdot 159.0 \text{ pm} \cdot \cos(58° / 2) = 278 \text{ pm}$$

The *de Broglie wavelength* of a matter wave is:

$$\lambda = h / p$$

Therefore:

$$p = h / \lambda$$

continued…

and

$$E = p^2 / 2m$$

$$E = (6.626 \times 10^{-34}\ \text{J·s} / 278 \times 10^{-12}\ \text{m})^2 / (2 \cdot 1.67 \times 10^{-27}\ \text{kg})$$

$$E = (1.70 \times 10^{-21}\ \text{J}) / (1.6 \times 10^{-19}\ \text{J/eV})$$

$$E = 0.0106\ \text{eV}$$

32. A is correct.

The energy of each incident photon is transferred to an electron.

The electron's energy goes into overcoming the photocathode's 2.5 eV work function.

The remaining energy is then stopped by the stopping potential.

The highest amount of energy an electron can have comes from a photon with a wavelength of 360 nm.

The remaining energy is then:

$$\text{remaining energy} = hv - 2.5\ \text{eV}$$

$$\text{remaining energy} = hc / \lambda - 2.5\ \text{eV}$$

$$\text{remaining energy} = (6.626 \times 10^{-34}\ \text{J·s})\cdot(3.00 \times 10^{8}\ \text{m/s}) / (360 \times 10^{-9}\ \text{m}) - 2.5\text{eV}$$

$$\text{remaining energy} = [(5.52 \times 10^{-19}\ \text{J}) / (1.6 \times 10^{-19}\ \text{J/eV})] - 2.5\ \text{eV}$$

$$\text{remaining energy} = 3.45\ \text{eV} - 2.5\ \text{eV} = 0.95\ \text{eV}$$

The electron has a charge of e.

Therefore, this energy can be stopped by a voltage of 0.95 volts.

33. D is correct.

The *de Broglie wavelength* is given by:

$$\lambda = h / p$$

$$\lambda = (6.626 \times 10^{-34}\ \text{J·s}) / (1.95 \times 10^{-27}\ \text{kg·m/s})$$

$$\lambda = 340\ \text{nm}$$

34. C is correct.

The *Balmer formula* for hydrogen for the emission of radiation from the nth level down to the mth (i.e., $n > m$) is:

$$1 / \lambda = (1 / 91.2\ \text{nm})\cdot(1 / m^2 - 1 / n^2)$$

If $n = 9$ and $m = 6$, then:

$$1 / \lambda = (1 / 91.2\ \text{nm})\cdot(1 / 6^2 - 1 / 9^2)$$

$$1 / \lambda = (1 / 91.2\ \text{nm})\cdot(0.01543)$$

continued…

Thus:

$v = c / \lambda$

$v = (3.00 \times 10^8 \text{ m/s})\cdot(1 / 91.2 \text{ nm})\cdot(0.01543)$

$v = 5.08 \times 10^{13} \text{ /s}$

$v = 5.08 \times 10^{13} \text{ Hz}$

35. A is correct.

$E = hv$

$E = (6.626 \times 10^{-34} \text{ J}\cdot\text{s})\cdot(110 \text{ GHz})$

$E = (6.626 \times 10^{-34} \text{ J}\cdot\text{s})\cdot(110 \times 10^9 / \text{s})$

$E = 7.29 \times 10^{-23} \text{ J}$

36. C is correct.

The visible spectrum ranges from 400 nm to 700 nm.

The Balmer formula for H for the emission of radiation from the nth level down to the mth (i.e., n > m) is:

$1 / \lambda = (1 / 91.2 \text{ nm})\cdot(1 / m^2 - 1 / n^2)$

Consider emission from any level down to the m = 1 level.

The lowest energy is from the n = 2 level, and its wavelength is:

$91.2 \text{ nm} \cdot (4 / 3) = 121 \text{ nm}$, which is not visible.

Therefore, no visible lines are radiating down to the m = 1 level.

Consider emission from any level down to the m = 3 level.

The highest energy comes from n = ∞ and its wavelength is:

$91.2 \text{ nm} \cdot 9 = 820 \text{ nm}$, which is not visible.

Therefore, no visible lines radiate down to the m = 3 level, nor are there visible lines radiating down to any m level higher than 3.

Consider radiation from various levels down to the m = 2 level.

From n = 3, the wavelength is:

$91.2 \text{ nm} / (1 / 4 - 1 / 9) = 91.2 \text{ nm} \cdot 7.2 = 656.6 \text{ nm}$

From n = 4, the wavelength is:

$91.2 \text{ nm} / (1 / 4 - 1 / 16) = 91.2 \text{ nm} \cdot 5.333 = 486.4 \text{ nm}$

continued…

From n = 5, the wavelength is:

91.2 nm / (1 / 4 – 1 / 25) = 91.2 nm · 4.762 = 434.3 nm

From n = 6, the wavelength is:

91.2 nm / (1 / 4 – 1 / 36) = 91.2 nm · 4.5 = 410.4 nm

From n = 7, the wavelength is:

91.2 nm / (1 / 4 – 1 / 49) = 91.2 nm · 4.355 = 397.2 nm, which is not visible.

Therefore, there are 4 visible lines:

m = 2 to n = 3   656.6 nm

m = 2 to n = 4   486.4 nm

m = 2 to n = 5   434.3 nm

m = 2 to n = 6   410.4 nm

37. B is correct.

$E = hv = hc / \lambda$

$\lambda = hc / E$

$\lambda = (6.626 \times 10^{-34}\ J \cdot s) \cdot (3.00 \times 10^8\ m/s) / (4.20\ eV)$

$\lambda = (4.73 \times 10^{-26}\ J \cdot m/eV) / (1.60 \times 10^{-19}\ J/eV)$

$\lambda = 2.96 \times 10^{-7}\ m$

$\lambda = 296\ nm$

38. C is correct.

The energy of each incident photon is transferred to an electron, which must then overcome the material's work function to be ejected.

Therefore, if a wavelength of light is just able to eject an electron, then:

$hv = hc / \lambda$ = Work function

$(6.626 \times 10^{-34}\ J \cdot s) \cdot (3.00 \times 10^8\ m/s) / (500 \times 10^{-9}\ m)$ = Work function

$(3.98 \times 10^{-19}\ J) / (1.6 \times 10^{-19}\ J/eV)$ = Work function

2.48 eV = Work function

39. C is correct.

In the Bohr theory, there is a fixed number of de Broglie wavelengths of the electron in orbit.

This number is the principal quantum number:

$$n = 2\pi r / \lambda$$

The de Broglie wavelength is given by:

$$\lambda = h / p$$

Therefore:

$$n = (2\pi / h)rp$$

alternatively,

$$n^2 = (2\pi / h)^2 r^2 p^2$$

The attractive force of the proton keeps the electron in a circular orbit.

The attractive force is proportional to $1 / r^2$, and if that force keeps the electron in orbit, $1 / r^2$ must be proportional to v^2 / r.

Therefore, v^2 (or p^2) is proportional to $1 / r$.

Since r^2p^2 is proportional to n^2, and p^2 is proportional to $1 / r$, then r (i.e., $r^2 \cdot 1 / r$) is proportional to n^2.

40. B is correct.

Heisenberg's Uncertainty Principle states that:

$$\Delta p \Delta x \geq h / 2\pi$$

Therefore:

$$m\Delta v \Delta x \geq h / 2\pi$$

$$\Delta v \geq (6.626 \times 10^{-34}\ \text{J·s}) / (2\pi \cdot 0.053\ \text{nm} \cdot 1.67 \times 10^{-27}\ \text{kg})$$

$$\Delta v \geq 1.19 \times 10^3\ \text{m/s}$$

41. C is correct.

The Balmer formula for hydrogen for the emission of radiation from the nth level down to the mth (i.e., n > m) is:

$$1 / \lambda = (1 / 91.2\ \text{nm})\cdot(1 / m^2 - 1 / n^2)$$

$$1 / \lambda = (1 / 91.2\ \text{nm})\cdot(1 / 9^2 - 1 / 11^2)$$

$$\lambda = (91.2\ \text{nm}) / (0.00408)$$

$$\lambda = 22{,}300\ \text{nm}$$

42. E is correct.

Heisenberg's Uncertainty Principle states that:

$$\Delta E \Delta t \geq h/2\pi$$

Therefore:

$$\Delta t \geq (h / 2\pi) / \Delta E$$

$$\Delta t \geq (6.626 \times 10^{-34}\ \text{J·s} / 2\pi) / (10^{-18}\ \text{J})$$

$$\Delta t \geq 1.05 \times 10^{-16}\ \text{s}$$

43. C is correct.

The energy of each incident photon is transferred to an electron, which must then overcome the material's work function to be ejected.

Therefore, if a wavelength of light can eject an electron with energy 2.58 eV, then:

$$hv = hc / \lambda = \text{Work function} + 2.58\ \text{eV}$$

$$(6.626 \times 10^{-34}\ \text{J·s})·(3.00 \times 10^{8}\ \text{m/s}) / (240 \times 10^{-9}\ \text{m}) = \text{Work function} + 2.58\ \text{eV}$$

$$\text{Work function} = [(8.28 \times 10^{-19}\ \text{J}) / (1.6 \times 10^{-19}\ \text{J/eV})] - 2.58\text{eV}$$

$$\text{Work function} = 2.60\ \text{eV}$$

Notes for active learning

Notes for active learning

Special Relativity – Detailed Explanations

1. A is correct.

Alpha sees Beta moving away at $+0.7c$, and Beta sees Gamma moving away at $+0.7c$.

To find the speed that Alpha sees Gamma moving at, apply the relativistic velocity addition expression:

$$\beta = (\beta_1 + \beta_2) / (1 + \beta_1\beta_2)$$

$$\beta = (0.7 + 0.7) / [1 + (0.7)\cdot(0.70)]$$

$$\beta = 0.94$$

$$v = 0.94c$$

2. E is correct.

The train is not moving in the vertical direction.

Therefore, there is no Lorentz Contraction of lengths in that direction.

3. D is correct.

As measured in the frame of Earth, the total distance is:

$$d = 2\cdot(4.367 \text{ years})\cdot c$$

The time is therefore:

$$t = d / v$$

$$t = (2 \cdot 4.367 \text{ years} \cdot c) / (0.5c)$$

$$t = 17.468 \text{ years}$$

4. E is correct.

From the point of view of an observer in the lab, the muon's time appears to be slowed by relativistic time dilation.

The lifetime for the muon in a frame in which it is at rest ($\Delta x = 0$) is $\Delta t = 2.2\mu\text{s}$.

The Lorentz transformation to the frame in which the muon is moving is:

$$\Delta t' = \gamma\,(\Delta t - v\,\Delta x / c^2)$$

where $\gamma = 1 / \sqrt{(1 - \beta^2)}$ and $\beta = v / c$

$$\Delta t' = \gamma\,\Delta x$$

$$\gamma = 1 / \sqrt{(1 - 0.99^2)}$$

$$\gamma = 7.09$$

$$\Delta t' = 15.6\ \mu\text{s}$$

5. D is correct.

The speed of the scout, as measured by the asteroid, is the relativistic addition of the speed of the spaceship as measured from the asteroid, plus the speed of the scout as measured from the spaceship.

$\beta = (\beta_1 + \beta_2) / (1 + \beta_1\beta_2)$

$\beta = (0.6 + 0.4) / [1 + (0.6)\cdot(0.4)]$

$\beta = 0.81$

$v = 0.81c$

6. D is correct.

The angle depends on the ratio of the elevation of the bed to the length of the bed along the floor.

The elevation of the bed does not appear different because the motion of the spaceship is zero in that direction.

The length along the floor appears shorter due to Lorentz Contraction when viewed from the observer's frame.

Therefore, the ratio appears to be larger, and the angle appears to have increased.

7. A is correct.

In the frame of the spaceship, the separation of the two events, Δx, is zero, and the time is dilated according to the Lorentz transformation:

$\Delta t' = \gamma \Delta t$

$\gamma = 1 / \sqrt{(1 - \beta^2)}$

$\beta = v / c$

If $v = 0.9640c$:

$\gamma = 1 / \sqrt{(1 - 0.964^2)}$

$\gamma = 3.76$

$\Delta t = (10.5 \text{ years}) / (3.76)$

$\Delta t = 2.79 \text{ years}$

8. C is correct.

Conservation of energy requires:

$E_p + E_n = E_{\text{final}}$

In the center of the mass frame, the proton and neutron start with insignificant kinetic energy, so their initial energies are approximately their rest energies.

The final energy is the rest energy of the deuterium plus the released energy:

$m_pc^2 + m_nc^2 = m_dc^2 + E_{\text{released}}$

continued…

So:

$$m_dc^2 = m_pc^2 + m_nc^2 - E_{\text{released}}$$

The proton rest mass energy is $m_pc^2 = 938.28$ MeV, and the neutron rest mass energy is:

$$m_nc^2 = 939.57 \text{ MeV}$$

The rest energy of the deuterium is then:

$$m_dc^2 = [(938.28 \text{ MeV}) + (939.57 \text{ MeV}) - (3.6 \times 10^{-13} \text{ J})] / [(1.6 \times 10^{-19} \text{ J/eV})\cdot(10^6 \text{ eV/MeV})]$$

$$m_dc^2 = (1877.85 \text{ MeV}) - (2.25 \text{ MeV})$$

$$m_dc^2 = 1875.6 \text{ MeV}$$

Converting to SI units:

$$m_d = (1875.6 \text{ MeV})\cdot(1.6 \times 10^{-19} \text{ J/eV}) / c^2$$

$$m_d = [(1875.6 \text{ MeV})\cdot(1.6 \times 10^{-19} \text{ J/eV})\cdot(10^6 \text{ eV/MeV})] / (3.0 \times 10^8 \text{ m/s})^2$$

$$m_d = 3.3 \times 10^{-27} \text{ kg.}$$

9. B is correct.

The time interval between events increases in the moving frame from the value measured in the rest frame because of time dilation.

Let primed variables represent quantities measured in the moving frame, and unprimed variables represent quantities measured in the rest frame.

Then, the expression for time dilation is:

$$\Delta t' = \gamma \Delta t$$

where $\gamma = 1 / \sqrt{(1 - \beta^2)}$ and $\beta = v^2 / c^2$

If some factor increases time intervals, rates are decreased by the same factor.

Let r be the heart rate measured in the rest frame, and r' be the heart rate measured in the moving frame.

Because of time dilation, the rate in the moving frame is reduced according to:

$$r' = r / \gamma$$

If $v = 0.99c$, then $\gamma = 7.09$.

Thus:

$$r' = 12 \text{ beats/min}$$

10. C is correct.

As measured by the first spaceship, the speed of Earth is $0.31c$, and the speed of the second spaceship as measured by the Earth is also $0.31c$.

The speed of the second spaceship, as measured by the first, is the relativistic sum of the two speeds.

$$\beta = (\beta_1 + \beta_2) / (1 + \beta_1\beta_2)$$

$$\beta = (0.31 + 0.31) / [1 + (0.31)\cdot(0.31)]$$

$$\beta = 0.566$$

$$v = 0.57\ c$$

11. E is correct.

The kinetic energy, total energy, and linear momentum have no limit.

They increase monotonically with increasing γ.

Although the speed of a particle has an upper limit of c, the value of γ goes to infinity as v approaches c.

12. B is correct.

This problem can be solved using Lorentz transformation, time dilation, or length contraction.

Using length contraction, the distance between the beacons as measured by the spaceship is contracted:

$$l' = l / \gamma$$

where $\beta = v / c$ and $\gamma = 1 / \sqrt{(1 - \beta^2)}$

$$\gamma = 1 / \sqrt{[1 - (0.5)^2]} = 1.155$$

$$l' = (49 \times 10^6\ \text{m}) / 1.155$$

$$l' = 42.42 \times 10^6\ \text{m}$$

The time interval for passage across this distance as measured in the spaceship frame is:

$$t' = l' / v = l' / (0.5\ c)$$

$$t' = (42.42 \times 10^6\ \text{m}) / [(0.5)\cdot(3 \times 10^8\ \text{m/s})]$$

$$t' = 283\ \text{ms}$$

13. A is correct.

The lifetime of the pion is greater as measured by clocks on Earth due to time dilation.

The lifetime measured on Earth is:

$$\Delta t' = \gamma\Delta t$$

where $\gamma = 1 / \sqrt{(1 - \beta^2)}$ and $\beta = v^2 / c^2$

continued…

If $\beta = 0.23$, then $\gamma = 1.028$, so the lifetime as measured on Earth is:

$$\Delta t' = 2.62 \times 10^{-8} \text{ s}$$

The distance traveled in that time interval is:

$$d' = v\Delta t'$$

$$d' = 0.23(3 \times 10^{8} \text{ m/s})\cdot(2.62 \times 10^{-8} \text{ s})$$

$$d' = 1.81 \text{ m}$$

14. B is correct.

The problem can be solved in the rest frame of the planets or the rest frame of the astronaut.

In the rest frame of the planets, the light from the nuclear explosion from planet Y arrives at the astronaut when:

$$c\Delta t = (1 \text{ hour})\ c + v\Delta t$$

$$\Delta t = (1 \text{ hour})\ c / (c - v)$$

If $v = 0.6c$, then:

$$\Delta t = 2.5 \text{ hours}$$

From planet Z, the light arrives to the astronaut at:

$$c\Delta t = (1 \text{ hour})\ c - v\Delta t$$

$$\Delta t = (1 \text{ hour})\ c / (c + v)$$

$$\Delta t = 0.625 \text{ hours}$$

This is 1.875 hours sooner as measured in the frame of the planets.

In the planets' frame, the astronaut's clock appears to be moving slower because of time dilation.

Therefore, the time interval measured by the astronaut is:

$$\Delta t' = (1.875 \text{ hours}) / \gamma$$

$$\gamma = 1 / \sqrt{(1 - \beta^2)}$$

$$\gamma = 1.25$$

$$\Delta t' = 1.5 \text{ hours} = 90 \text{ minutes}$$

Alternatively

Consider the problem in the rest frame of the astronaut.

In that frame, the two explosions do not occur simultaneously but at a time interval given by:

$$\Delta t = \gamma\ (\Delta t' + v\ \Delta x' / c^2)$$

where $\Delta t'$, the time interval as measured in the frame of the planets, is 0 and $\Delta x'$ is 2 light-hours

continued…

Therefore:

$\Delta t = 1.25 \cdot (0.6c) \cdot (2 \text{ hours} \cdot c) / c^2$

$\Delta t = 1.5 \text{ hours} = 90 \text{ minutes}$

15. D is correct.

The speed of the Klingon ship, as measured by the Enterprise, is the relativistic addition of the speed of the Klingon ship as measured from Earth, plus the speed of Earth as measured from the Enterprise.

The formula for the relativistic addition of speeds is:

$\beta = (\beta_1 + \beta_2) / (1 + \beta_1\beta_2)$

$\beta = (0.9 + 0.8) / [1 + (0.9) \cdot (0.8)]$

$\beta = 0.988$

$v = 0.988\ c$

16. E is correct.

In Sofia's inertial frames of reference, one before the spaceship increased its speed and one after, the spaceship is at rest.

Therefore, the measurements of time and length must give the same result.

Otherwise, the laws of physics would be different in these two inertial frames of reference.

During the acceleration of the spaceship, there are some unusual effects caused by the acceleration, and those effects are the subject of General Relativity.

17. C is correct.

In the frame of the beacons:

$t = d / v$

$t = (40 \times 10^6 \text{ m}) / [(0.30) \cdot (3 \times 10^8 \text{ m/s})]$

$t = 44.4 \times 10^{-2} \text{ s}$

$t = 444 \text{ ms}$

18. B is correct.

Kinetic energy is total energy minus rest energy:

$\text{K} = \text{E} - mc^2$

$\text{K} = \gamma mc^2 - mc^2$

$\text{K} = (\gamma - 1)mc^2$

where $\gamma = 1 / \sqrt{(1 - \beta^2)}$ and $\beta = v / c$

continued…

$\gamma = 1 / (\sqrt{[1 - (0.737)^2]}$

$\gamma = 1.480$

So:

K = (1 – 1.480)·(511 keV)

K = 245 keV

19. D is correct.

The space time interval, Δs, is given by:

$(\Delta s)^2 = (c\Delta t)^2 - (\Delta x)^2$

For these events,

$\Delta s = 90c$ years and $\Delta t = 100$ years

Solving for Δx:

$(\Delta x)^2 = (100c \text{ years})^2 - (90c \text{ years})^2$

$(\Delta x)^2 = 1900c^2 \text{ years}^2$

$\Delta x = 44$ light-years

20. A is correct.

Lorentz contraction and time dilation are proportional to:

$\gamma = 1 / \sqrt{(1 - \beta^2)} \qquad \beta = v / c$

Therefore, if $\gamma = 2$:

$\beta^2 = 0.75$

$\beta = 0.866$

$v = 0.866c$

21. A is correct.

As measured from the inertial frame of Earth, time on the spaceship is dilated, and a clock on the moving spaceship moves slower than a stationary clock on the Earth.

This is true when the spaceship is traveling away from Earth and traveling towards the Earth.

Therefore, the elapsed time on the spaceship and the consequential advance of the age of the twin on the spaceship is less than the elapsed time and age of the twin on Earth.

What is the difference between the two inertial frames? As viewed from the frame of Earth, nothing unusual happens while the spaceship is decelerating and re-accelerating as it changes directions and heads back to Earth.

As viewed from the spaceship frame, time on Earth advances quickly while the spaceship changes directions.

continued…

It is during this brief period that the spaceship-bound twin observes the Earth-bound twin aging rapidly.

If the twin calculates the apparent rate of time change on Earth on the decelerating and then accelerating spaceship, the result reflects the General Relativistic effect of a differing gravitational potential on time.

During the deceleration and acceleration, the twin on the spaceship is in a frame equivalent to a stationary frame with a uniform gravitational field.

In that frame, the Earth, situated at a very different gravitational potential than the spaceship, is experiencing time moving very quickly due to the perceived gravitational potential difference, according to the weak equivalence principle.

22. E is correct.

Using subscript 1 and 2 for the two initial masses and subscript 3 for the final combined mass, conservation of momentum gives:

$$\gamma_1 m v_1 - \gamma_2 m v_2 = \gamma_3 m_3 v_3$$

Conservation of energy gives:

$$\gamma_1 mc^2 + \gamma_2 mc^2 = \gamma_3 m_3 c^2$$

Using these two equations to solve for v_3:

$$v_3 = (\gamma_1 v_1 - \gamma_2 v_2) / (\gamma_1 + \gamma_2)$$

$$\gamma_1 = 1 / \sqrt{(1 - 0.650^2)}$$

$$\gamma_1 = 1.316$$

$$\gamma_2 = 1 / \sqrt{(1 - 0.850^2)}$$

$$\gamma_2 = 1.898$$

$$v_3 = [(1.316 \cdot 0.650c) - (1.898 \cdot 0.850c)] / (1.316 + 1.898)$$

$$v_3 = -0.234c$$

$$\gamma_3 = 1 / \sqrt{(1 - 0.277^2)}$$

$$\gamma_3 = 1.029$$

Using the conservation of energy to solve for m_3:

$$m_3 = m(\gamma_1 + \gamma_2) / \gamma_3$$

$$m_3 = (100 \text{ kg}) \cdot (1.316 + 1.898) / 1.029$$

$$m_3 = 312 \text{ kg}$$

23. D is correct.

The relativistic expression for momentum is:

$$p = \gamma mv = \gamma m\beta c$$

where $\beta = v / c$ and $\gamma = 1/ \sqrt{(1 - \beta^2)}$

Therefore:

$$\beta = 0.60$$

$$\gamma = 1 / \sqrt{(1 - 0.60^2)}$$

$$\gamma = 1.25$$

$$p = (1.25)\cdot(1.67 \times 10^{-27}\text{ kg})\cdot(0.60)\cdot(3 \times 10^8)$$

$$p = 3.8 \times 10^{-19}\text{ kg·m/s}$$

24. B is correct.

The total energy of the rock, $\gamma\, mc^2$, is the sum of its rest mass plus the added energy:

$$E_{\text{Total}} = \gamma\, mc^2$$

$$E_{\text{Total}} = \text{mc}^2 + E_{\text{Added}}$$

$$E_{\text{Added}} = (\gamma - 1)\, mc^2$$

$$\gamma = 1 / \sqrt{(1 - \beta^2)}$$

$$\beta = v / c$$

Given that $v = 0.866c$

$$\beta = 0.866$$

$$\gamma = 2.00$$

$$E_{\text{Added}} = (2.00 - 1)\cdot(1.0\text{ kg})\cdot(3.0 \times 10^8\text{ m/s})^2$$

$$E_{\text{Added}} = 9.00 \times 10^{16}\text{ kg·m/s}$$

$$E_{\text{Added}} = 9.00 \times 10^{16}\text{ J}$$

25. C is correct.

The speed of the secondary rocket, as measured in the frame of Earth, is the relativistic addition of the speed of the secondary rocket as measured from the spaceship, plus the speed of the spaceship as measured from Earth. The formula for the *relativistic addition of speeds* is:

$$\beta = (\beta_1 + \beta_2) / (1 + \beta_1\beta_2)$$

$$\beta = (0.5 + 0.5) / [1 + (0.5)\cdot(0.5)]$$

$$\beta = 0.80$$

$$v = 0.80\, c$$

26. A is correct.

The observer moving with the rocket is making the time measurements at the same location:

$$\Delta x_B = 0$$

The *Lorentz transformation* is, therefore, simply:

$$T_A = \gamma T_B$$

Since γ is greater than one,

$$T_A > T_B$$

This is *time dilation*— the shortest time interval for two events occurs in the frame in which the two events occur at the same location, a frame stationary with respect to the two events.

If those events are clicking a clock, the clock appears to move the fastest in its rest frame.

27. B is correct.

As measured by an observer on Earth, the spaceship will have traveled for:

$$t = d / v$$

$$t = (4.367\ c\ \text{years}) / 0.50c$$

$$t = 8.734\ \text{years}$$

To an observer on the spaceship, $\Delta x' = 0$, and the time $\Delta t'$ is dilated by the Lorentz transformation:

$$\Delta t = \gamma\ \Delta t'$$

$$\gamma = 1 / \sqrt{(1 - \beta^2)}$$

$$\beta = v / c$$

$$\gamma = 1 / \sqrt{(1 - 0.50^2)}$$

$$\gamma = 1.15$$

$$\Delta t' = (8.734\ \text{years}) / 1.15$$

$$\Delta t' = 7.6\ \text{years}$$

28. D is correct.

The decay time Δt in the rest frame of the particle (where $\Delta x = 0$) is dilated to a shorter elapsed time.

Since,

$$\Delta x = 0$$

The Lorentz transformation for the two events is:

$$\Delta t' = \gamma \Delta t$$

continued...

$\gamma = 1 / \sqrt{(1 - \beta^2)}$

$\beta = v / c = 0.997$

$\gamma = 12.9$

$\Delta t = 37.0\ \mu s / 12.9$

$\Delta t = 2.9\ \mu s$

29. C is correct.

The total energy of a particle is:

$E_{Total} = \gamma\, mc^2$

When the kinetic energy is correctly defined as the difference between the total energy and the rest mass energy, then:

$E_{Total} = mc^2 + E_{Kinetic}$

$E_{Kinetic} = (\gamma - 1)\, mc^2$

$\gamma = 1 / \sqrt{(1 - \beta^2)}$

$\beta = v / c$

Given that $v = 0.5c$

$\beta = 0.5$

$\gamma = 1.1547$

$E_{Kinetic} = 0.1547\, mc^2$

$E_{Kinetic} = (mv^2 / 2) \cdot (0.1547 \cdot 2) \cdot (c / v)^2$

$E_{Kinetic} = (mv^2 / 2) \cdot (1.24)$

Therefore,

by using $mv^2 / 2$ as the kinetic energy, the calculation is off by:

$mv^2 / 2 = 0.81 \times E_{Kinetic}$, which is a 19% error.

30. B is correct.

The separation $\Delta x'$ of the two stars in a frame moving with respect to the two stars is affected by Lorentz Contraction.

In the spaceship's frame, the measurement of $\Delta x'$ was made with $\Delta t' = 0$.

The Lorentz transformation is therefore:

$$\Delta x = \gamma\, \Delta x'$$

$$\Delta x' = \Delta x / \gamma$$

$$\gamma = 1 / \sqrt{(1 - \beta^2)} \qquad \beta = v / c$$

Since $\Delta x = 90$ light-years and $\Delta x' = 68.7$ light-years:

$$\gamma = 90 / 68.7$$

Solving for β:

$$\beta = \sqrt{[1 - (1 / \gamma^2)]}$$

$$\beta = 0.646$$

Therefore:

$$v = 0.646c$$

31. D is correct.

One of the postulates of Special Relativity is that the laws of physics are the same in inertial frames.

If the laws of physics are the same in all inertial frames, there is no way to differentiate frames.

32. D is correct.

Using subscript 1 and 2 for the two initial masses and subscript 3 for the final combined mass, conservation of momentum gives:

$$\gamma_1 m v_1 - \gamma_2 m v_2 = \gamma_3 m_3 v_3$$

Conservation of energy gives:

$$\gamma_1 mc^2 + \gamma_2 mc^2 = \gamma_3 m_3 c^2$$

Using these two equations to solve for $v_{\square}$:

$$v_3 = (\gamma_1 v_1 - \gamma_2 v_2) / (\gamma_1 + \gamma_2)$$

$$\gamma_1 = 1 / \sqrt{(1 - 0.550^2)} = 1.197$$

$$\gamma_2 = 1 / \sqrt{(1 - 0.750^2)} = 1.512$$

$$v_3 = [(1.197 \cdot 0.550c) - (1.512 \cdot 0.750c)] / (1.197 + 1.512)$$

$$v_3 = -0.1756c$$

(The direction is in the opposite direction of the first satellite.)

33. A is correct.

In the moving frame, the time interval between events is increased from the value measured in the rest frame because of time dilation.

Consider two heartbeats as two events on the spaceship.

They have a value of $\Delta x = 0$ and $\Delta t = (1 / 70)$ minutes.

In the frame of Earth, because Δx on the spaceship is 0, the Lorentz transformation for $\Delta t'$ is simply:

$$\Delta t' = \gamma \, \Delta t$$

where:

$$\gamma = 1 / \sqrt{(1 - \beta^2)}$$

$$\beta = v / c$$

$$\gamma = 1 / \sqrt{(1 - 0.50^2)}$$

$$\gamma = 1.15$$

$$\Delta t' = 1.15 \cdot (1 / 70) \text{ minutes}$$

$$\Delta t' = 0.0164 \text{ minutes}$$

The heart rate is, therefore, the inverse of this interval, 61 beats per minute.

34. B is correct.

The accelerator's length is contracted as measured in the frame of the particle:

$$l' = l / \gamma$$

where $\beta = v / c$ and $\gamma = 1 / \sqrt{(1 - \beta^2)}$

$$l' = (453 \text{ m}) [\sqrt{(1 - (0.875)^2}]$$

$$l' = 219 \text{ m}$$

35. C is correct.

In the spaceship's frame, the two events have a value of $\Delta x = 0$.

The time difference in the spaceship's frame appears to be dilated so that the time difference in the frame of Earth is given by:

$$\Delta t' = \gamma \, \Delta t$$

$$\gamma = 1 / \sqrt{(1 - \beta^2)} \qquad \beta = v / c$$

Since $\Delta t = 5.78$ years and $\Delta t' = 10$ years:

$$\gamma = 10 / 5.78$$

continued…

Solving for β:

$$\beta = \sqrt{(1 - 1 / \gamma^2)}$$

$$\beta = 0.816$$

Therefore:

$$v = 0.816c$$

36. C is correct.

The length of a moving object appears to be shortened by Lorentz contraction.

In the stationary observer's frame on the ground, the length measurement involves two events, with $\Delta t' = 0$ and $\Delta x'$ being the length of the moving train as measured by the stationary observer.

In the frame of the moving train, those two events have an Δx value equal to the length of the train as measured in the rest frame of the train. (Note that the value of Δt is not zero.)

Since $\Delta t'$ is zero, the Lorentz transformation for Δx is simply:

$$\Delta x = \gamma\, \Delta x'$$

Since γ is greater than 1, $\Delta x'$, the length measured by the stationary observer on the ground is less than Δx, the length measured in the rest frame of the train.

37. C is correct.

At first glance, this is a complicated problem since the spaceship's velocity is unknown and cannot be readily determined.

However, the space-time interval, Δs^2, of the two events (i.e., the spaceship crossing the orbit of Jupiter and Mars) can be determined.

The space-time interval is the same in all frames.

$$\Delta s^2 = (c\Delta t)^2 + (\Delta x)^2$$

$$\Delta s'^2 = (c\Delta t')^2 + (\Delta x')^2$$

In the spaceship's frame, the distance between the two measurements is $\Delta x = 0$, and the time between the two measurements is Δt.

In the Earth's frame:

$$\Delta x' = (778 \times 10^9 \text{ m}) - (228 \times 10^9 \text{ m})$$

$$\Delta x' = 550 \times 10^9 \text{ m}$$

It is given that:

$$\Delta t' = 50.0 \text{ minutes}$$

continued…

Solving for Δt by equating the space-time interval in the two frames:

$$(c\Delta t)^2 = (c\Delta t')^2 + (\Delta x')^2$$

$$\Delta t^2 = \Delta t'^2 + (\Delta x'/c)^2$$

$$\Delta t^2 = (50 \text{ minutes} \cdot 60 \text{ s/minute})^2 + (550 \times 10^9 \text{ m}) / (3.0 \times 10^8 \text{ m/s})$$

$$\Delta t^2 = (3000 \text{ s})^2 + (1833 \text{ s})^2$$

$$\Delta t^2 = (3515 \text{ s})^2$$

$$\Delta t = 58.6 \text{ minutes}$$

38. D is correct.

In the frame of the two beacons, in a time Δt, a radar signal goes from one beacon to the other beacon and back to the moving spaceship, while the moving spaceship travels from the first beacon towards the second beacon.

The total distance traveled by the spaceship plus the radar signal in Δt is twice the distance between the beacons:

$$2 \cdot (65 \times 10^6 \text{ m}) = (0.90c + c)\Delta t$$

$$\Delta t = (68.4 \times 10^6 \text{ m}) / (3 \times 10^8 \text{ m})$$

$$\Delta t = 228 \text{ ms}$$

$$\Delta t = 228 \text{ ms}$$

As measured in the frame of the moving spaceship, the time interval is transformed into the spaceship's frame for which a stationary clock has $\Delta x' = 0$ (note that Δx is not zero).

Therefore,

The Lorentz transformation is:

$$\Delta t = \gamma \, \Delta t'$$

$$\Delta t' = \Delta t / \gamma$$

$$\gamma = 1 / \sqrt{(1 - \beta^2)}$$

$$\beta = v/c = 0.9$$

$$\gamma = 2.29$$

$$\Delta t' = 228 \text{ ms} / 2.29$$

$$\Delta t' = 100 \text{ ms}$$

39. B is correct.

The total energy of the electron, $\gamma\ mc^2$, is the sum of its rest mass plus the potential difference through which it is accelerated, V, times its charge, e:

$$\gamma\ mc^2 = mc^2 + eX$$

$$V = mc\ (\gamma - 1)$$

where $\gamma = 1 / \sqrt{(1 - \beta^2)}$ and $\beta = v / c$

$$\gamma = 1 / \sqrt{[1 - (0.648)^2]}$$

$$\gamma = 1.313$$

$$V = 511\ \text{keV} \cdot 0.313$$

$$V = 160\ \text{keV}$$

40. C is correct.

In the particle's frame, the two events (creation and annihilation) have a value of $\Delta x = 0$.

The time difference in the laboratory's frame appears to be dilated by:

$$\Delta t' = \gamma\ \Delta t$$

$$\gamma = 1 / \sqrt{(1 - \beta^2)} \qquad \beta = v / c$$

and:

$$\Delta x' = v\Delta t'$$

Since $\Delta t = 1.52 \times 10^{-6}$ s and $\Delta x' = 342$m:

$$342\ \text{m} = v\gamma\ (1.52 \times 10^{-6}\ \text{s})$$

Dividing both sides by c:

$$(342\ \text{m}) / (3.00 \times 10^8\ \text{m/s}) = (v/c)\ \gamma\ (1.52 \times 10^{-6}\ \text{s})$$

$$0.75 = \beta\ \gamma$$

$$0.75 = \sqrt{[\beta^2 / (1 - \beta^2)]}$$

$$0.75 = \sqrt{[1 / \beta^2 - 1)]}$$

Solving for B:

$$\beta = 0.6$$

$$v = 0.6c$$

41. C is correct.

The relativistic kinetic energy formula is derived from the conservation of energy:

$$E_{total} = \gamma\, mc^2$$

This is true for all values of γ (or velocity).

The kinetic energy formula acts as a definition of kinetic energy by dividing the total energy into a rest mass component and a kinetic component:

$$E_{total} = mc^2 + E_{kinetic}$$

$$E_{kinetic} = (\gamma - 1)mc^2$$

For $v < c$, a Taylor expansion for $(\gamma - 1)$ gives the Newtonian formula for kinetic energy:

$E_{kinetic} = mv^2 / 2$, which is true only for non-relativistic velocities.

42. B is correct.

The separation $\Delta x'$ of Earth and the star in a frame moving with respect to the spaceship is affected by Lorentz contraction.

Since the separation $\Delta x'$ is measured when $\Delta t'$ is zero, the Lorentz transformation is simply:

$$\Delta x = \gamma \Delta x'$$

$$\gamma = 1 / \sqrt{(1 - \beta^2)} \qquad \beta = v / c$$

With $v = 0.800c$ and $\Delta x = 4.30$ light-years:

$$\gamma = 1.67$$

$\Delta x' = 2.58$ light-years

43. B is correct.

As measured in the frame of the asteroid, the speed of the missile is the relativistic addition of the speed of the spaceship as measured from the asteroid, plus the speed of the missile as measured from the spaceship.

The two velocities are in the opposite direction.

The formula for the relativistic addition of two speeds in the same directions is:

$$\beta = (\beta_1 + \beta_2) / (1 + \beta_1\beta_2)$$

$$\beta = (0.8 + 0.5) / [1 + (0.8)\cdot(0.5)]$$

$$\beta = 0.93$$

$$v = 0.93c$$

44. A is correct.

The total energy of a particle of mass m moving with velocity v is given by:

$E = \gamma\, mc^2$

$\gamma = 1 / \sqrt{(1 - \beta^2)}$

$\gamma = v / c$

$\gamma = 0.950$

$\gamma = 1 / \sqrt{(1 - 0.950^2)}$

$\gamma = 3.20$

$E = (3.20)\cdot(9.11 \times 10^{-31}\ \text{kg})\cdot(3.0 \times 10^8\ \text{m/s})^2$

$E = 2.6 \times 10^{-13}\ \text{J}$

45. D is correct.

In the spaceship's frame, the two events (i.e., two time measurements on the same clock) have a value of Δx = 0.

The time difference in the spaceship's frame appears to be dilated so that the time difference in the frame of Earth is given by:

$\Delta t' = \gamma\, \Delta t$

$\gamma = 1 / \sqrt{(1 - \beta^2)}$ $\beta = v / c$

Since $\Delta t = 1.0$ s and $v = 0.960c$:

$\gamma = 3.6$

$\Delta t = 3.6$ s

46. A is correct.

In the frame of the first particle, the speed of the second particle is the relativistic sum of the speed of Earth (as measured from the frame of the first particle - note that Earth appears to be moving to the right), plus the speed of the second particle (as measured from the frame of Earth – note that it appears to be moving to the right).

The formula for the relativistic addition of two speeds in the same directions is:

$\beta = (\beta_1 + \beta_2) / (1 + \beta_1\beta_2)$

$\beta = (0.741 + 0.543) / [1 + (0.741)\cdot(0.543)]$

$\beta = 0.916$

$v = 0.916c$

47. A is correct.

Relativistic momentum is given by:

$$p = \gamma m v = \gamma m \beta c$$

where $\beta = v / c$ and $\gamma = 1 / \sqrt{(1 - \beta^2)}$

Newtonian momentum is given by:

$$p = mv$$

The *percentage difference* is:

$$(\gamma mv - mv) / \gamma mv = [(1 - (1 / \gamma)]$$

For $v = 0.86c$:

$$\gamma = 1 / \sqrt{(1 - 0.86^2)}$$

$$\gamma = 1.96$$

$$1 - (1 / \gamma) = 0.49 = 49\%$$

48. B is correct.

The speed of the secondary rocket as measured in the frame of the space platform is the relativistic addition of the speed of the secondary rocket as measured in the frame of the spaceship, plus the speed of the spaceship as measured in the frame of the space platform.

The two velocities are in the opposite direction.

The formula for the relativistic addition of two speeds in opposite directions is:

$$\beta = (\beta_1 - \beta_2) / (1 - \beta_1\beta_2)$$

$$\beta = (0.1 - 0.56) / [1 - (0.1)\cdot(0.56)]$$

$$\beta = 0.487$$

$$v = 0.487c$$

49. D is correct.

The relativistic expression for momentum is:

$$p = \gamma m v = \gamma m \beta c$$

where $\beta = v / c$ and $\gamma = 1 / \sqrt{(1 - \beta^2)}$

Therefore:

$$\Delta p = \gamma_1 v_1 m - \gamma_2 v_2 m$$

where subscript 1 refers to before the slowing and subscript 2 refers to after the slowing

continued…

$\Delta p = [(\gamma_1 v_1 / c) - (\gamma_2 v_2 / c)]mc^2/c$

$v_1 = 0.998c$

$\beta_1 = 0.998$

$\gamma_1 = 1 / \sqrt{(1 - 0.998^2)}$

$\gamma_1 = 15.82$

$v_2 = 0.500c$

$\beta_2 = 0.500$

$\gamma_2 = 1 / \sqrt{(1 - 0.500^2)}$

$\gamma_2 = 1.155$

$\Delta p = [(15.82 \cdot 0.998) - (1.155 \cdot 0.500)] \cdot (511 \text{ keV}) / c$

$\Delta p = (7772 / c \text{ keV})$

$\Delta p = 7.77 \text{ MeV/c}$

50. D is correct.

The speed of light has the same value for any observer, regardless of the state of motion of the source or the observer. This is a fundamental principle of Special Relativity.

Notes for active learning

Notes for active learning

Answer Keys

and

Detailed Explanations

Diagnostic Tests

Diagnostic Test 1 – Answer Key and Explanations

Answer Key

1	D	Kinematics & dynamics	26	C	Electric circuits
2	A	Force, motion, gravitation	27	D	Light & optics
3	A	Fluids & solids	28	C	Atomic & nuclear structure
4	C	Work & energy	29	C	Heat & thermodynamics
5	A	Waves & periodic motion	30	B	Special relativity
6	B	Sound	31	B	Equilibrium & momentum
7	C	Fluids & solids	32	A	Work & energy
8	A	Electrostatics & electromagnetism	33	D	Waves & periodic motion
9	C	Electric circuits	34	B	Heat & thermodynamics
10	E	Light & optics	35	E	Light & optics
11	B	Atomic & nuclear structure	36	B	Electric circuits
12	C	Equilibrium & momentum	37	B	Electric circuits
13	B	Work & energy	38	E	Fluids & solids
14	D	Waves & periodic motion	39	C	Sound
15	B	Sound	40	D	Waves & periodic motion
16	D	Fluids & solids	41	E	Work & energy
17	D	Electrostatics & electromagnetism	42	A	Equilibrium & momentum
18	A	Electric circuits	43	C	Force, motion, gravitation
19	C	Light & optics	44	E	Kinematics & dynamics
20	D	Rotational motion	45	A	Heat & thermodynamics
21	B	Heat & thermodynamics	46	C	Work & energy
22	E	Force, motion, gravitation	47	A	Quantum mechanics
23	B	Sound	48	D	Electrostatics & electromagnetism
24	C	Fluids & solids	49	B	Light & optics
25	A	Electrostatics & electromagnetism	50	C	Electrostatics & electromagnetism

51	**C**	Equilibrium & momentum	**76**	**E**	Fluids & solids
52	**E**	Waves & periodic motion	**77**	**A**	Electrostatics & electromagnetism
53	**D**	Fluids & solids	**78**	**A**	Electric circuits
54	**A**	Electric circuits	**79**	**C**	Light & optics
55	**B**	Atomic & nuclear structure	**80**	**D**	Atomic & nuclear structure
56	**D**	Heat & thermodynamics	**81**	**E**	Heat & thermodynamics
57	**E**	Kinematics & dynamics	**82**	**C**	Force, motion, gravitation
58	**B**	Force, motion, gravitation	**83**	**A**	Sound
59	**D**	Equilibrium & momentum	**84**	**B**	Fluids & solids
60	**E**	Atomic & nuclear structure	**85**	**B**	Electrostatics & electromagnetism
61	**A**	Kinematics & dynamics	**86**	**E**	Special relativity
62	**D**	Force, motion, gravitation	**87**	**B**	Light & optics
63	**C**	Heat & thermodynamics	**88**	**C**	Atomic & nuclear structure
64	**B**	Work & energy	**89**	**B**	Kinematics & dynamics
65	**A**	Waves & periodic motion	**90**	**E**	Force, motion, gravitation
66	**A**	Sound	**91**	**D**	Equilibrium & momentum
67	**A**	Fluids & solids	**92**	**A**	Work & energy
68	**D**	Electrostatics & electromagnetism	**93**	**C**	Waves & periodic motion
69	**C**	Electric circuits	**94**	**D**	Heat & thermodynamics
70	**B**	Light & optics	**95**	**E**	Light & optics
71	**A**	Atomic & nuclear structure	**96**	**E**	Electric circuits
72	**E**	Equilibrium & momentum	**97**	**D**	Force, motion, gravitation
73	**D**	Work & energy	**98**	**D**	Quantum mechanics
74	**E**	Rotational motion	**99**	**D**	Sound
75	**D**	Sound	**100**	**B**	Waves & periodic motion

1. D is correct.

When an object is thrown into the air, the acceleration vector equals gravity (for objects in free fall).

The velocity vector changes direction when the object starts to come down.

2. A is correct.

As shown in the diagram, mg (i.e., the force of the mass due to gravity) is broken down into two components.

The $mg \sin \theta$ component is parallel to the slope of the road.

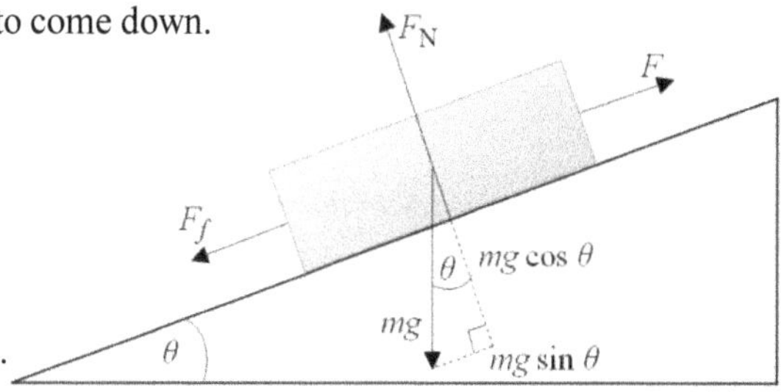

3. A is correct.

Archimedes' principle: the buoyant force acting on an object of volume V completely submerged in a liquid of a specific density has a magnitude that is expressed as:

$$F_B = \rho(V_{displaced})g$$

A change in the density of the liquid, and thus, the volume affects the buoyant force.

density = mass / volume

F_B does not include the density of the object.

Differences in the density of a liquid resulting from depth can be ignored.

4. C is correct.

$$P = W / t$$

$$W = Fd$$

$$P = (Fd) / t$$

$$F = mg$$

$$P = (mgd \sin \theta) / t$$

$$P = [(54 \text{ kg})\cdot(9.8 \text{ m/s}^2)\cdot(10 \text{ m}) \sin 30°] / (4 \text{ s})$$

$$P = [(54 \text{ kg})\cdot(9.8 \text{ m/s}^2)\cdot(10 \text{ m})\cdot(0.5)] / (4 \text{ s})$$

$$P = 2{,}646 \text{ J} / (4 \text{ s})$$

$$P = 661 \text{ J/s}$$

$$P = 661 \text{ W}$$

Convert watts into horsepower:

$$1 \text{ hp} = 745 \text{ W}$$

$$P = (661 \text{ W})\cdot(1 \text{ hp} / 745 \text{ W})$$

$$P = 0.89 \text{ hp}$$

5. A is correct.

The *amplitude* of a wave is a measure of the energy of the wave.

If energy dissipates, the amplitude is reduced.

6. B is correct.

To determine the *frequency of the fundamental*:

$f_n = nf_1$

where f_1 = fundamental

$f_1 = f_3 / 3$

$f_1 = 783\text{ Hz} / 3$

$f_1 = 261\text{ Hz}$

7. C is correct.

To calculate the *buoyant force* due to the water:

$F_B = 7.86\text{ N} - 6.92\text{ N}$

$F_B = 0.94\text{ N}$

Volume of displaced water is equal to volume of object:

$F_B = \rho g\text{V}$

$\text{V} = F_B / \rho g$

$\text{V} = 0.94\text{ N} / (1{,}000\text{ kg/m}^3)\cdot(9.8\text{ m/s}^2)$

$\text{V} = 9.6 \times 10^{-5}\text{ m}^3$

Mass of the object:

$m = W / g$

$m = 7.86\text{ N} / 9.8\text{ m/s}^2$

$m = 0.8\text{ kg}$

ρ = mass / volume

$\rho = 0.8\text{ kg} / 9.6 \times 10^{-5}\text{ m}^3$

$\rho = 8{,}333\text{ kg/m}^3$

8. A is correct.

Height determines the energy per mass required to position the water at a particular point.

This is analogous to an electric potential, as a measure of energy per charge required to position the charge at a particular point.

9. C is correct.

As the copper sheet is quickly passed through the magnetic field, eddy currents form.

According to Lenz's Law, the eddy currents rotate in such a way to produce magnetic fields that oppose the changing magnetic flux.

This opposing magnetic force acts to impede the motion of the copper sheet.

10. E is correct.

$n_1 \sin \theta_1 = n_2 \sin \theta_2$

$1.33 \sin 42° = 1 \sin \theta_2$

$1.33 \times (0.67) = 1 \sin \theta_2$

$\sin \theta_2 = 0.89$

$\theta_2 = \sin^{-1}(0.89)$

$\theta_2 = 63°$

11. B is correct.

Half-life is the time it takes for ½ of the quantity to decay.

1st half-life: $3{,}200\ \mu g / 2 = 1{,}600\ \mu g$

2nd half-life: $1{,}600\ \mu g / 2 = 800\ \mu g$

If two periods took 24.6 years, then one cycle takes 12.3 years.

Mathematically solving for half-life:

$A_{final} = A_{initial} \times (½)^{t/h}$

where t = time and h = half-life

$800\ \mu g = (3{,}200\ \mu g)\cdot(½)^{24.6\ yr / h}$

$(800\ \mu g) / (3{,}200\ \mu g) = (½)^{24.6\ yr / h}$

$0.25 = (½)^{24.6\ yr / h}$

$\ln(0.25) = (24.6\ yr / h) \ln(½)$

$h = 12.3\ yr$

12. C is correct.

If the object's velocity is constant, then the net force is zero.

In the *y*-direction:

$$0 = F_N + F_g$$

In the *x*-direction:

$$0 = F_{friction} + F$$

$$F_N + F_g = F_{friction} + F$$

$$-F = F_{friction}$$

Since kinetic friction exerts a force opposing the object's motion, there must be an equal and opposing force propelling it forward for the net force to be zero.

13. B is correct.

According to the *conservation of mechanical energy principle*, total mechanical energy in a system remains constant if the only forces acting are conservative forces.

14. D is correct.

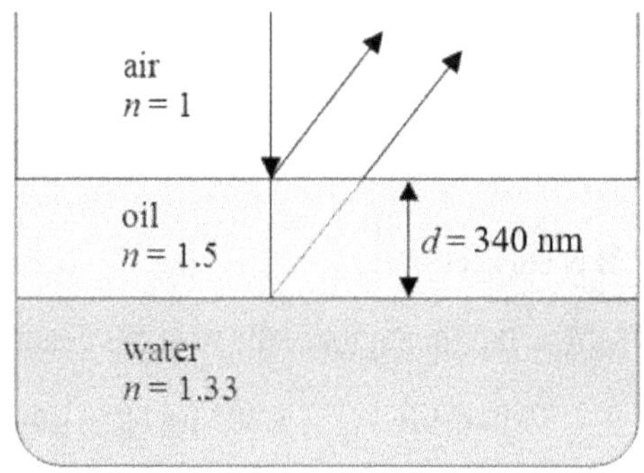

This is a thin film interference problem.

When light strikes the surface of the oil, some is transmitted, and some light will be reflected off the surface.

This process is repeated when the light reaches the oil-water interface.

However, the light reflected from the oil-water interface may combine with the light reflected initially off the oil before constructive or destructive interference.

The strongly reflected light is formed through constructive interference; the rays that destructively interfere with the initially reflected light cannot be seen.

Use the thin film constructive interference equation:

$$2n_{oil}d = (m + ½)\lambda \text{ (m = 0, 1, 2...)}$$

where m is the order of the reflected light

$$\lambda = (2n_{oil}d) / (m + ½)$$

Solve for range:

$$400 \text{ nm} \leq \lambda \leq 800 \text{ nm}$$

continued…

Use m = 0

$\lambda_0 = (2)\cdot(1.5)\cdot(340 \times 10^{-9}\text{ m}) / (0 + ½)$

$\lambda_0 = 1{,}020 \times 10^{-9}\text{ m} = 1{,}020\text{ nm}$

λ_0 is out of range

Use m =1

$\lambda_1 = (2)\cdot(1.5)\cdot(340 \times 10^{-9}\text{ m}) / (1 + ½)$

$\lambda_1 = 680 \times 10^{-9}\text{ m} = 680\text{ nm}$

λ_1 is in range

Use m = 2

$\lambda_2 = (2)\cdot(1.5)\cdot(340 \times 10^{-9}\text{ m}) / (2 + ½)$

$\lambda_2 = 408 \times 10^{-9}\text{ m}$

$\lambda_2 = 408\text{ nm}$

λ_2 is in range

If m is a value greater than 2, it produces a wavelength outside the 400 nm to 800 nm range.

Thus, the two most strongly reflected wavelengths are:

$\lambda_1 = 680\text{ nm}$

$\lambda_2 = 408\text{ nm}$

15. B is correct.

The harmonic wavelength λ_n occurs when:

$\lambda_n = (2L) / n$

where L is the length of the string, and n is the harmonic (n = 1, 2, 3…)

The fourth harmonic wavelength occurs at n = 4.

$\lambda_4 = (2L) / 4$

$\lambda_4 = (2 \times 1\text{ m}) / 4$

$\lambda_4 = 0.5\text{ m}$

16. D is correct.

The elastic modulus is calculated as stress divided by strain.

$E = \sigma / \varepsilon$

Stress is measured in N/m^2, the strain is unitless, and work in Joules, so the other choices are incorrect based on different units.

17. D is correct.

Electric field at a distance:

$$E = kQ / d^2$$

Solve for $Q_1 = 18\ \mu C$, where d is half the distance:

$$E_1 = (9 \times 10^9\ N \cdot m^2 \cdot C^{-2}) \cdot (18 \times 10^{-6}\ C) / (0.075\ m)^2$$

$$E_1 = 28.8 \times 10^6\ N/C$$

Solve for Q_2:

$$E_2 = (9 \times 10^9\ N \cdot m^2 \cdot C^{-2}) \cdot (-6 \times 10^{-6}\ C) / (0.075\ m)^2$$

$$E_2 = -9.6 \times 10^6\ N/C$$

Note that the negative sign indicates that the electric field goes into the charge because the charge is negative.

Because the electric fields generated by both charges point in the same direction, sum E_1 and E_2 to find the strength halfway between the charges:

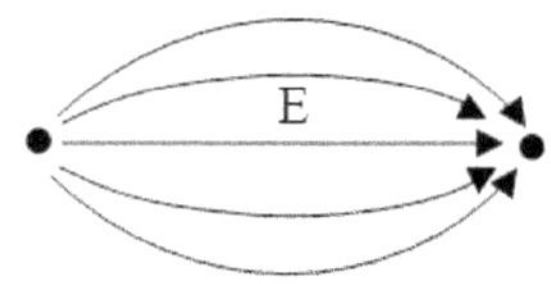

$$E_{total} = E_1 + E_2$$

$$E_{total} = (28.8 \times 10^6\ N/C) + (9.6 \times 10^6\ N/C)$$

$E_{total} = 38.4 \times 10^6$ N/C towards the negative charge.

18. A is correct.

$$P = I^2R$$

$$P = (4I)^2R$$

$$P = (16)I^2R$$

Power increased by a factor of 16

19. C is correct.

$$1 / f = 1 / d_o + 1 / d_i$$

$$1 / d_i = 1 / f - 1 / d_o$$

Since $d_o > f$,

$1 / f - 1 / d_o$ is always positive, so the image is real.

$$m = -d_i / d_o$$

Since i is positive, m = –(+i / o) is negative, so the image is inverted.

If the resulting magnification is positive, the image is upright.

If the magnification is negative, the image is inverted (upside down).

20. D is correct.

Apply the law of conservation of energy. There is no potential energy in this situation – all of the energy is kinetic (rotational kinetic energy in the initial state, translational kinetic energy in the final state).

$$E_f = E_i$$

$$K_f = K_i$$

$$\frac{1}{2}m_{car}v^2 = \frac{1}{2}I\omega^2$$

The moment of inertia of a disk is:

$$\frac{1}{2}mR^2$$

where m is the mass of the disk, and R is its radius

Then:

$$\frac{1}{2}m_{car}v^2 = \frac{1}{2}(\frac{1}{2}m_{wheel}R^2)\omega^2$$

Solving for velocity:

$$v = R\omega\sqrt{(m_{wheel} / 2m_{car})}$$

The angular speed in rev/s must be converted to rad/s:

$$200.0 \text{ rev/s}\cdot(2\pi \text{ rad/rev}) = 1{,}256.6 \text{ rad/s}$$

Finally:

$$v = (0.50 \text{ m})\cdot(1{,}256.6 \text{ rad/s})\cdot\sqrt{[370.0 \text{ kg} / (2\cdot1{,}500.0 \text{ kg})]}$$

$$v = 221 \text{ m/s}$$

21. B is correct.

Solve for *heat required*:

$$Q = cm\Delta T$$

$$Q = (92 \text{ cal/kg}\cdot°\text{C})\cdot(0.5 \text{ kg})\cdot(70 °\text{C} - 20 °\text{C})$$

$$Q = 2{,}300 \text{ cal}$$

Convert to Joules:

$$Q = (2{,}300 \text{ cal} / 1 \text{ J})\cdot(4.186 \text{ J/cal})$$

$$Q = 9{,}628 \text{ J}$$

22. E is correct.

The *force of gravity* depends only on the m and the distance between their centers.

$$F_{grav} = GM_1M_2 / d^2$$

If one m decreases by a factor of 2, then F_{grav} decreases by a factor of 2.

23. B is correct.

The *Doppler effect* for a receding source:

$$f_{observed} = [v_{sound} / (v_{sound} + v_{source})]f_{source}$$

Thus, a *receding source* has a lower pitch (higher frequency) and decreases further as it falls due to the increase in the source velocity.

24. C is correct.

As the block just enters the water, the total pressure is the sum of atmospheric pressure and the pressure produced by submersion.

$$P_{total} = P_{atmosphere} + \rho gh$$

When the block just enters the water:

$$P_{total} = P_{atmosphere} + 0$$

Graph C depicts the scenario as the block will experience an initial pressure of $P_{atmosphere}$. This will linearly increase (from ρgh) as the block is lowered further into the water.

25. A is correct.

$$F = qE$$

where q is the charge and E is the electric field

They both act in the same direction because they have a positive charge and are moving opposite the negative charge generated by the plates.

Since E is constant, and q is twice as large for the α particle (He has two protons and two neutrons), the charge is multiplied by +2.

$$F \propto q$$

F is twice as large.

26. C is correct.

Resistance = resistivity × (length / area)

$$A = \pi r^2$$

$$R = (1.68 \times 10^{-8}\ \Omega\cdot\text{m})\cdot[(57\ \text{m}) / \pi(5.7 \times 10^{-3}\ \text{m})^2]$$

$$R = 9.4 \times 10^{-3}\ \Omega$$

$$V = IR$$

$$I = V / R$$

$$I = 70\ \text{V} / 9.4 \times 10^{-3}\ \Omega$$

$$I = 7{,}447\ \text{A}$$

27. D is correct.

Virtual images can be seen (image in a plane mirror) but cannot be projected onto a screen.

28. C is correct.

Beta-decay (plus): the parent nuclide ejects a positron and neutrino.

A proton converts to a neutron in the process, so the mass number remains the same, but the atomic number decreases by 1.

$${}^{A}_{Z}X \rightarrow {}^{A}_{Z-1}Y + {}^{0}_{+1}e^{+} + {}^{0}_{0}v$$

29. C is correct.

Isothermal means that the temperature is constant.

Therefore, if there is no temperature change, U (internal energy) experiences no change.

$$\Delta U = 0$$

30. B is correct.

As measured in the rest frame of a second object, the speed of an object is the same as the speed of the second, as measured in the rest frame of the first.

Note that the question did not ask for the speed of Beta.

An observer on Alpha will observe that speed to be $0.94c$ (ignoring the effects of the curvature of the universe).

31. B is correct.

In projectile motion, the acceleration due to gravity acts in the vertical component.

Thus, acceleration due to gravity remains a nonzero constant.

32. A is correct.

Power = current × voltage

$P = I\Delta V$

$P = (5\text{ A})\cdot(25\text{ V})$

$P = 125\text{ W}$

Energy = power × time

Energy = (125 W)·(60 s)

Energy = 7,500 J

At 30% efficiency, the energy converted into PE:

7,500 J × 30% = 2,250 J

continued…

Set PE equal to *mgh*:

$$2{,}250\text{ J} = mgh$$

$$2{,}250\text{ J} = (50\text{ kg})\cdot(10\text{ m/s}^2)h$$

$$2{,}250\text{ J} = (500\text{ kg m/s}^2)h$$

$$2{,}250\text{ J} = (500\text{ J/m})h$$

$$h = 2{,}250\text{ J} / (500\text{ J/m})$$

$$h = 4.5\text{ m}$$

33. D is correct.

Gravitational PE is converted into KE as the pendulum swings back from its maximum height.

To calculate the pendulum's speed as it passes through its lowest point:

$$mgh = \tfrac{1}{2}mv^2$$

cancel m from both sides of the expression

$$gh_0 = \tfrac{1}{2}v_0^2$$

$$v_0 = \sqrt{(2gh_0)}$$

If h_0 is doubled the velocity is:

$$v = \sqrt{[2g(2h_0)]}$$

$$v = \sqrt{(2gh_0)} \times \sqrt{2}$$

$$v = v_0\sqrt{2}$$

The velocity increases by a factor of $\sqrt{2}$.

34. B is correct.

$$\Delta L / L = \alpha\Delta T$$

where α is coefficient of linear expansion

$$\Delta L = \alpha\Delta TL$$

Using $3L$ for L:

$$\Delta L = \alpha\Delta T(3L)$$

$$\Delta L = 3(\alpha\Delta TL)$$

ΔL is 3 times larger

35. E is correct.

$n_1 \sin \theta_1 = n_2 \sin \theta_2$

When the critical angle is reached:

$\theta_2 = 90°$

$\sin \theta_C = (n_2 / n_1)$

For this statement to be valid:

$n_1 > n_2$

36. B is correct.

The graph of resistivity *vs.* temperature for a semiconductor material has a negative sloping line. Thus, as temperature increases, resistivity decreases.

37. B is correct.

In a circuit with capacitors connected in parallel, the charge in each branch of the circuit is not constant (like the current in different branches is different).

The equivalent capacitance of the two capacitors in series, C_2 and C_3, is lower than the capacitance of each capacitor ($1 / C_{eq} = 1 / C_2 + 1 / C_3$).

The charge across these capacitors is lower than across C_1 due to the relation $V = Q / C$, and the voltage across C_1 and the voltage across the branch containing C_2 and C_3 are the same.

For a capacitor:

$U = (1/2)\ C\ V^2$

Note: U is the symbol for potential energy.

Since C is the same for each capacitor, but V is higher in C_1 than in C_2 or C_3, the PE is higher.

38. E is correct.

$F_B = \rho g V$

The buoyant force is directly proportional to volume.

Volume of a sphere is given as:

$V = 4 / 3\pi r^3$

If radius doubles:

$V = 4 / 3\pi(2r)^3$

$V = (8)\cdot(4 / 3\pi r^3)$

Therefore,

when r doubles, F_B increase by a factor of 8 times.

39. C is correct.

Speed of sound in a gas:

$$V_{sound} = \sqrt{[(\gamma RT) / M]}$$

where γ = adiabatic constant, R = gas constant, T = temperature and M = molar mass of a gas

$$V_{sound\ H} = \sqrt{[(\gamma RT) / M]}$$

If M is doubled, the speed of sound in He compared to in H is:

$$V_{sound\ He} = \sqrt{[(\gamma RT) / 2\ M]}$$

$$V_{sound\ He} = (1 / \sqrt{2})\cdot\sqrt{[(\gamma RT) / M]}$$

$$V_{sound\ He} = (0.707)\cdot\sqrt{[(\gamma RT) / M]}$$

Comparing the two with respect to He:

$$V_{sound\ He} = (0.707)\cdot(V_{sound\ H})$$

$$(1 / 0.707)V_{sound\ He} = V_{sound\ H}$$

$$V_{sound\ H} = 1.41\ V_{sound\ He}$$

40. D is correct.

The amplitude of a wave is a measure of the energy of the wave.

If energy increases, the amplitude increases.

41. E is correct.

I: PE = work

II: PE = $(mg)h$

III: PE = (weight)h

All the statements are measurements of the potential energy of the box.

42. A is correct.

Conservation of momentum:

$$mv_{im} + Mv_{iM} = mv_{fm} + Mv_{fM}$$

$$v_{fM} = [m(v_{im} - v_{fm}) + Mv_{im}] / M$$

$$v_{fM} = [(3.3\text{ kg})\cdot(8.5\text{ m/s} - (-3\text{ m/s})) + (6.5\text{ kg})\cdot(0\text{ m/s})] / 6.5\text{ kg}$$

$$v_{fM} = (3.3\text{ kg})\cdot(11.5\text{ m/s}) / 6.5\text{ kg}$$

$$v_{fM} = 5.8\text{ m/s}$$

43. C is correct.

$(m_1 + m_2)a = m_1g$

$(70 \text{ kg} + 10 \text{ kg})a = (70 \text{ kg})\cdot(10 \text{ m/s}^2)$

$a = 8.8 \text{ m/s}^2$

Another method to solve this problem:

Sum of forces for each system equals zero:

(A) $0 = T + ma$

$T = -(10 \text{ kg})a$

(B) $0 = T - ma - mg$

$0 = T - (70 \text{ kg})a - (70 \text{ kg})\cdot(10 \text{ m/s}^2)$

$0 = T - (70 \text{ kg})a - 700 \text{ N}$

Substituting (A) into (B):

$0 = [(-10 \text{ kg})a] - (70 \text{ kg})a - 700 \text{ N}$

$0 = (-80 \text{ kg})a - 700 \text{ N}$

$700 \text{ N} / 80 \text{ kg} = a$

$a = 8.8 \text{ m/s}^2$

44. E is correct.

$v_f = v_i + at$

$a = (v_f - v_i) / t$

$a = (16 \text{ m/s} - 4 \text{ m/s}) / 2 \text{ s}$

$a = 6 \text{ m/s}^2$

$d = d_i + v_it + ½(6 \text{ m/s}^2)\cdot(2 \text{ s})^2$

$d = 0 + (4 \text{ m/s})\cdot(2 \text{ s}) + ½(6 \text{ m/s}^2)\cdot(2 \text{ s})^2$

$d = (8 \text{ m}) + ½(6 \text{ m/s}^2)\cdot(4 \text{ s}^2)$

$d = 20 \text{ m}$

45. A is correct.

$Q = m_s c_p \Delta T$

$Q = (70 \text{ g})\cdot(0.11 \text{ cal/g}\cdot°\text{C})\cdot(450 \text{ °C} - 100 \text{ °C})$

$Q = 2{,}695 \text{ cal}$

$Q = m_w c_p \Delta T$

$m_w = Q / c_p \Delta T$

$m_w = 2{,}695 \text{ cal} / [(1 \text{ cal/g}\cdot°\text{C})\cdot(100 \text{ °C} - 25 \text{ °C})]$

$m_w = 36 \text{ g}$

46. C is correct.

$PE_{spring} = KE_{mass}$

$½kx^2 = ½mv^2$

$x = \sqrt{[(m / k) / v^2]}$

$m / k = 0.031 \text{ kg}\cdot\text{m/N}$

$x = \sqrt{[(0.031 \text{ kg}\cdot\text{m/N})\cdot(30 \text{ m/s})^2]}$

$x = 5.3 \text{ m}$

47. A is correct.

The *Balmer formula for Hydrogen* is:

$1 / \lambda = (1 / 91.2 \text{ nm})\cdot(1 / m^2 - 1 / n^2)$

From the n = 3 level, the transition to n = 1 has the greatest energy and, therefore, the shortest wavelength.

The Balmer formula for n = 3, m = 1 is:

$1 / \lambda = (1 / 91.2 \text{ nm})\cdot(1 / 1^2 - 1 / 3^2)$

$1 / \lambda = (1 / 91.2 \text{ nm})\cdot(8 / 9)$

$\lambda = 102.6 \text{ nm}$

48. D is correct.

According to *Coulomb's Law*, like charges repel with equal and opposite force:

$F_1 = F_2 = kq_1q_2 / r^2$

49. B is correct.

$1 / f = 1 / d_i + 1 / d_o$

$1 / 10 \text{ m} = 1 / d_i + 1 / 20 \text{ m}$

$1 / 10 \text{ m} - 1 / 20 \text{ m} = 1 / d_i$

$1 / 20 \text{ m} = 1 / d_i$

$d_i = 20$ m, positive so it must be in front of the mirror

The image is 20 m in front of the mirror – same position as the object (i.e., in front), only inverted.

50. C is correct.

$E = kQ / r^2$

$E_2 = k(2Q) / r^2$

$E_2 = 2E$

51. C is correct.

Note that the collision of the bullet and the block is inelastic because the bullet and block stuck after the collision, so the momentum is conserved, but kinetic energy is not.

Find the velocity of the bullet-block system through the conservation of energy when it compressed the spring.

Then the velocity of the bullet can be found using conservation of momentum in the bullet-block collision.

To determine the final velocity of the system, use *conservation of energy*:

$KE_{system} = PE_{system}$

$½m_{system}v_{system}^2 = ½kx^2$

$m_{system}v_{system}^2 = kx^2$

$v_{system}^2 = kx^2 / m_{system}$

$v_{system} = \sqrt{(kx^2 / m_{system})}$

$v_{system} = \sqrt{[(2{,}400 \text{ N/m})·(0.034 \text{ m})^2 / (4 \text{ kg} + 0.009 \text{ kg})]}$

$v_{system} = 0.832 \text{ m/s}$

To determine the initial velocity of the bullet, use *conservation of momentum*:

$p = mv$

$m_{bullet}v_{bullet} = m_{final: block+bullet}v_{final: block+bullet}$

$v_{bullet} = m_{final: block+bullet}v_{final: block+bullet} / m_{bullet}$

$v_{bullet} = (4.009 \text{ kg} \times 0.832 \text{ m/s}) / (0.009 \text{ kg})$

$v_{bullet} \approx 370 \text{ m/s}$

52. E is correct.

The *period of a simple pendulum* only depends on the length:

$$T = 2\pi\sqrt{(L / g)}$$

When L is tripled:

$$T_2 = 2\pi\sqrt{(3L / g)}$$

$$T_2 = 2\pi\sqrt{3}\sqrt{(L / g)}$$

$$T_2 / T_1 = [2\pi\sqrt{3}\sqrt{(L / g)}] / [2\pi\sqrt{(L / g)}]$$

$$T_2 / T_1 = \sqrt{3}$$

53. D is correct.

Static fluid pressure only depends on fluid density, height, and acceleration.

$$P = \rho gh$$

$$P = (1{,}000 \text{ kg/m}^3)\cdot (9.8 \text{ m/s}^2)\cdot(15 \text{ m} + 30 \text{ m})$$

$$P = 4.4 \times 10^5 \text{ N/m}^2$$

54. A is correct.

The *resistance of a wire* is given by:

$$R = \rho L / A$$

where ρ = resistivity, L = length and A = cross-sectional area

Resistance of thin wire:

$$R_0 = \rho L / A_0$$

Resistance of thicker wire:

$$A = 2A_0$$

$$R = \rho L / 2A_0$$

$$R = \tfrac{1}{2}R_0$$

The thicker wire has ½ the resistance of the thinner wire.

55. B is correct.

The *orbital angular momentum* is given by:

$$L = \sqrt{[\ell(\ell + 1)]}\, h$$

Allowable angular momentum quantum numbers: $\ell = 0, 1, 2 \ldots n - 1$

56. D is correct.

$Q = mc_p\Delta T$

$c_p = Q / m\Delta T$

$c_p = (14 \text{ J}) / (0.185 \text{ kg})\cdot(10 \text{ °C})$

$c_p = 7.6 \text{ J/kg·C}$

$1 \text{ °C} = 1 \text{ K}$

$c_p = 7.6 \text{ J/kg·K}$

57. E is correct.

$v_f = v_0 + at$

The force causing acceleration acting on the ball after it has been thrown is gravity.

Thus, acceleration due to gravity is –10 m/s^2 because it acts opposite the upward direction.

The acceleration due to gravity points downwards, giving a negative value.

58. B is correct.

$a = \Delta v / \Delta t$

$a = (0.9 \text{ m/s}) / (8 \text{ s})$

$a = 0.1125 \text{ m/s}^2$

$F = ma$

Total mass:

$m = F_{net} / a$

$m = 900 \text{ N} / 0.1125 \text{ m/s}^2$

$m = 8{,}000 \text{ kg}$

The mass of the rocket:

$m_{rocket} = m - m_{spacecraft}$

$m_{rocket} = (8{,}000 \text{ kg} - 3{,}500 \text{ kg})$

$m_{rocket} = 4{,}500 \text{ kg}$

59. D is correct.

An *impulse* is a change in momentum.

If the ball's initial direction is positive, then it is negative after the bounce.

$$\Delta p = p_2 - p_1$$

$$\Delta p = mv_i - mv_f$$

$$\Delta p = [(0.3\ \text{kg})\cdot(7\ \text{m/s})] - [(0.3\ \text{kg})\cdot(-5\ \text{m/s})]$$

$$\Delta p = (2.1\ \text{kg}\cdot\text{m/s}) - (-1.5\ \text{kg}\cdot\text{m/s})$$

$$\Delta p = 3.6\ \text{kg}\cdot\text{m/s}$$

60. E is correct.

The *Bohr model* of the atom has two major failures:

1. It could not explain why some spectral emission lines were brighter than others.
2. It defines a radius and momentum for the electron, which violates the Heisenberg uncertainty principle.

61. A is correct.

Involves proportions: what change in h (distance) is required for t to double?

$$t = (v_f - v_i) / a$$

$$t = v_f / g$$

If t is doubled, then v_f is also doubled:

$$2t = 2(v_f / g)$$

$$2t = (2v_f) / g$$

The height h is related to the velocity v_f through the equation for conservation of energy:

$$mgh = ½mv_f^2$$

cancel m from both sides and rearrange for h:

$$h = v_f^2 / 2g$$

Since v_f was doubled when t was doubled, the new height is:

$$h = (2v_f)^2 / 2g$$

$$h = (4)v_f^2 / 2g$$

If t is doubled, h increases by a factor of $2^2 = 4$.

62. D is correct.

Take the system to comprise the lower block and Rope 2.

This system has two forces: the tension due to Rope 1 (pointing up) and gravity (pointing down).

The net force is:

$$F_{net} = T - m_{system}\,g$$

The mass of the system is the sum of the mass of the lower block and Rope 2:

$$m_{system} = m_{block} + m_{Rope2}$$

By Newton's second law:

$$F_{net} = m_{system}\,a$$

Combined:

$$T - (m_{block} + m_{Rope2})g = (m_{block} + m_{Rope2})a$$

Or:

$$T = (m_{block} + m_{Rope2})\cdot(g + a)$$

$$T = (1.0\text{ kg} + 0.35\text{ kg})\cdot(9.8\text{ m/s}^2 + 5.5\text{ m/s}^2)$$

$$T = (1.35\text{ kg})\cdot(15.3\text{ m/s}^2)$$

$$T = 20.7\text{ N} \approx 21\text{ N}$$

63. C is correct.

$$Q = 9\text{ kJ}$$

$$W = -5\text{ kJ}$$

By the first law of thermodynamics:

$$\Delta U = Q + W$$

$$\Delta U = 9\text{ kJ} - 5\text{ kJ}$$

$$\Delta U = 4\text{ kJ}$$

64. B is correct.

The force of gravity and the normal force on the table sum to zero.

The spring provides the centripetal force:

$F_{spring} = mv^2 / r$

$20\ \text{N} = (5\ \text{kg})v^2 / (4\ \text{m})$

$80\ \text{N·m} = (5\ \text{kg})v^2$

$v^2 = 16\ \text{m}^2/\text{s}^2$

$v = 4\ \text{m/s}$

One revolution is equivalent to the circumference:

$C = 2\pi r$

$C = 2\pi(4\ \text{m})$

$C = 25.1\ \text{m}$

$t = C / v$

$t = (25.1\ \text{m}) / (4\ \text{m/s})$

$t = 6.3\ \text{s}$

65. A is correct.

Waves transport energy from one region to another but do not transport matter.

66. A is correct.

An organ pipe produces sounds of wavelength:

$\lambda_n = (2L / \text{n})$

where n is harmonic number: n = 1, 2, 3…

The lowest tone is the fundamental, where n = 1.

$\lambda_1 = (2L / 1)$

$\lambda_1 = 2L$

The next lowest tone is the second harmonic where n = 2.

$\lambda_2 = (2L / 2)$

$\lambda_2 = L$

The next lowest tone is the third harmonic where n = 3.

$\lambda_3 = (2L / 3)$

$\lambda_3 = 2L/3$

67. A is correct.

Young's Modulus (E) is given as:

$$E = FL / A\Delta L$$

where $F = mg$

$$E = mgL / A\Delta L$$

Rearranging for m:

$$m = EA\Delta L / gL_i$$

$$A = \pi r^2$$

$$m = E\pi r^2\Delta L / gL_i$$

$$m = [(2 \times 10^{11}\ N/m^2)\cdot\pi(0.9 \times 10^{-3}\ m)^2\cdot(1.8 \times 10^{-3}\ m)] / (9.8\ m/s^2)\cdot(4.2\ m)$$

$$m = 916.8\ kg\cdot m^2 / 41.2\ m^2/s^2$$

$$m = 22\ kg$$

68. D is correct.

Parallel to the field direction.

Faraday's Law (induced emf in a coil with changing magnetic flux):

$$V_{generated} = -N\Delta(BA) / \Delta t$$

where N = number of turns, B = magnetic field, A = area of coil and t = time

Positioning the loop such that the area vector is parallel to the magnetic field maximizes the flux through the loop (BA) and maximizes $V_{generated}$.

69. C is correct.

The nuclei have the same charge because each has one proton.

Therefore, the forces are the same.

70. B is correct.

A diverging thin lens always produces images that are virtual, erect, and reduced in size.

A converging thin lens produces real or virtual images, erect or inverted and reduced or magnified.

71. A is correct.

Beta Decay (plus): the parent nuclide ejects a positron and neutrino.

A proton converts to a neutron in the process, so the mass number remains the same, but the atomic number decreases by 1.

$${}^{A}_{Z}X \rightarrow {}^{A}_{Z-1}Y + {}^{0}_{+1}e^{+} + {}^{0}_{0}v_e$$

72. E is correct.

Conservation of momentum for inelastic collision:

$$m_Av_A = (m_A + m_B)v$$

$$v = m_Av_A / (m_A + m_B)$$

Calculate velocity:

$$v = (2 \text{ kg})\cdot(2 \text{ m/s}) / (2 \text{ kg} + 5 \text{ kg})$$

$$v = 0.6 \text{ m/s}$$

73. D is correct.

Potential energy is equal to the work performed:

$$\text{PE} = \text{W}$$

$$\text{PE} = mgh$$

$$mgh = \text{W}$$

$$m = \text{W} / gh$$

$$m = 400 \text{ J} / (9.8 \text{ m/s}^2)\cdot(4 \text{ m})$$

$$m = 10.2 \text{ kg}$$

74. E is correct.

Since the motor spins at a constant speed, there is no tangential component for the linear acceleration.

The linear acceleration is just the centripetal acceleration:

$$a = r\omega^2$$

Angular speed is in rpm, but needs to be in rad/s:

$$2695.0 \text{ rpm}\cdot(1 \text{ min/60 s})\cdot(2\pi \text{ rad/rev}) = 282.22 \text{ rad/s}$$

$$a = (0.07165 \text{ m})\cdot(282.22 \text{ rad/s})^2$$

$$a = 5{,}707 \text{ m/s}^2$$

75. D is correct.

The diagram represents the Doppler effect for a source with the velocity with respect to the observer (positioned to the right).

The siren on the truck changes pitches because, as the source approaches, the pitch is higher.

As the source is receding, the pitch is lower.

76. E is correct.

$P = F / A$

$F = PA$

$F = (3\text{ atm})\cdot(1.01 \times 10^5\text{ Pa/1 atm})\cdot(0.2\text{ m})^2$

$F = 12{,}120\text{ N} = 1.2 \times 10^4\text{ N}$

77. A is correct.

The force from a magnetic field on a charged particle equals the electric field force on the charged particle.

For an electromagnetic wave:

$B = E / c$

$B = (1{,}200\text{ V/m}) / (3 \times 10^8\text{ m/s})$

$B = 4 \times 10^{-6}\text{ T}$

78. A is correct.

$V_{rms} = V_{max} / \sqrt{2}$

$V_{rms} = 200\text{ V} / 1.41$

$V_{rms} = 142\text{ V}$

79. C is correct.

For a diverging lens, the image is always located on the same side as the object.

The equation is:

$1 / f = 1 / d_i + 1 / d_o$

For a diverging lens, the focal length is negative by convention.

$-1 / 3\text{ m} = 1 / d_i + 1 / 4\text{ m}$

$1 / d_i = -1 / 3\text{ m} - 1 / 4\text{ m}$

$1 / d_i = -4 / 12\text{ m} - 3 / 12\text{ m}$

$1 / d_i = -7 / 12\text{ m}$

$d_i = -12 / 7\text{ m}$

For a diverging lens, the image distance is always negative.

The focal length is negative for a concave or diverging lens because the focus used in the ray diagram is located on the lens's left side (x- and y-axis coordinate system).

The image distance is negative because it is also formed on the left-hand side.

80. D is correct.

The transition from level 3 to 2 produces a wavelength λ so that wavelength has the energy of:

$$E_{3,2} = hf$$

$$f = c / \lambda$$

$$E_{3,2} = h(c / \lambda)$$

$$E_{3,2} = (1 / \lambda)\cdot(hc)$$

Then the energy from 2 to 1 is:

$$E_{2,1} = 2E_{3,2}$$

Then the energy from 3 to 1 is:

$$E_{3,1} = 3E_{3,2}$$

Because the energy and wavelength are inversely related, the transition from 2 to 1 must have ½ λ and the transition from 3 to 1 must have 1/3 λ.

81. E is correct.

$$\text{PE} = mgh$$

Energy used in one repetition:

$$\text{PE} = (2)\cdot(3\text{ kg})\cdot(9.8\text{ m/s}^2)\cdot(0.5\text{ m})$$

$$\text{PE} = 29.4\text{ J}$$

Convert 19 kcal to joules:

$$\text{PE} = (19\text{ kcal} / 1)\cdot(1{,}000\text{ cal} / \text{kcal})\cdot(4.186\text{ J} / \text{cal})$$

$$\text{PE} = 79{,}534\text{ J}$$

Divide by 29.4 J to find the # of repetitions

$$\# = (79{,}534\text{ J}) / (29.4\text{ J})$$

$$\# = 2{,}705\text{ repetitions}$$

82. C is correct.

Find force due to gravity on 18 kg block:

$F_{g1} = mg \sin \theta$

$F_{g1} = (18 \text{ kg})\cdot(9.8 \text{ m/s}^2) \sin 20°$

$F_{g1} = 60.3 \text{ N}$

Find force due to gravity on 21 kg block:

$F_{g2} = m_2 g$

$F_{g2} = (21 \text{ kg})\cdot(9.8 \text{ m/s}^2)$

$F_{g2} = 205.8 \text{ N}$

Sum forces (note opposite signs so the F_{g1} values is subtracted):

$F_{\text{tot}} = F_{g2} - F_{g1}$

$F_{\text{tot}} = (205.8 \text{ N} - 60.3 \text{ N})$

$F_{\text{tot}} = 145.5 \text{ N}$

Find the acceleration of system:

$F = m_{\text{tot}} a$

$a = F / m_{\text{tot}}$

$a = 145.5 \text{ N} / (18 \text{ kg} + 21 \text{ kg})$

$a = 3.7 \text{ m/s}^2$

83. A is correct.

The time it takes for the point on the string to move from $+x$ to $-x$ is half a period, or $T / 2$.

$\lambda = vT,$

where v is the velocity of the wave

$T = \lambda / v$

$T = (8 \text{ m}) / (4 \text{ m/s})$

$T = 2 \text{ s}$

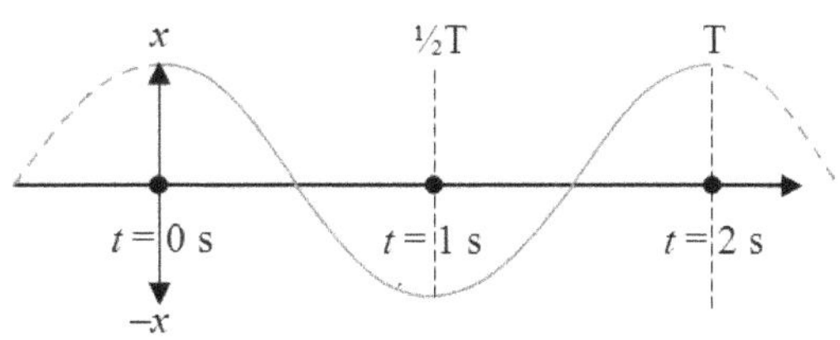

In general, calculating the frequency of a repeating event is accomplished by counting the number of times that event occurs within a specific period, then dividing the count by the period.

$t = ½T$

$t = 1 \text{ s}$

84. B is correct.

The *equation of continuity*:

$f = Av$ remains constant.

$A_2V_2 = A_1V_1$

$V_2 = A_1V_1 / A_2$

$V_2 = \pi(16\text{ cm})^2\cdot(V_1) / \pi(4\text{ cm})^2$

$V_2 = 16V_1$

If the radius decreases by a factor of 4 then the velocity increases by a factor of 16.

85. B is correct.

Coulomb's Law equation:

$F = kQ_1Q_2 / d$

Because of the multiplication of Q_1 and Q_2 by Coulomb's Law, each ball will experience the same force regardless of the magnitude of its charge.

$F_1 = k(2Q_1)Q_2 / d$

$F_2 = k(2Q_1)Q_2 / d$

$F_1 = F_2$

86. E is correct.

The *principle of relativity* requires that the laws of physics be the same in all inertial reference frames.

There can be no difference in observations between one taken in a frame at rest and one taken in a uniformly moving frame.

87. B is correct.

Linear magnification of lens:

$m = -d_i / d_o$

The negative in the image distance is because real images are inverted.

If the image distance value is negative, the image is virtual, m is positive, and the image is erect:

m	Image	Inverted / Erect
+	virtual	erect
–	real	inverted

88. C is correct.

All the masses of the elements are determined relative to 12C (Carbon-12), which is defined as an exact number 12 amu. Elements exist as a variety of isotopes, and two major isotopes of carbon are ^{12}C and ^{13}C.

Each carbon atom has the same number of protons and electrons – 6.

^{12}C has 6 neutrons, ^{13}C has 7 neutrons.

89. B is correct.

Kinetic energy is equal to potential energy at the top of the flight.

$$PE = KE$$

$$mgh = ½mv^2$$

cancel m from the expression

$$gh = ½v^2$$

$$h = ½v^2 / g$$

$$h = v^2 / 2g$$

Thus, height does not depend on mass.

Objects thrown at the same velocity travel to equal heights, regardless of mass.

90. E is correct.

According to Newton's First Law: "An object at rest, or in uniform motion, will remain that way unless acted upon by an outside force."

Because no outside forces act on the puck, no force is required to keep the puck in uniform motion.

91. D is correct.

Final velocity of an object with respect to time:

$$v_f = v_i + at$$

Solve for acceleration:

$$0 = 16 \text{ m/s} + a(0.091 \text{ s})$$

$$a = -175.8 \text{ m/s}^2$$

The acceleration is negative because it acts opposite the velocity vector.

Calculate the force using the magnitude of acceleration:

$$F = ma$$

$$F = (0.24 \text{ kg})\cdot(175.8 \text{ m/s}^2)$$

$$F = 42 \text{ N}$$

92. A is correct.

$d_{tot} = d_1 + d_2 + d_3$

$d_{tot} = (5 \text{ km} + 7.3 \text{ km} + 3.4 \text{ km})$

$d_{tot} = 15.7 \text{ km}$

$d_{tot} = 15.7 \times 10^3 \text{ m}$

Force to work relationship:

$W = Fd$

$F = W / d$

$F = (2.6 \times 10^6 \text{ J}) / (15.7 \times 10^3 \text{ m})$

$F = 1.7 \times 10^2 \text{ N}$

93. C is correct.

A redshift is created when a light source moves away from the observer, and thereby, its wavelength is perceived to be longer due to the Doppler effect.

A blueshift occurs when the light source moves towards the observer, and the perceived wavelength is shorter.

94. D is correct.

Stefan-Boltzmann Law:

$P = A\varepsilon\sigma T^4$

Temperature is doubled:

$P = A\varepsilon\sigma(2T)^4$

$P = A\varepsilon\sigma(16)T^4$

The power increases by a factor of 16.

$P_2 = P(16)$

$P_2 = (15 \text{ W})\cdot(16)$

$P_2 = 240 \text{ W}$

95. E is correct.

Angular magnification equation:

$$M_\alpha = NP / f$$

where NP = near point

$$M_\alpha = (250 \text{ mm} / 50 \text{ mm})$$

$$M_\alpha = 5$$

96. E is correct.

Faraday's Law states that a voltage will be induced in a coil exposed to a magnetic field:

$$V = -N\Delta(BA) / \Delta t$$

All three choices are correct because options I and III would increase the rate of change of the magnetic field (B) and increase the voltage.

Option II is correct because it would increase the number of turns (N) and increase the induced voltage.

97. D is correct.

Use a rotating coordinate system in which the normal force points along the positive y-axis and "up the incline" is in the negative x-direction.

In this coordinate system, there is no motion or acceleration in the y-direction; this means the normal force F_N must be equal to $mg \cos \theta$, so the y-forces balance.

Next, consider forces and components of forces in the x-direction.

There is the frictional force to the right ("down the incline," since friction opposes the motion and is stopping the car), and there is a component of the gravitational force to the right $mg \sin \theta$.

$$F_{net} = F_{net,x} = F_{friction} + F_{gravity,x}$$

$$F_{friction} = \mu_k F_N = \mu_k (mg \cos \theta)$$

$$F_{net,x} = \mu_k (mg \cos \theta) + mg \sin \theta$$

$$F_{net,x} = mg (\mu_k \cos \theta + \sin \theta)$$

where downhill is positive

Note: use μ_k because the tires are locked up, meaning the tires are sliding over the road surface.

If the tires were gripping the surface (a condition called "smooth-rolling"), the static friction would be used because, in that case, the tires are *not* sliding against the surface.

98. D is correct.

The uncertainty principle states that the uncertainty in position and momentum must be at least:

$\Delta x \Delta p \approx h / 2\pi$

$\Delta p \approx h / (2\pi\Delta x)$

If:

$\Delta x = 5.0 \times 10^{-15}$ m

Then:

$\Delta p \approx 6.626 \times 10^{-34}$ J·s / $(2\pi \cdot 5.0 \times 10^{-15}$ m)

$\Delta p \approx 2.11 \times 10^{-20}$ kg·m/s

The uncertainty in the (non-relativistic) kinetic energy of the proton is then:

$\Delta KE \approx (\Delta p)^2 / (2m)$

$\Delta KE \approx [h / (2\pi\Delta x)]^2 / (2m)$

$\Delta KE \approx [(6.626 \times 10^{-34}$ J·s$) / (2\pi \cdot 5.0 \times 10^{-15}$ m$)]^2 / (2 \cdot 1.67 \times 10^{-27}$ kg)

$\Delta KE \approx (1.33 \times 10^{-13}$ J)

Converting to MeV:

$\Delta KE \approx (1.33 \times 10^{-13}$ J$) / (1.6 \times 10^{-13}$ J/MeV)

$\Delta KE \approx (0.83$ MeV)

99. D is correct.

Standing wave: relationship of wavelength to length and harmonic number.

$\lambda = 2L / n$

$\lambda = 2(0.2$ m$) / 5$

$\lambda = 0.08$ m

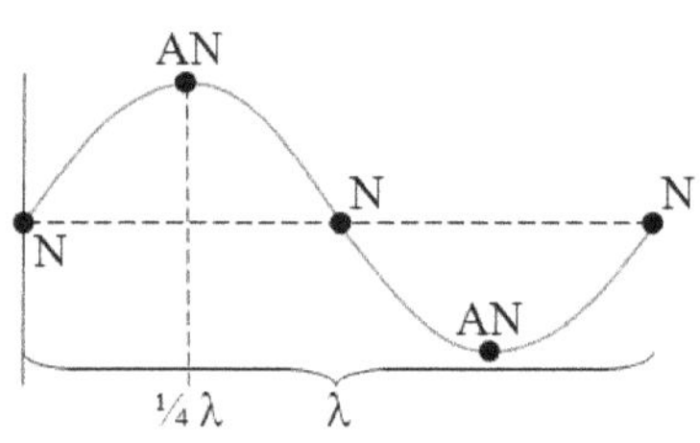

In a wave, the length between a node and antinode is ¼λ:

$d = \frac{1}{4}\lambda$

$d = \frac{1}{4}(0.08$ m)

$d = 0.02$ m = 20 mm

100. B is correct.

Notice that every time the position graph reads a maximum, it must be changing direction.

Therefore, velocity must be zero at these points.

Graphs B and C have zero velocity at the maximum of the position graph.

Graph B is the correct answer because graph C displays a negative velocity while the position graph increases, which is not possible.

Graph B is correct because it has a positive velocity as the position increases and a negative velocity as the position decreases.

Notes for active learning

Diagnostic Test 2 – Answer Key and Detailed Explanations

Answer Key

1	**D**	Work & energy	**26**	**B**	Sound
2	**D**	Equilibrium & momentum	**27**	**C**	Fluids & solids
3	**A**	Rotational motion	**28**	**A**	Electrostatics & electromagnetism
4	**E**	Kinematics & dynamics	**29**	**C**	Electric circuits
5	**D**	Heat & thermodynamics	**30**	**A**	Light & optics
6	**A**	Work & energy	**31**	**C**	Atomic & nuclear structure
7	**D**	Sound	**32**	**B**	Equilibrium & momentum
8	**E**	Electrostatics & electromagnetism	**33**	**E**	Work & energy
9	**C**	Light & optics	**34**	**B**	Waves & periodic motion
10	**B**	Kinematics & dynamics	**35**	**C**	Sound
11	**C**	Equilibrium & momentum	**36**	**E**	Fluids & solids
12	**E**	Waves & periodic motion	**37**	**B**	Electrostatics & electromagnetism
13	**A**	Fluids & solids	**38**	**C**	Electric circuits
14	**B**	Kinematics & dynamics	**39**	**D**	Light & optics
15	**C**	Atomic & nuclear structure	**40**	**E**	Atomic & nuclear structure
16	**E**	Heat & thermodynamics	**41**	**A**	Heat & thermodynamics
17	**C**	Special relativity	**42**	**B**	Force, motion, gravitation
18	**B**	Quantum mechanics	**43**	**B**	Sound
19	**A**	Equilibrium & momentum	**44**	**B**	Fluids & solids
20	**C**	Atomic & nuclear structure	**45**	**A**	Electrostatics & electromagnetism
21	**D**	Kinematics & dynamics	**46**	**C**	Electric circuits
22	**E**	Force, motion, gravitation	**47**	**C**	Light & optics
23	**C**	Heat & thermodynamics	**48**	**E**	Atomic & nuclear structure
24	**B**	Work & energy	**49**	**A**	Kinematics & dynamics
25	**D**	Waves & periodic motion	**50**	**D**	Force, motion, gravitation

51	B	Equilibrium & momentum	76	B	Heat & thermodynamics
52	C	Work & energy	77	E	Kinematics & dynamics
53	E	Waves & periodic motion	78	E	Force, motion, gravitation
54	D	Heat & thermodynamics	79	D	Equilibrium & momentum
55	D	Light & optics	80	E	Atomic & nuclear structure
56	E	Electric circuits	81	B	Kinematics & dynamics
57	A	Electrostatics & electromagnetism	82	C	Force, motion, gravitation
58	B	Fluids & solids	83	E	Heat & thermodynamics
59	D	Sound	84	A	Work & energy
60	C	Waves & periodic motion	85	B	Waves & periodic motion
61	D	Work & energy	86	C	Sound
62	C	Equilibrium & momentum	87	A	Waves & periodic motion
63	A	Force, motion, gravitation	88	A	Special relativity
64	B	Kinematics & dynamics	89	E	Electric circuits
65	A	Heat & thermodynamics	90	D	Light & optics
66	C	Kinematics & dynamics	91	B	Atomic & nuclear structure
67	B	Rotational motion	92	D	Equilibrium & momentum
68	C	Electrostatics & electromagnetism	93	C	Work & energy
69	D	Light & optics	94	A	Waves & periodic motion
70	C	Heat & thermodynamics	95	C	Sound
71	D	Equilibrium & momentum	96	E	Fluids & solids
72	B	Waves & periodic motion	97	C	Electrostatics & electromagnetism
73	A	Fluids & solids	98	B	Electric circuits
74	D	Electric circuits	99	B	Quantum mechanics
75	E	Atomic & nuclear structure	100	A	Atomic & nuclear structure

1. D is correct.

Potential energy:

$$PE = mgh$$

$$PE = (2\ \text{kg})\cdot(9.8\ \text{m/s}^2)\cdot(4\ \text{m})$$

$$PE \approx 80\ \text{J}$$

2. D is correct.

The diver has a maximum velocity at point D because the PE from point B is converted into KE.

If velocity is maximized, then momentum is maximized:

$$p = mv$$

3. A is correct.

The kinetic energy stored in a rotating object is:

$$K = \tfrac{1}{2} I\omega^2$$

With this, the angular speed as a function of energy is:

$$\omega^2 = 2K / I$$

The moment of inertia of a disk is:

$$I = \tfrac{1}{2}mr^2$$

Giving:

$$\omega^2 = 4K / mr^2$$

$$\omega = \sqrt{(4K / mr^2)}$$

$$\omega = \sqrt{[4\cdot(3.2 \times 10^7\ \text{J})] / [(400.0\ \text{kg})\cdot(0.60\ \text{m})^2]}$$

$$\omega = 943\ \text{rad/s}$$

4. E is correct.

Constant velocity means balanced forces are acting on the box.

Therefore, the horizontal forces must be equal.

5. D is correct.

Isometric refers to constant volume.

Since the container does not expand when heat energy is added, it undergoes an isometric process.

6. A is correct.

A watt is a J/s

J = N·m

N = kg·m/s^2

J = (kg·m/s^2)m

W = J/s

W = N·m/s

W = kg·m^2/s^3

7. D is correct.

Sound speed and frequency are related:

$f = v / \lambda$

where λ is twice the length of the organ pipe

If sound speed is 3% slower, then v decreases by a factor of 0.97, and f decreases by a factor of 0.97 because f is directly proportional to v.

Thus, f decreases by 3%.

8. E is correct.

Particle 1 is negative because it is deflected in the magnetic field but goes in the opposite direction of the right-hand rule.

Particle 2 is neutral because it is unaffected by the magnetic field.

Particle 3 is positive because it is deflected in the magnetic field and obeys the right-hand rule.

9. C is correct.

Among the choices listed, radio waves have the lowest frequency.

$f_{radio} < f_{microwave} < f_{infrared} < f_{X\text{-}ray} < f_{\gamma\ rays}$

10. B is correct.

If both marbles are thrown with speed v_0 then both marbles have equal KE_0.

When they reach the ground, their final KE is:

$KE_0 + PE_0 = KE_{final}$

Because the height of the cliff is the same, the PE_0 of both marbles is equal.

Thus, both marbles have equal KE_{final} and equal speed at the bottom.

11. C is correct.

Moment of *inertia solid cylinder*:

$$I_1 = ½mr^2$$

Moment of *inertia thin-loop*:

$$I_2 = mr^2$$

If both have the same mass and radius, then:

$$I_1 = ½I_2$$

The thin loop has a greater moment of inertia.

12. E is correct.

Calculate the *cable tension*:

$$T = mg$$

$$T = (2{,}500 \text{ kg})\cdot(9.8 \text{ m/s}^2)$$

$$T = 24{,}500 \text{ N}$$

Calculate the *wave velocity*:

$$v = \sqrt{[T / (m / L)]}$$

$$v = \sqrt{(24{,}500 \text{ N} / 0.65 \text{ kg/m})}$$

$$v = 194 \text{ m/s}$$

13. A is correct.

Volumetric flow rate:

$$V_f = Av$$

$$A = \pi d^2 / 4$$

So, for the original pipe:

$$V_f = (\pi d^2 / 4) \cdot v$$

If the diameter is doubled:

$$V_f = [\pi (2d)^2 / 4]\, v$$

$$V_f = (\pi / 4)\cdot(4d^2)\, v$$

$$V_f = 4\, [(\pi d^2 / 4)\, v]$$

where the term in square brackets is identical to the original flow rate

Thus, the flow rate increases by a factor of 4.

14. B is correct.

They are traveling at the same speed.

15. C is correct.

Protons are positively charged and repel each other in the nucleus due to the forces from electrostatic repulsion.

The strong nuclear force keeps the protons together because it overcomes the electrostatic repulsion and thus binds the nucleus.

16. E is correct.

$$10\ °C = 10\ K$$

$$Q = mc\Delta T$$

$$Q = (0.3\ kg)·(128\ J/kg·K)·(10\ K)$$

$$Q = 384\ J$$

17. C is correct.

The *relativistic expression for momentum* is:

$$p = \gamma mv = \gamma m\beta c$$

where as usual $\beta = v / c$ and $\gamma = 1 / \sqrt{(1 - \beta^2)}$

The two particles are to have the same momentum:

$$\gamma_1 m_1 \beta_1 c = \gamma_2 m_2 \beta_2 c$$

$$(m_1\beta_1) / \sqrt{(1 - \beta_1^2)} = (m_2\beta_2) / \sqrt{(1 - \beta_2^2)}$$

$$\beta_1^2 / (1 - \beta_1^2) = (m_2 / m_1)^2 ·[\beta_2^2 / (1 - \beta_2^2)]$$

For clarity, the right-hand side is set equal to a temporary variable x.

$$x = (m_2 / m_1)^2·[\beta_2^2 / (1 - \beta_2^2)]$$

$$x = [(1.30\ kg) / (0.723\ kg)]^2·[0.515^2 / (1 - 0.515)^2]$$

$$x = 1.1670$$

Furthermore:

$$\beta_1^2 / (1 - \beta_1^2) = x$$

Solving this for β_1:

$$\beta_1 = \sqrt{[x / (1 + x)]}$$

$$\beta_1 = \sqrt{(1.1670 / 2.1670)}$$

$$\beta_1 = 0.734$$

$$v_1 = 0.734c$$

18. B is correct.

The electron absorbs the entire energy of the photon, losing 2.4 eV as it escapes from the photocathode and an additional 1.1 eV of kinetic energy after it escapes from the photocathode.

This allows it to be stopped only with a potential exceeding 1.1 volts.

It therefore absorbs 2.4 eV + 1.1 eV = 3.5 eV from the incident photon.

A photon with energy 3.5 eV has a wavelength given by:

$$hv = hc / \lambda = \mathrm{E}$$

$$\lambda = hc / \mathrm{E} = (6.626 \times 10^{-34}\ \mathrm{J{\cdot}s}){\cdot}(3.00 \times 10^{8}\ \mathrm{m/s}) / (3.5\ \mathrm{eV} \cdot 1.60 \times 10^{-19}\ \mathrm{J/eV})$$

$$\lambda = 355\ \mathrm{nm}$$

19. A is correct.

The collision is inelastic because the bullet and block stick after impact, so kinetic energy is not conserved.

Use energy conservation after the collision to find the velocity of the bullet and block system.

$$\mathrm{KE} = \mathrm{PE}$$

$$\tfrac{1}{2}(m_{\mathrm{bullet}} + m_{\mathrm{block}})v_{\mathrm{system}}^2 = (m_{\mathrm{bullet}} + m_{\mathrm{block}})gh$$

$$v_{\mathrm{system}} = \sqrt{(2gh)}$$

$$v_{\mathrm{system}} = \sqrt{[2(9.8\ \mathrm{m/s^2}){\cdot}(0.4\ \mathrm{m})]}$$

$$v_{\mathrm{system}} = 2.8\ \mathrm{m/s}$$

Conservation of momentum to find the velocity of the bullet:

$$(m_{\mathrm{bullet}}){\cdot}(v_{\mathrm{bullet}}) = (m_{\mathrm{bullet}} + m_{\mathrm{block}})v_{\mathrm{system}}$$

$$v_{\mathrm{bullet}} = (0.01\ \mathrm{kg} + 1.5\ \mathrm{kg}){\cdot}(2.8\ \mathrm{m/s}) / (0.01\ \mathrm{kg})$$

$$v_{\mathrm{bullet}} = 423\ \mathrm{m/s}$$

20. C is correct.

Heavy nuclei tend to have more neutrons than protons because neutrons contribute to the nuclear strong force but do not contribute to the electrostatic repulsion (i.e., protons) because of their neutral charge.

Thus, an excess of neutrons helps keep the nuclei together without adding repulsion forces within the nuclei.

21. D is correct.

$v_f = v_0 + at$

$5\text{ m/s} = (0\text{ m/s}) + a(1\text{ s})$

$a = 5\text{ m/s}^2$

$x = x_0 + v_0t + ½at^2$

$x = (0\text{ m}) + (0\text{ m/s})\cdot(3\text{ s}) + ½(5\text{ m/s}^2)\cdot(3\text{ s})^2$

$x = 22.5\text{ m}$

22. E is correct.

The component of the block's weight (w) perpendicular to the inclined plane equals the normal force (N) on the block.

N is given by the expression $w \cos \theta$, where θ is the angle of incline.

As the angle increases toward 90°, $\cos \theta$ decreases.

Therefore, N also decreases.

23. C is correct.

Linear expansion:

$\Delta L = L_i\alpha\Delta T$

Total linear expansion of two rods:

$\Delta L_{total} = \Delta L_A + \Delta L_S$

$\Delta L_{total} = L_{iA}\alpha_A\Delta T + L_{iS}\alpha_S\Delta T$

$\Delta L_{total} = \Delta T(L_{iA}\alpha_A + L_{iS}\alpha_S)$

$\Delta L_{total} = (90\text{ °C} - 15\text{ °C})\cdot[(0.1\text{ m})\cdot(2.4 \times 10^{-5}\text{ K}^{-1}) + (0.8\text{ m})\cdot(1.2 \times 10^{-5}\text{ K}^{-1})]$

$\Delta L_{total} = 9 \times 10^{-4}\text{ m}$

Convert to mm:

$\Delta L_{total} = (9 \times 10^{-4}\text{ m}/1)\cdot(1{,}000\text{ mm}/1\text{ m})$

$\Delta L_{total} = 0.9\text{ mm}$

24. B is correct.

The object's KE and momentum become zero after impact.

The total energy of the system (including heat, sound, etc.) remains constant.

25. D is correct.

Frequency is the number of cycles per second.

f = cycles / sec

$2f$ = 2 cycles / sec

By doubling the frequency, the number of cycles per second is doubled.

The speed increases by a factor of two to accomplish this in the same amount of time.

26. B is correct.

For a pipe that has one end closed and one end open, the diagram for the 3rd overtone looks as follows:

Note: the closed end of the pipe must have an antinode, and the open end will have a node.

Because the open end cannot have an antinode, the even-numbered harmonics cannot occur in a pipe like this.

That means that the 3rd overtone corresponds to the *seventh* harmonic in this case.

The fundamental frequency for this pipe is 1/7 the frequency of the 3rd overtone.

From the diagram, the number of antinodes (points on the waveform that have zero displacements) can be counted – there are four of them.

27. C is correct.

The solid circle and wire circle have the same diameters.

However, the wire has two surface areas exposed to the water (i.e., inner and outer circumference), where the difference in circumference length is assumed to be negligible.

The solid circle only has the outer edge exposed to the water's surface (i.e., circumference).

$L_{wire} = 2(2\pi r)$ = the relevant length for the wire circle

$2\pi r$ = the relevant length for the solid circle

Solid circle on water:

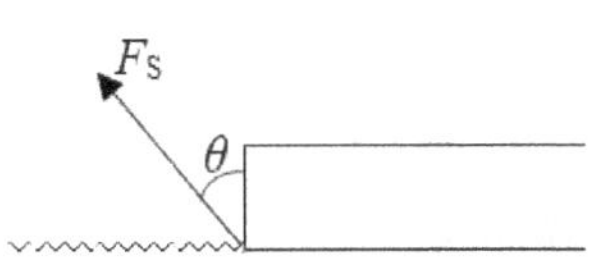

$F_{s1} = y \cos \theta \, L$

Wire on water:

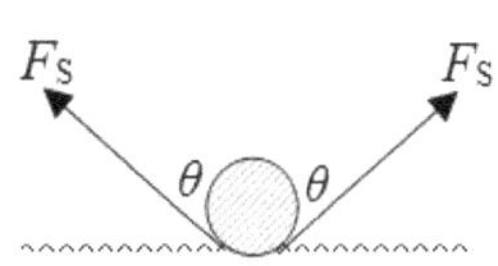

$F_{s2} = 2y \cos \theta \, L$

$F_{s2} = 2F_{s1}$

Balance weight *vs.* surface tension where F_s point up and F_w points down.

continued…

For a *solid circle*:

$$F_{w1} = F_{s1}$$

For a *wire circle*:

$$F_{w2} = F_{s2}$$

$$F_{w2} = 2F_{s1}$$

$$F_{w2} = 2F_{w1}$$

The wire circle can have double the mass without sinking.

28. A is correct.

The *volt* is expressed in Joules per Coulomb:

$$1\text{ V} = 1\text{ J} / 1\text{ C}$$

Thus, 120 V is 120 J per Coulomb of charge.

29. C is correct.

Current:

$$6\text{ mA} = 0.006\text{ A}$$

$$0.006\text{ A} = 0.006\text{ Coulombs/sec}$$

During 1 minute (60 sec), calculate the charge that has flowed:

$$0.006\text{ C/s} \times 60\text{ s} = 0.36\text{ C}$$

Each electron carries: 1.602×10^{-19} C

Calculate the # of electrons:

$$(0.36\text{ C})\cdot(1\text{ electron} / 1.602 \times 10^{-19}\text{ C}) = 2.25 \times 10^{18}\text{ electrons}$$

30. A is correct.

A *convex mirror* always produces an upright, virtual, and smaller image regardless of object location.

31. C is correct.

According to the *de Broglie hypothesis*, the idea of matter waves describes the wave-like behavior of particles that are moving.

32. B is correct.

Assume that the system is the massless spring and the two masses.

Although KE is not conserved, KE and spring PE are conserved because KE is converted into PE.

Momentum is conserved only if the system is isolated.

The fixed wall provides an external force on the back of the spring; therefore, the system is not isolated, and momentum is not conserved.

This results in the changing speed of the massive objects.

33. E is correct.

The potential energy of the pebble at the top of its flight is equal to its kinetic energy right as it is being thrown.

$$KE = PE$$

$$\frac{1}{2}mv_0^2 = mgh_0$$

cancel m from both sides of the expression

$$v_0 = \sqrt{(2gh_0)}$$

When the height doubles:

$$v = \sqrt{(2g2h_0)}$$

$$v = \sqrt{2}\ \sqrt{2gh_0}$$

$$v = \sqrt{2} \times v_0$$

34. B is correct.

The period of a simple pendulum only depends on the length:

$$T = 2\pi\sqrt{(L / g)}$$

Double the length:

$$T = 2\pi\sqrt{(2L / g)}$$

When the length is doubled, the period increases by a factor of $\sqrt{2}$.

35. C is correct.

Open pipes can produce even and odd # harmonics.

Closed pipes only produce odd # harmonics.

36. E is correct.

Bernoulli's Equation:

$$P_1 + \frac{1}{2}\rho v_1^2 + \rho g h_1 = P_2 + \frac{1}{2}\rho v_2^2 + \rho g h_2$$

When simplifying Bernoulli's Equation, the top opening and bottom opening experience atmospheric pressure, so P_1 and P_2 cancel.

Likewise,

$v_1 << v_2$, so it can be assumed to be negligible and go to zero.

Finally, $h_2 = o$ because it is at the bottom of the tank and is the reference for further heights in the problem

This results in the simplified expression:

$$\rho g h_1 = \frac{1}{2}\rho v_2^2$$

cancel ρ from both sides of the expression

$$g h_1 = \frac{1}{2} v_2^2$$

$$v_2 = \sqrt{(2 g h_1)}$$

$$v_2 = \sqrt{[(2)\cdot(9.8\ \text{m/s}^2)\cdot(0.5\ \text{m})]}$$

$$v_2 = 3.1\ \text{m/s}$$

37. B is correct.

The electric field strength of parallel-plate capacitors:

Energy = voltage / distance

$$E = V / d$$

If V remains constant and d increases, E decreases

38. C is correct.

The force on a charged particle due to an electric field:

$$F = qE_0$$

The charge on the deuteron (1 proton and 1 neutron) has a charge equal to and opposite that of an electron.

The forces on the two particles have the same magnitude but opposite directions.

39. D is correct.

When a wave enters a different medium, its speed and wavelength change.

Electromagnetic waves transport energy in a vacuum at approximately 3.00×10^8 m/s (speed of light or c).

The frequency remains constant and is related to wavelength by:

$$\lambda = c / f$$

$$\lambda = (3 \times 10^8 \text{ m/s}) / (1.8 \times 10^{14} \text{ Hz})$$

$$\lambda = 1.667 \times 10^{-6} \text{ m} = 1{,}667 \text{ nm}$$

40. E is correct.

Wavelength relates to photon energy:

$$E = hc / \lambda$$

$$E = [(6.626 \times 10^{-34} \text{ J·s})·(3 \times 10^8 \text{ m/s})] / (580 \times 10^{-9} \text{ m})$$

$$E = 3.4 \times 10^{-19} \text{ J}$$

41. A is correct.

Graham's Law:

$$\text{rate}_1 / \text{rate}_2 = \sqrt{(m_2 / m_1)}$$

42. B is correct.

$$\text{W} = mg$$

On a slope, the force on an object due to the acceleration of gravity:

$$F = mg \sin \theta$$

$$F = \text{W} \sin \theta$$

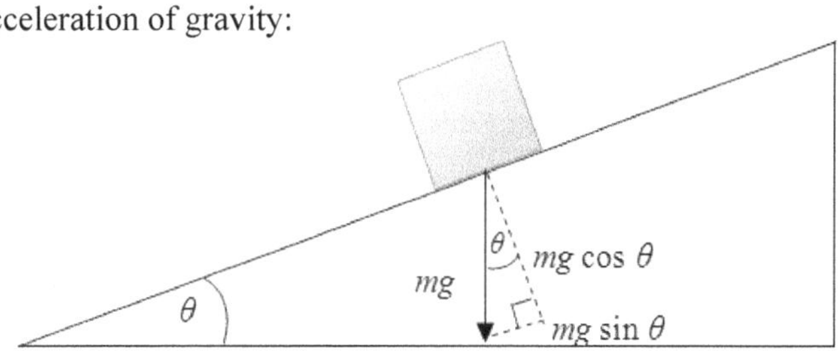

43. B is correct.

Velocity of sound in ideal gas:

$$v_0 = \sqrt{(\text{YRT} / \text{M})}$$

where Y = adiabatic constant, R = gas constant, T = temperature and M = molecular mass

If T is doubled:

$$v_1 = \sqrt{(\text{YR2T} / \text{M})}$$

$$v_1 = \sqrt{2}\, v_0$$

The velocity increases by $\sqrt{2}$

44. B is correct.

The flow rate must be constant along the flow because water is incompressible.

Volumetric Flow Rate is:

$\upsilon = Av$

Volumetric Flow Rate at different points in a pipe with different areas:

$A_1v_1 = A_2v_2$

Thus, the Volumetric Flow Rate at one point must match that of the other.

$A_2v_2 = 0.09\ m^3/s$

45. A is correct.

From the right-hand rule for a negative charge, the magnetic field causes the electron to curve by path X.

If the charge were positive, the charge would curve by path Z.

46. C is correct.

For the magnetic field around a current-carrying wire, the magnetic field strength is:

$B = (\mu_0 I) / (2\pi r)$

where B is the magnetic field strength, μ_0 is the magnetic constant, I is the current in the wire, and r is the distance of the particle from the wire

According to the equation for magnetic field strength, there are two ways to increase B:

1) increase the current I through the wire, or

2) decrease the distance r from the wire

Changing the particle's charge does not affect the field, nor does changing the particle's speed.

The field becomes weaker as the particle moves away from the wire (r increases).

47. C is correct.

Relationship between lens power and focal length:

$P = 1 / f$, with f expressed in meters

$f = 1 / P$

$f = 1/4$ m

The lens is converging because the focal length of ¼ m is more than 0.

48. E is correct.

An alpha particle is more massive than a beta particle and thus has more inertia.

An alpha particle deflects less in a magnetic field because its extra inertia requires more force to change its path.

49. A is correct.

The marble and sponge balls should fall at the same rate.

Air resistance causes more friction, increasing the time for the sponge ball to reach the ground.

50. D is correct.

Newton's Third Law of Motion: the gravitational pull of the Earth on the Moon is equal to the gravitational pull of the Moon on the Earth.

51. B is correct.

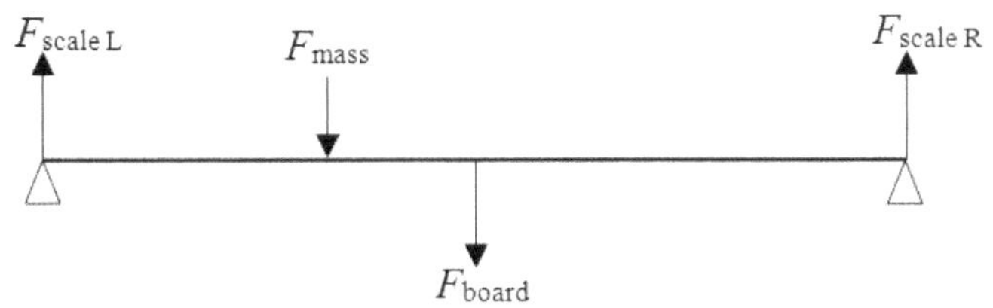

where $F_{\text{scale L}}$ and $F_{\text{scale R}}$ are forces from the scales

$F_{\text{mass}} = (4 \text{ kg})\cdot(9.8 \text{ m/s}^2)$

$F_{\text{mass}} = 39.2 \text{ N}$

$F_{\text{board}} = (16 \text{ kg})\cdot(9.8 \text{ m/s}^2)$

$F_{\text{board}} = 156.8 \text{ N}$

F_{board} acts at $L / 2$ because the board is uniform.

Movement of a force (torque) is:

$\tau = L \times F$

The board is in equilibrium, so the sum of torques = 0.

Taking torque about left side of the board:

$\Sigma\tau = 0$

$0 = (0)\cdot(F_{\text{scale L}}) + (L / 3)\cdot(-F_{\text{mass}}) + (L / 2)\cdot(-F_{\text{board}}) + (L)\cdot(F_{\text{scale R}})$

Note: F_{mass} and F_{board} are negative because they turn the board clockwise, which is designated negative.

$0 = (L / 3)\cdot(-39.2 \text{ N}) + (L / 2)\cdot(-156.8 \text{ N}) + (L)\cdot(F_{\text{scale R}})$

Divide by L:

$0 = (-39.2 \text{ N}) / 3 + (-156.8 \text{ N}) / 2 + (F_{\text{scale R}})$

$0 = -13.1 \text{ N} - 78.4 \text{ N} + F_{\text{scale R}}$

$F_{\text{scale R}} = (13.1 \text{ N} + 78.4 \text{ N})$

$F_{\text{scale R}} = 91.5 \text{ N}$

52. C is correct.

The KE of the car must be entirely dissipated by work due to friction.

The friction is static and has a constant normal force, so the frictional force is constant in both scenarios.

So:

$$\frac{1}{2}mv^2 = F_f \times d$$

where d is the distance of the skid

$$\frac{1}{2}mv_0^2 = F_f \times d_0$$

When the velocity is doubled,

$$d = \frac{1}{2}m(2v_0)^2 / F_f$$

$$d = \frac{1}{2}mv_0^2\,(4) / F_f$$

$$d = 4d_0$$

If the velocity is doubled, the car will skid 4 times further.

53. E is correct.

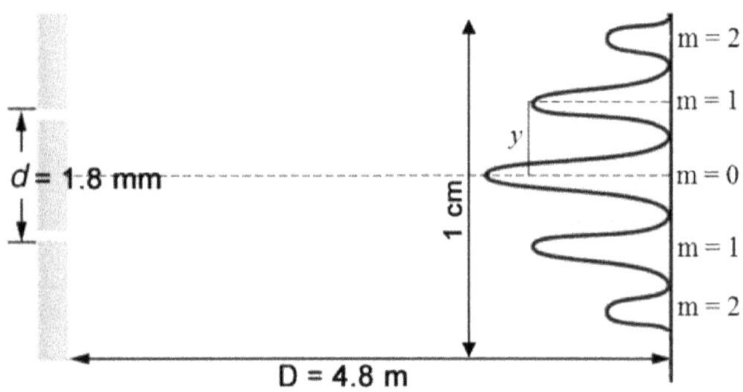

Double slit interference equation:

$$y = (m\lambda D) / d$$

$$\lambda = (yd) / (mD)$$

Because 5 bright spots are visible in 1 cm, the first order fringe m = 1 will be separated by a distance y from the central fringe m = 0.

$$y = (1\text{ cm}) / (5)$$

$$y = 0.2\text{ cm}$$

$$\lambda = (yd) / (mD)$$

Solve: $\lambda = (0.002\text{ m})\cdot(0.0018\text{ m}) / (1)\cdot(4.8\text{ m})$

$$\lambda = 7.5 \times 10^{-7}\text{ m}$$

$$\lambda = 750\text{ nm}$$

54. D is correct.

Heat required to melt a solid:

$Q_1 = mL_f$

$Q_1 = (0.4 \text{ kg})\cdot(334 \times 10^3 \text{ J/kg})$

$Q_1 = 133.6 \text{ kJ}$

Heat required to raise temperature of water from 0 °C to 60 °C:

$Q_2 = mc\Delta\text{T}$

$Q_2 = (0.4 \text{ kg})\cdot(4.186 \times 10^3 \text{ J/Kg}\cdot{}^\circ\text{C})\cdot(60\ {}^\circ\text{C})$

$Q_2 = 100.5 \text{ kJ}$

Total heat added:

$Q_1 + Q_2 = Q_{\text{total}}$

$Q_{\text{total}} = (133.6 \text{ kJ} + 100.5 \text{ kJ})$

$Q_{\text{total}} = 234.1 \text{ kJ}$

55. D is correct.

Lens equation:

$1 / f = 1 / d_i + 1 / d_o$

$1 / 20 \text{ cm} = 1 / d_i + 1 / 10 \text{ cm}$

$-1 / 20 \text{ cm} = 1 / d_i$

$d_i = -20 \text{ cm}$

The negative sign indicates the image should be on the object side of the mirror. By the sign convention for mirrors, a negative value places the image behind the mirror.

56. E is correct.

For a DC circuit, the slope of the current *vs.* power graph represents 1 / potential.

57. A is correct.

$\text{W} = Q\Delta\text{V}$ (for work or energy with charges)

The charge transferred is:

$Q = 10^{-10} \text{ C}$

A positive charge moving from 8,000 V to –8,000 V is a negative sign.

$\text{W} = (10^{-10} \text{ C})\cdot(-8{,}000 \text{ V} - 8{,}000 \text{ V})$

$\text{W} = (10^{-10} \text{ C})\cdot(-1.6 \times 10^4 \text{ V})$

$\text{W} = -1.6 \times 10^{-6} \text{ J}$

58. B is correct.

Archimedes Principle:

$$\rho_{object} / \rho_{fluid} = W_{object} / W_{fluid}$$

$$\rho_{object} / \rho_{fluid} = (11.3 \text{ g/cm}^3) / (13.6 \text{ g/cm}^3)$$

$$\rho_{object} / \rho_{fluid} = 0.83$$

83% of the lead ball is below the surface by weight.

The ball has 17% above the surface because the density is assumed to be consistent throughout the sphere.

The weight is directly correlated to volume.

59. D is correct.

For an approaching sound source, the Doppler equation becomes:

$$f_{observed} = [v / (v - v_{source})]\cdot(f_{source})$$

$$f_{observed} = [350 \text{ m/s} / (350 \text{ m/s} - 50 \text{ m/s})]\cdot(420 \text{ Hz})$$

$$f_{observed} = (1.17)\cdot(420 \text{ Hz})$$

$$f_{observed} = 490 \text{ Hz}$$

60. C is correct.

The period of a simple pendulum only depends on the length:

$$T = 2\pi\sqrt{(L / g)}$$

where L is the length of the pendulum and g is the acceleration due to gravity

Since the gravity on the Moon is 1/6 of that on Earth, the period of the pendulum on the Moon is:

$$T_M = 2\pi\sqrt{(L / (1/6)g)}$$

$$T_M = 2\pi\sqrt{(6L / g)}$$

$$T_M = (\sqrt{6})\cdot 2\pi\cdot\sqrt{(L / g)}$$

Since $T = 2\pi\sqrt{(L / g)}$, the period of the pendulum on the Moon can be rewritten as:

$$T_M = (\sqrt{6})T$$

$$T_M = (\sqrt{6})\cdot(3 \text{ s})$$

$$T_M = 7.3 \text{ s}$$

61. D is correct.

The gravitational force and the spring force add to zero; they are equal in magnitude (Newton's Second Law of Motion).

The magnitude of the force is:

$F_{grav} = mg$

$F_{grav} = (1.2 \text{ kg})\cdot(10 \text{ m/s}^2)$

$F_{grav} = 12 \text{ N}$

Use the spring equation:

$F_{spring} = kx$

$x = F_{spring} / k$

$x = (12 \text{ N}) / (3 \text{ N/m})$

$x = 4 \text{ m}$

Add the amount that the spring stretches to the resting length of 0.25 m.

Total length = (0.25 m + 4 m)

Total length = 4.25 m

62. C is correct.

The horizontal component of velocity is:

$v_x = v \cos \theta$

$v_x = (20 \text{ m/s}) \cos 25°$

$v_x = 18.1 \text{ m/s}$

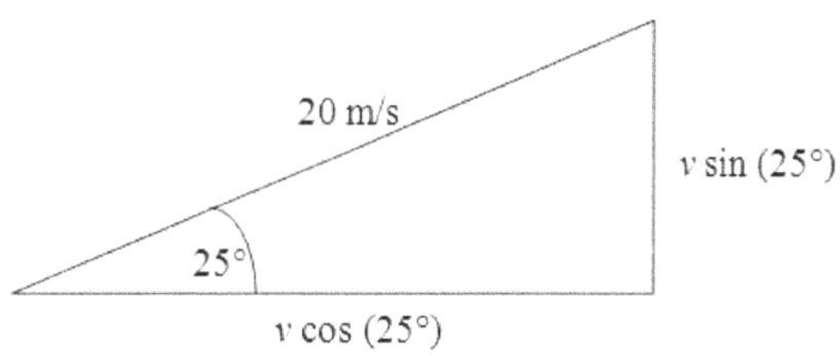

63. A is correct.

Force due to friction is expressed as:

$F_f = \mu F_N$

where μ = coefficient of friction and F_N = the normal force on the block

$F_N = W \cos \theta$

As θ decreases, $\cos \theta$ becomes larger.

F_N increases, and substituting F_N back into the equation for frictional force, the value of F_f increases.

Weight is the force of the mass due to gravity, both constant.

Therefore, weight remains constant.

64. B is correct.

If the velocity of an object is constant, then the instantaneous velocity at an arbitrary time equals the average velocity.

If the velocity increases or decreases at a constant rate, then the instantaneous velocity at an arbitrary time will not equal the average velocity.

65. A is correct.

Stefan-Boltzmann Law:

$$P = \varepsilon A\sigma T^4$$

$$P = (1)\cdot(1.25\ m^2)\cdot(5.67 \times 10^{-8}\ W/m^2K^4)\cdot(100 + 273\ K)^4$$

$$P = 1{,}371.9\ W \approx 1.4\ kW$$

66. C is correct.

$$t = d_x / v_x$$

$$t = (1{,}000\ m) / (50\ m/s)$$

$$t = 20\ s$$

Check that the projectile is still moving upwards during the duration of the allotted time:

$$v_{fy} = v_{iy} + at$$

$$v_{fy} = 240\ m/s + (-9.8\ m/s^2)t$$

$$v_{fy} = 240\ m/s + (-9.8\ m/s^2)\cdot(20\ s)$$

$$v_{fy} = 240\ m/s + (-196\ m/s)$$

$v_{fy} = 44$ m/s, positive value so projectile is still moving upwards

$$d_y = v_{iy}t + \tfrac{1}{2}at^2$$

$$d_y = (240\ m/s)\cdot(20\ s) + \tfrac{1}{2}(-9.8\ m/s^2)\cdot(20\ s)^2$$

$$d_y = (4{,}800\ m) + \tfrac{1}{2}(-9.8\ m/s^2)\cdot(400\ s^2)$$

$$d_y = (4{,}800\ m) - (1{,}960\ m)$$

$$d_y = 2{,}840\ m$$

67. B is correct.

An object having a moment of inertia *I* rotating with an angular speed of *ω* has angular momentum:

$L = I\omega$

The moment of inertia of a solid right circular cylinder about the axis indicated is $\frac{1}{2}mR^2$:

$L = (\frac{1}{2}mR^2)\omega$

$L = 0.5(15.0\ \text{kg})\cdot(1.4\ \text{m})^2\cdot(2.7\ \text{rad/s})$

$L = 39.69\ \text{kg m}^2/\text{s}$

$L = 40\ \text{kg m}^2/\text{s}$

68. C is correct.

Coulomb's law:

$F = kq_1q_2 / r^2$

If the charge and the separation distance are doubled:

$F = k(2q_1)\cdot(2q_2) / (2r)^2$

$F = 4kq_1q_2 / 4r^2$

$F = kq_1q_2 / r^2$ (same as the first expression)

69. D is correct.

For a diverging lens, regardless of the object position, the image is:

- virtual
- upright
- reduced

70. C is correct.

Work during an isothermal process:

$W = nRT \times \ln(V_f / V_i)$

$W = (4\ \text{mol})\cdot(8.134\ \text{J/mol}\cdot\text{K})\cdot(650\ \text{K}) \times \ln(0.33\ \text{m}^3 / 0.025\ \text{m}^3)$

$W = 56\ \text{kJ}$

71. D is correct.

Momentum:

$$p = mv$$

Sum momentum:

$$p_{tot} = m_1v_1 + m_2v_2 + m_3v_3$$

Objects moving left have negative velocity.

$$p_{tot} = 7 \text{ kg} \times 6 \text{ m/s} + 12 \text{ kg} \times (-3 \text{ m/s}) + 4 \text{ kg} \times 2 \text{ m/s}$$

$$p_{tot} = 42 \text{ kg·m/s} + (-36 \text{ kg·m/s}) + 8 \text{ kg·m/s}$$

$$p_{tot} = 14 \text{ kg·m/s}$$

72. B is correct.

Sound velocity in an ideal gas:

$$v_{sound} = \sqrt{(\text{y}RT / M)}$$

where y = adiabatic constant, R = gas constant, T = temperature and M = molecular mass of gas

Increasing the temperature increases the velocity of sound in air.

73. A is correct.

The Bernoulli Equation originated as a conservation of energy relationship for flowing fluids:

$$P_1 + ½\rho v_1^2 + \rho gh_1 = P_2 + ½\rho v_2^2 + \rho gh_2$$

where P_1 = pressure energy, $½\rho v_1^2$ = kinetic energy (volumetric) and ρgh_1 = potential energy (volumetric)

The Bernoulli Equation is a statement of the conservation of pressure energy, kinetic energy, and potential energy of a flowing fluid.

74. D is correct.

Resistance in a wire:

$$R = (\rho L) / A$$

where ρ = resistivity, L = length of wire and A = cross-sectional area of wire

If the first wire has resistance R_1:

$$R_1 = \rho L_1 / A_1$$

The second wire has:

$$L_2 = 2L_1$$

$$A_2 = 2A_1$$

continued…

$R_2 = \rho 2L_1 / 2A_1$

$R_2 = \rho L_1 / A_1$

$R_2 = R_1$

The resistances are equal.

75. E is correct.

Positron: ${}^{0}_{1}\beta^{+}$

The atomic mass remains the same, but the atomic number decreases.

76. B is correct.

The *Carnot cycle* is an example of a reversible process with two isothermal and two adiabatic processes and is the most efficient heat engine cycle.

77. E is correct.

$a = (v_2 - v_1) / \Delta t$

$a = [12 \text{ m/s} - (-6 \text{ m/s})] / 10 \text{ s}$

$a = 1.8 \text{ m/s}^2$

78. E is correct.

Force of gravity on a mass:

$F = mg$

Force of gravity on the 1 kg book:

$F_{\text{book}} = (1 \text{ kg})\cdot(9.8 \text{ m/s}^2)$

$F_{\text{book}} = 9.8 \text{ N}$

Force of gravity on the 10 kg brick:

$F_{\text{brick}} = (10 \text{ kg})\cdot(9.8 \text{ m/s}^2)$

$F_{\text{brick}} = 98 \text{ N}$

The force on the brick is 10 times that on the book.

79. D is correct.

Find acceleration:

$v_f = v_0 + at$

$106 \text{ m/s} = (0 \text{ m/s}) + a(0.85 \times 10^{-3} \text{ s})$

$106 \text{ m/s} = a(0.85 \times 10^{-3} \text{ s})$

$a = (106 \text{ m/s}) / (0.85 \times 10^{-3} \text{ s})$

$a = 1.25 \times 10^5 \text{ m/s}^2$

Find average force:

$F = ma$

$F = (0.6 \text{ kg})\cdot(1.25 \times 10^5 \text{ m/s}^2)$

$F = 7{,}500 \text{ N}$

80. E is correct.

During beta minus decay, the parent nuclide ejects an electron and electron antineutrino.

A neutron converts to a proton in the process, so the mass number remains the same, but the atomic number increases by 1.

$${}^{A}_{Z}X \rightarrow {}^{A}_{Z+1}Y + {}^{0}_{-1}e^{-} + {}^{0}_{0}\nu_e$$

81. B is correct.

The equation for displacement given an initial velocity and constant acceleration is:

$d = v_i t + ½at^2$

$a = (v_f - v_i) / t$

$a = (1.8 \text{ m/s} - 0.4 \text{ m/s}) / 4 \text{ s}$

$a = 0.35 \text{ m/s}^2$

Substituting $a = 0.35 \text{ m/s}^2$ into the equation for displacement,

$d = v_i t + ½at^2$

$d = (0.4 \text{ m/s})\cdot(4 \text{ s}) + ½(0.35 \text{ m/s}^2)\cdot(4 \text{ s})^2$

$d = 4.4 \text{ m}$

82. C is correct.

The object experiencing the force increases in speed due to the acceleration from the force:

$$F = ma$$

However, because the force decreases, *the rate of increase of its speed* (not the speed itself) decreases.

83. E is correct.

$$Q = mc\Delta T$$

$$c = Q / m\Delta T$$

$$c = (150 \text{ kcal}) / [(3 \text{ kg})\cdot(200 \text{ °C})]$$

$$c = 0.25 \text{ kcal/kg}\cdot\text{°C}$$

84. A is correct.

$$W = F \times d \cos \theta$$

$$F = W / d$$

$$F = 40 \text{ J} / 4 \text{ m}$$

$$F = 10 \text{ N}$$

85. B is correct.

Equation for the period T of a simple pendulum:

$$T = 2\pi\sqrt{(L / g)}$$

Rearranging the equation for g:

$$g = (2\pi / T)^2L$$

$$g = (4\pi^2)L / T^2$$

$$g = 4\pi^2(0.58 \text{ m}) / (2.5 \text{ s})^2$$

$$g = 3.7 \text{ m/s}^2$$

86. C is correct.

Electromagnetic waves are created by accelerating electric charges (change in velocity or speed).

87. A is correct.

Speed of the wave = wavelength × frequency

$$v = \lambda f$$

$$v = (0.5\ \text{m})\cdot(800\ \text{Hz})$$

$$v = 400\ \text{m/s}$$

Distance = velocity × time

The distance covered in one second is:

$$d = vt$$

$$d = (400\ \text{m/s})\cdot(1\ \text{s})$$

$$d = 400\ \text{m}$$

88. A is correct.

In Elena's inertial frames of reference, one before the spaceship increased its speed and one after, the spaceship is at rest.

Therefore, the measurements of length must give the same result. Otherwise, the laws of physics would be different in these two inertial frames of reference.

During the acceleration of the spaceship, there are some unusual effects caused by the acceleration.

Those effects are the subject of General Relativity.

89. E is correct.

Connecting two identical batteries in parallel yields the same voltage but twice the total available charge.

90. D is correct.

The power of the combination of the lenses is the sum of the powers.

$$5\ \text{D} + 3\ \text{D} = 8\ \text{D}$$

91. B is correct.

A nucleon is a proton or neutron.

An alpha particle consists of two neutrons and two protons.

A hydrogen atom only has one proton, so it cannot emit an alpha particle.

92. D is correct.

$m_1v_1 = m_2v_2$

$(1{,}450\text{ kg})v_1 = (90\text{ kg})\cdot(30\text{ m/s})$

$v_1 = (90\text{ kg})\cdot(30\text{ m/s}) / (1{,}450\text{ kg})$

$v_1 = 1.9\text{ m/s}$

93. C is correct.

work = force × distance

$W = Fd$

Machines reduce the amount of force required, but the distance through which the force acts increases. Machines do not reduce the amount of work required.

If 900 J of energy is expended to lift the block to a certain height, then, even if a perfectly efficient pulley were used, the engineer expends a total of 900 J of work to lift the block to the same height using the pulley.

94. A is correct.

amplitude = ½(total displacement)

amplitude = ½(0.4 m)

amplitude = 0.2 m

The amplitude is the displacement from the position of equilibrium.

The amplitude is 0.2 m because the mass travels 0.2 m to the left of equilibrium and 0.2 m to the right of equilibrium in one oscillation.

95. C is correct.

One wave is completed at the 3rd node.

From the figure, the given wave has 1.5 λ:

$L = 1.5\ \lambda$

$\lambda = 2/3\ L$

$\lambda = (2/3)\cdot(\text{total length})$

$\lambda = (2/3)\cdot(0.6\text{ m})$

$\lambda = 0.4\text{ m}$

$v = f\lambda$

$1\text{ Hz} = 1\text{ s}^{-1}$

$v = (900\text{ s}^{-1})\cdot(0.4\text{ m})$

$v = 360\text{ m/s}$

96. E is correct.

density = mass / volume

$\rho = m / \text{V}$

$\rho = \text{kg/m}^3$

97. C is correct.

The charge q is equally attracted to both Qs because q is equidistant from each.

$q_{\text{net}} = 0$

$F_{1\to2} = kq_1q_2 / r^2$

$F_{2\to1} = -kq_1q_2 / r^2$

One force is attractive to the first Q, but the other is attractive to the second Q and, therefore, points in the opposite direction.

The sign of the second force is reversed.

$F_{\text{net}} = 0$

98. B is correct.

A simple LC circuit with 0 resistance acts as an oscillator with charge flowing back and forth across the capacitor and inductor.

99. B is correct.

The Compton effect measures the reduction in wavelength of an x-ray beam as it scatters off a sample.

The reduction in wavelength can be explained if the incident x-ray beam is considered as individual particles, each with momentum and energy.

Electrons scatter those particles in the sample.

Hence, the Compton effect demonstrates the particle nature of electromagnetic radiation.

While the energy content and momenta of the individual x-rays are included in the scattering calculation, they are not directly demonstrated in the Compton effect.

100. A is correct.

The orbital 2*d* does not exist.

The order of the electron orbital energy levels (least to greatest), is:

1*s*, 2*s*, 2*p*, 3*s*, 3*p*, 4*s*, 3*d*, 4*p*, 5*s*, 4*d*, 5*p*, 6*s*, 4*f*, 5*d*, 6*p*, 7*s*, 5*f*, 6*d*, 7*p*

Notes for active learning

Notes for active learning

Diagnostic Test 3 – Answer Key and Detailed Explanations

Answer Key

1	**D**	Heat & thermodynamics	**26**	**E**	Work & energy
2	**A**	Force, motion, gravitation	**27**	**B**	Sound
3	**B**	Sound	**28**	**C**	Rotational motion
4	**A**	Fluids & solids	**29**	**D**	Light & optics
5	**E**	Electrostatics & electromagnetism	**30**	**E**	Kinematics & dynamics
6	**B**	Electric circuits	**31**	**A**	Equilibrium & momentum
7	**C**	Light & optics	**32**	**B**	Waves & periodic motion
8	**D**	Atomic & nuclear structure	**33**	**E**	Fluids & solids
9	**B**	Rotational motion	**34**	**D**	Electric circuits
10	**B**	Special relativity	**35**	**C**	Atomic & nuclear structure
11	**B**	Equilibrium & momentum	**36**	**D**	Heat & thermodynamics
12	**C**	Work & energy	**37**	**C**	Kinematics & dynamics
13	**E**	Waves & periodic motion	**38**	**C**	Special relativity
14	**A**	Heat & thermodynamics	**39**	**B**	Equilibrium & momentum
15	**A**	Light & optics	**40**	**C**	Atomic & nuclear structure
16	**B**	Electric circuits	**41**	**D**	Kinematics & dynamics
17	**D**	Electrostatics & electromagnetism	**42**	**C**	Force, motion, gravitation
18	**E**	Fluids & solids	**43**	**D**	Heat & thermodynamics
19	**C**	Sound	**44**	**A**	Work & energy
20	**E**	Waves & periodic motion	**45**	**C**	Waves & periodic motion
21	**D**	Work & energy	**46**	**B**	Sound
22	**C**	Equilibrium & momentum	**47**	**A**	Fluids & solids
23	**B**	Quantum mechanics	**48**	**C**	Electrostatics & electromagnetism
24	**A**	Kinematics & dynamics	**49**	**E**	Electric circuits
25	**C**	Heat & thermodynamics	**50**	**C**	Light & optics

51	E	Atomic & nuclear structure	76	C	Electric circuits
52	D	Equilibrium & momentum	77	A	Electrostatics & electromagnetism
53	B	Work & energy	78	C	Fluids & solids
54	A	Waves & periodic motion	79	E	Sound
55	B	Sound	80	C	Waves & periodic motion
56	D	Fluids & solids	81	A	Work & energy
57	A	Electrostatics & electromagnetism	82	E	Equilibrium & momentum
58	C	Electric circuits	83	E	Force, motion, gravitation
59	E	Light & optics	84	D	Kinematics & dynamics
60	A	Atomic & nuclear structure	85	C	Heat & thermodynamics
61	A	Heat & thermodynamics	86	E	Work & energy
62	E	Force, motion, gravitation	87	D	Sound
63	C	Sound	88	D	Electrostatics & electromagnetism
64	A	Fluids & solids	89	A	Light & optics
65	B	Electrostatics & electromagnetism	90	D	Kinematics & dynamics
66	D	Electric circuits	91	D	Quantum mechanics
67	B	Light & optics	92	B	Waves & periodic motion
68	C	Atomic & nuclear structure	93	D	Fluids & solids
69	B	Kinematics & dynamics	94	E	Electric circuits
70	E	Force, motion, gravitation	95	B	Atomic & nuclear structure
71	D	Equilibrium & momentum	96	C	Heat & thermodynamics
72	B	Work & energy	97	A	Kinematics & dynamics
73	A	Waves & periodic motion	98	E	Force, motion, gravitation
74	E	Heat & thermodynamics	99	A	Equilibrium & momentum
75	C	Light & optics	100	B	Atomic & nuclear structure

1. D is correct.

Because the path *ca* is adiabatic, no heat is added or lost by the system during this process.

Adiabatic process:

$$Q = 0 \text{ kJ}$$

2. A is correct.

$$\Sigma F_x = 0$$

$$0 = F_{Ax} + F_{Bx} + F_{Cx} + F_x$$

$$0 = (30 \text{ N})\cdot(\sin 35°) - (40 \text{ N})\cdot(\cos 25°) + (50 \text{ N})\cdot(\cos 40°) + F_x$$

$$0 = (30 \text{ N})\cdot(0.57) - (40 \text{ N})\cdot(0.91) + (50 \text{ N})\cdot(0.77) + F_x$$

$$0 = (17 \text{ N} - 36 \text{ N} + 39 \text{ N}) + F_x$$

$$F_x = -20 \text{ N}$$

3. B is correct.

The 4th harmonic has four nodes (i.e., three more than the fundamental), and the ends are antinodes.

The harmonic to wavelength relationship for an organ pipe is calculated by:

$$\lambda_n = 2L / n$$

where L is the length of the pipe and n is the harmonic number (1, 2, 3,…)

For n = 4:

$$\lambda_4 = (2)\cdot(0.2 \text{ m}) / 4$$

$$\lambda_4 = 0.1 \text{ m}$$

4. A is correct.

Volumetric flow rate = velocity × area

$$V = vA$$

Water is incompressible, so volumetric flow rate is constant.

$$(vA)_{in} = (vA)_{out}$$

$$(0.02 \text{ m/s})\cdot(\pi/4)\cdot(0.15 \text{ m})^2 = v_{out}(\pi/4)\cdot(0.003 \text{ m})^2$$

$$v_{out} = [(0.02 \text{ m/s})\cdot(0.15 \text{ m})^2] / (0.003 \text{ m})^2$$

$$v_{out} = 50 \text{ m/s}$$

5. E is correct.

Microwaves and blue light are electromagnetic radiation.

The difference between them is with respect to their frequency, wavelength, and corresponding energy.

$$c = \lambda f$$

$$E = hf$$

6. B is correct.

The internal resistance is added to the external resistance of the circuit.

If the external resistance is large, the current is small, and the internal resistance can be ignored.

7. C is correct.

A plane mirror does not magnify the image and always produces an erect, virtual image.

$$\mathrm{m} = 1$$

$$1 = d_i / d_o$$

$$1 = h_i / h_o$$

8. D is correct.

An *alpha particle* is essentially a helium nucleus and consists of two protons and two neutrons.

Alpha decay: the parent nuclide sheds two protons and two neutrons.

$${}^{A}_{Z}\mathrm{X} \rightarrow {}^{A-4}_{Z-2}\mathrm{Y} + {}^{4}_{2}\alpha$$

9. B is correct.

The angular speed changes according to the kinematic relation:

$$\Delta\omega = \alpha\Delta t$$

$$\Delta t = \Delta\omega / \alpha$$

The torque is given, and the moment of inertia of a right circular cylinder is $½mR^2$.

Find the angular acceleration from the dynamic relation:

$$\tau = I\alpha$$

$$\alpha = \tau / I = \tau / (½mR^2)$$

$$\alpha = 2\tau / mR^2$$

Combining the two results:

$$\Delta t = mR^2\Delta\omega / 2\tau$$

$$\Delta t = (10.0\ \mathrm{kg})\cdot(3.00\ \mathrm{m})^2\cdot(8.13\ \mathrm{rad/s}) / [2(110.0\ \mathrm{N\ m})]$$

$$\Delta t = 3.33\ \mathrm{s}$$

10. B is correct.

The speed of the missile as measured in the frame of the asteroid is the relativistic addition of the speed of the spaceship as measured from the asteroid, plus the speed of the missile as measured from the spaceship.

The two velocities are in the opposite direction.

The formula for the relativistic addition of two speeds in opposite directions is:

$\beta = (\beta_1 - \beta_2) / (1 - \beta_1\beta_2)$

$\beta = (0.8 - 0.5) / [1 - (0.8)(0.5)]$

$\beta = 0.50$

$v = 0.50c$

11. B is correct.

Impulse:

$J = F\Delta t$

Increase in Δt = decrease in F

Decreased force means that it is less likely to break.

Impulse and change in momentum are directly related.

$F\Delta t = m\Delta v$

Because the wine glass goes from terminal velocity to zero, the mass is constant, and the impulse is constant and equal in both cases.

Thus, the only way to decrease force is to increase Δt, which is the stopping time.

12. C is correct.

Work equation:

$W = Fd \cos \theta$

θ is perpendicular to direction of velocity

$\cos 90° = 0$

$Fd \cos \theta = 0$ J, no work is done against the force of gravity.

Since the velocity is constant, there is no net horizontal force, so no net work is done by horizontal forces.

13. E is correct.

velocity = wavelength × frequency

$v = \lambda f$

For velocity to remain constant, as frequency decreases, the wavelength must increase.

14. A is correct.

$\Delta L / L_i = \alpha_L \Delta T$

$\alpha_L = \Delta L / L_i \Delta T$

$\alpha_L = (2.0005 \text{ m} - 2.0000 \text{ m}) / (2.0000 \text{ m})\cdot(40 \text{ °C} - 20 \text{ °C})$

$\alpha_L = 0.0005 \text{ m} / 40 \text{ m}\cdot\text{°C}$

$\alpha_L = 1.25 \times 10^{-5} \text{ K}^{-1}$

15. A is correct.

$1 / f = 1 / d_i + 1 / d_o$

$1 / 12 \text{ m} = 1 / d_i + 1 / 6 \text{ m}$

$1 / 12 \text{ m} - 2 / 12 \text{ m} = 1 / d_i$

$-1 / 12 \text{ m} = 1 / d_i$

$d_i = -12 \text{ m}$

A negative sign indicates that the image is behind the mirror.

16. B is correct.

The total resistance for a circuit with parallel resistors decreases as more resistors are added.

A decrease in the total resistance increases the total current in the circuit.

17. D is correct.

$E = kQ / r^2$

E is directly proportional to Q and inversely proportional to r^2:

$2Q = 2E$

The only solution that doubles the magnitude of E is to double Q.

18. E is correct.

If an object's entire volume is suspended in a liquid, the object's density is equal to that of the liquid.

If 20% of the floating buoy is above the surface of the liquid, 80% of its volume is suspended in the liquid.

Since 80% of the buoy's volume is in the liquid, it is only 80% as dense as the liquid.

19. C is correct.

The wave of the fundamental frequency for an open pipe is shown in the figure.

Therefore:

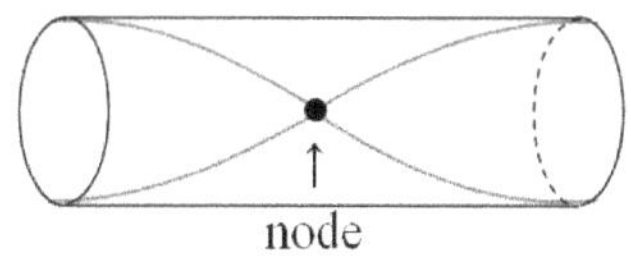

$\lambda = 2L$

The two endpoints are antinodes, and the midpoint is a node.

20. E is correct.

Wave interference occurs for all types of waves, both mechanical and electromagnetic.

21. D is correct.

$F = -kx$

$-k = F / x$

$-k = 30\ \text{N} / -0.75\ \text{m}$

$k = 40\ \text{N/m}$

22. C is correct.

Since the two objects come to rest immediately, they have equal but opposite momenta before the collision:

$p_A = p_B$

$p = mv$

$m_A v_A = m_B v_B$

$v_A / v_B = m_B / m_A$

v_A is 5 times greater than v_B

$v_A / v_B = 5$

$5 = m_B / m_A$

take the reciprocal of each side

$m_A / m_B = 1 / 5$

The ratio of the mass of A to the mass of B is 1: 5

23. B is correct.

The de Broglie wavelength of a matter wave is:

$\lambda = h / p$

$\lambda = h / (m\gamma v)$

Since $v / c < 0.1$, this will equal about 1.005 and can be ignored

$\lambda = (6.626 \times 10^{-34}\ \text{J·s}) / [(9.11 \times 10^{-31}\ \text{kg})·(2.5 \times 10^{7}\ \text{m/s})]$

$\lambda = 29 \times 10^{-12}\ \text{m}$

$\lambda = 29\ \text{pm}$

24. A is correct.

$v_{Parallel} = v \sin \theta$

The component of velocity in the direction of the slope is:

$v_P = v \sin \theta$

where $\theta = 20°$

The velocity parallel to the slope:

$v_P = v \sin \theta$

$v_P = v \sin 20°$

$v_P = (16\ \text{m/s})·(0.34)$

$v_P = 5.47 \approx 5.5\ \text{m/s}$

25. C is correct.

In *isothermal* compression, heat is released to the surroundings equal to the work needed to compress the gas.

$Q_{released} = W$

$W = nRT \ln (V_2 / V_1)$

26. E is correct. All three answer choices are correct because they all describe PE:

$PE = mgh$

PE equals the work done by the pile driver:

$PE = W$

$PE = Fd$

$PE / d = F$

The force is directly proportional to PE, so all three choices are correct because they relate to the PE of the piledriver.

27. B is correct.

An organ pipe is open at both ends or closed at one end.

Open pipe:

$f_n = nv / 2L$

where n = 1, 2, 3…

Closed pipe:

$f_n = nv / 4L$

where n = 1, 3, 5…

Observing the equation for frequencies produced by an open or closed pipe, there is an infinite number of frequencies possible depending on the value of n.

28. C is correct.

The gravitational force is the only influence making the satellite move in a circular path, so the centripetal acceleration in orbit is the gravitational acceleration at that orbital radius.

$v^2 / r = g'$

$v = \sqrt{(g'r)}$

$v = \sqrt{[(2.3 \text{ m/s}^2)\cdot(34{,}000 \text{ m})]}$

$v = 280 \text{ m/s}$

29. D is correct.

A diopter is a unit of measurement of the optical power of a lens or a curved mirror.

The power equals the reciprocal of the focal length (i.e., $1 / f$). Therefore, the diopter is a unit of reciprocal length.

For example, a 3 diopter lens brings parallel rays of light to focus at 1/3 m.

$1 / d_o + 1 / d_i = 1 / f$

Setting object at infinity (∞):

$1 / \infty + 1 / d_i = 1 / f$

$0 + 1 / d_i = 1 / f$

$1 / d_i = 1 / f$

Power of the lens:

$P = 1 / f$

$3 = 1 / d_i$

$d_i = 1 / 3 \text{ m} = 0.33 \text{ m}$

30. E is correct.

Accelerating uniformly for 8 s:

$\Delta v = a\Delta t$

$v_2 - v_1 = a\Delta t$

$v_2 = v_1 + a\Delta t$

$v_2 = 0 \text{ m/s} + (2.5 \text{ m/s}^2)\cdot(8 \text{ s})$

$v_2 = 20 \text{ m/s}$

31. A is correct. Set North as the positive direction.

Conservation of momentum:

$(m_1 + m_2)v = m_1v_1 + m_2v_2$

$(m_1 + m_2)v = (2{,}500 \text{ kg})\cdot(7 \text{ m/s}) + (2{,}000 \text{ kg})\cdot(-14 \text{ m/s})$

$(m_1 + m_2)v = 17{,}500 \text{ kg·m/s} + (-28{,}000 \text{ kg·m/s})$

$(m_1 + m_2)v = -10{,}500 \text{ kg·m/s}$

$-10{,}500 \text{ kg·m/s} = (4{,}500 \text{ kg})v$

Velocity immediately after the inelastic collision:

$v = (-10{,}500 \text{ kg·m/s}) / (4{,}500 \text{ kg})$

$v = -2.3$ m/s, the minus sign indicates v points South.

32. B is correct.

$\text{T} = 2\pi\sqrt{(L / g)}$

If the clock is too slow, that means the period (T) is too large.

Decrease T by raising the weight, which effectively decreases L.

33. E is correct.

When the air above the paper is blown, the air moves faster, and the decrease in pressure causes the air below the paper to fill that void, pushing the paper upward in the process.

34. D is correct.

voltage = current × resistance

$\text{V} = IR$

$4 \text{ V} = I(20 \ \Omega)$

$I = 4 \text{ V} / 20 \ \Omega$

$I = 0.2 \text{ A}$

35. C is correct.

Becquerel (Bq) is the SI unit of radioactivity.

One Bq is the activity of a quantity of radioactive material in which one nucleus decays per second (s^{-1}).

36. D is correct.

As the temperature rises, the mercury and glass will expand by:

$$\Delta V = \alpha V_i \Delta T$$

Thus, in order to reach 450 ml the expansion of the glass and mercury must be taken into account:

$$V_G + \Delta V_G = V_M + \Delta V_M$$

$$450 \text{ ml} + \Delta V_G = 442 \text{ ml} + \Delta V_M$$

$$8 \text{ ml} = \Delta V_M - \Delta V_G$$

$$8 \text{ ml} = \alpha_M V_{iM} \Delta T - \alpha_G V_{iG} \Delta T$$

$$8 \text{ ml} = \Delta T(\alpha_M V_{iM} - \alpha V_{iG})$$

$$8 \text{ ml} = \Delta T[(18 \times 10^{-5} \text{ K}^{-1})\cdot(442 \text{ ml}) - (2 \times 10^{-5} \text{ K}^{-1})\cdot(450 \text{ ml})]$$

$$8 \text{ ml} = \Delta T(0.0706 \text{ K}^{-1} \text{ ml})$$

$$\Delta T = 113.4 \text{ °C}$$

$$T_f = T_i + \Delta T$$

$$T_f = 22 \text{ °C} + 113.4 \text{ °C}$$

$$T_f = 135 \text{ °C}$$

37. C is correct.

A constant positive acceleration increases velocity linearly upwards over time.

Graphs with a positive slope represents a constant positive acceleration on a velocity *vs.* time graph.

38. C is correct.

This is one of the two postulates of special relativity.

The other postulate of special relativity is that the speed of light is the same in all frames moving with constant velocity with respect to an inertial frame.

39. B is correct.

Torque can be written as:

$$\tau = I\alpha$$

where τ = torque, I = moment of inertia and α = angular acceleration

Thus, if torque is constant, the angular acceleration must be constant because the moment of inertia does not change.

40. C is correct.

Dosimeter badges are used to monitor cumulative radiation dose from gamma radiation.

Typically, they do not measure beta or alpha radiation as they cannot penetrate the badges due to their lower energies.

41. D is correct.

Let $y = 0$ be the ground, which is the final position of the package.

Then:

$$y_f = y_i + v_i\Delta t - \tfrac{1}{2}g(\Delta t)^2$$

$$0 = d + v_i\Delta t - \tfrac{1}{2}g(\Delta t)^2$$

$$d = v_i\Delta t + \tfrac{1}{2}g\Delta t^2$$

$$d = (-15\text{ m/s})\cdot(16\text{ s}) + \tfrac{1}{2}(10\text{ m/s}^2)\cdot(16\text{ s})^2$$

$$d = (-240\text{ m} + 1{,}280\text{ m})$$

$$d = 1{,}040\text{ m}$$

42. C is correct.

$$F = ma$$

$$m = F / a$$

$$m = 3{,}000\text{ N} / (2\text{ m/s}^2)$$

$$m = 1{,}500\text{ kg}$$

43. D is correct.

If metal A has a larger coefficient of thermal expansion, it will elongate more than metal B when the strip is heated.

Because the two strips are bound, the bimetallic strip will curve downward when heated as the configuration gives strip A more room for expansion.

44. A is correct.

$F = -kx$

The negative sign is only by convention.

The negative sign indicates that the spring resting force is opposite the stretch direction.

$x = F / k$

$x = mg / k$

$x = [(30 \text{ kg})\cdot(10 \text{ m/s}^2)] / 4{,}600 \text{ N/m}$

$x = 0.0652 \text{ m}$

$\text{PE} = \frac{1}{2}kx^2$

$\text{PE} = \frac{1}{2}(4{,}600 \text{ N/m})\cdot(0.0652 \text{ m})^2$

$\text{PE} = 9.8 \text{ J}$

An increase in PE due to the addition of the weight.

45. C is correct.

Frequency = # of cycles / time

$f = 2 \text{ cycles} / 1 \text{ s}$

$f = 2 \text{ s}^{-1}$

$v = \lambda f$

$v = (6 \text{ m})\cdot(2 \text{ s}^{-1})$

$v = 12 \text{ m/s}$

46. B is correct.

The *decibel* rating for sound is based on a logarithmic scale.

Sound intensity equation:

$\text{I (dB)} = 10 \log_{10}(\text{I} / \text{I}_0)$

where $\text{I}_0 = 10^{-12} \text{ W/m}^2$ and is the threshold of human hearing

Example:

sound intensity decreases from 10 W/m^2 to 1 W/m^2

$\text{I}_1 \text{ (dB)} = 10 \log_{10}(10 \text{ W/m}^2 / 10^{-12} \text{ W/m}^2)$

$\text{I}_1 = 130 \text{ dB}$

$\text{I}_2 \text{ (dB)} = 10 \log_{10}(1 \text{ W/m}^2 / 10^{-12} \text{ W/m}^2)$

continued…

$I_1 = 120$ dB

$I_1 - I_2 = 130$ dB – 120 dB

$I_1 - I_2 = 10$ dB

Thus, the sound intensity decreases by 10 dB if the intensity decreases by a factor of 10.

47. A is correct.

Buoyancy force:

$$F_B = \rho g V$$

where V is the volume of water displaced

When the two blocks are stacked on each other:

$$F_B = \rho g V_0 = F_1 + F_2$$

When the two blocks are in the water:

$$F_{B1} = \rho g V_1 = F_1$$

$$F_{B2} = \rho g V_2 = F_2$$

Total volume displaced in both cases:

$$V_0 = (1/\rho g)\cdot(F_1 + F_2)$$

$$V_1 + V_2 = (1/\rho g)\cdot(F_1 + F_2)$$

$$V_0 = V_1 + V_2$$

Same displacement of water in both cases, so the water level does not change.

48. C is correct.

To find # excess electrons, divide Q_1 by the charge of a single electron:

excess electrons = Q_1 / e

excess electrons = $(-1 \times 10^{-6}\,\text{C}) / (-1.60 \times 10^{-19}\,\text{C})$

excess electrons = 6.3×10^{12}

49. E is correct.

Capacitance only depends on the geometry (i.e., surface area) of a conductor and the permittivity of the dielectric used.

50. C is correct.

A flat plane mirror has no curvature, so the radius is considered to be infinite.

51. E is correct.

$m = m_0 y$

where m_0 = rest mass, y = Lorentz factor and m = relativistic mass

$m = m_0 / \sqrt{(1 - v^2 / c^2)}$

$m = (1.675 \times 10^{-24}\ \text{g}) / \sqrt{[1 - (54{,}300 \times 10^3\ \text{m/s})^2/(3.0 \times 10^8\ \text{m/s})^2]}$

$m = (1.675 \times 10^{-24}\ \text{g}) / \sqrt{[1 - (2.948 \times 10^{15}\ \text{m}^2/\text{s}^2)/(9.0 \times 10^{16}\ \text{m}^2/\text{s}^2)]}$

$m = (1.675 \times 10^{-24}\ \text{g}) / \sqrt{[1 - (0.032)]}$

$m = (1.675 \times 10^{-24}\ \text{g}) / 0.984$

$m = 1.702 \times 10^{-24}\ \text{g}$

Because the speed of a particle is so low compared to light, the relativistic mass is very close to the rest mass.

This value shows how relatively high the speeds must be before physics deviate significantly from classical calculations.

52. D is correct.

The moment of inertia:

I = mass × the square of the distance from the rotational axis

$I = mr^2$

Since $r_1 = r_2$,

I_1 / I_2 is simply the ratio of the masses.

Therefore, the moment of inertia I is directly proportional to the mass m.

Doubling the mass doubles the moment of inertia, giving a ratio of 2 : 1 comparing the more massive to the less massive sphere.

53. B is correct.

power = work / time

$P = W / t$

$P = (1{,}000\ \text{J}) / (40\ \text{s})$

$P = 25\ \text{W}$

54. A is correct.

A spring mass oscillator has a frequency:

$$\omega = \sqrt{(k / m_0)}$$

$$f = \omega / 2\pi$$

$$f = \sqrt{(k / m_0)} / 2\pi$$

Period:

$$T = 1 / f$$

$$T_0 = 2\pi / \sqrt{(k / m_0)}$$

If mass is doubled:

$$T = 2\pi / \sqrt{(k / 2\ m_0)}$$

$$T = 2\pi / \sqrt{(k / m_0)} \times \sqrt{2}$$

$$T = \sqrt{2}\ T_0$$

The period increases by a factor of $\sqrt{2}$.

55. B is correct.

For a pipe or tube that is closed at one end and open at the other end:

harmonic wavelengths λ_n have values of n, which are only odd (n = 1, 3, 5, …).

After the fundamental wavelength (n = 1), the next consecutive harmonic wavelength is the third harmonic (n = 3).

The harmonic wavelength can be determined from the equation:

$$\lambda = 4L / n$$

$$\lambda = 4(1.5\ m) / 3$$

$$\lambda = 2\ m$$

56. D is correct.

$$\rho = m / V$$

$$V = (4/3)\pi r^3$$

$$\rho = m / (4/3)\pi^3$$

$$\rho = (115\ kg) / (4/3)\pi \times (0.6\ m)^3$$

$$\rho = 127\ kg/m^3$$

57. A is correct.

When a potential difference exists across a wire (for example connecting a battery to both ends of the wire), electrons flow to move from high potential to low potential.

For a battery, the electrons flow along the wire towards the positive terminal to lower their potential.

58. C is correct.

$1\ A = 1\ C/s$

59. E is correct.

For a concave spherical mirror:

$r = 2f$

If the object is located at the radius of curvature:

$r = d_o$

$d_o = 2f$

Use the mirror equation:

$1 / f = 1 / d_o + 1/ d_i$

$1 / f = 1 / 2f + 1 / d_i$

$1 / 2f = 1 / d_i$

$d_i = 2f = d_o$, the image and object are located at the same point.

Use the magnification equation:

$m = -d_i / d_o$

$m = h_i / h_o$

$-(2f) / 2f = h_i / h_o$

$-1 = h_i / h_o$

$-h_i = h_o$

If the object is located at the radius of curvature, then the image distance is equal to the object distance, and the image is inverted because the height is the negative of the object height.

60. A is correct.

Alpha particles (α) cannot penetrate human skin and requires only heavy paper or clothing as minimum protective shielding.

Ionizing power and *penetrating power* are inversely related.

Alpha particles have the lowest penetrating power but have the highest ionizing power, and this is due to their high mass and charge.

61. A is correct.

An isobaric process is a process in which the pressure remains constant throughout.

From the diagram, the volume changes, but the pressure is constant, consistent with an isobaric process.

62. E is correct.

$F = ma$, on Earth

$20\ \text{N} = m(4\ \text{m/s}^2)$

$m = 20\ \text{N} / (4\ \text{m/s}^2)$

$m = 5\ \text{kg}$

$\text{W} = ma$, on Moon

$\text{W} = (5\ \text{kg})\cdot(1.62\ \text{m/s}^2)$

$\text{W} = 8.1\ \text{N}$

63. C is correct.

Frequency is the amount of cycles that a wave makes in a certain amount of time.

$f = \#\ \text{cycles / time}$

The figure shows the wave making 2 cycles in 4 seconds.

$f = 2\ \text{cycles} / 4\ \text{s}$

$f = 0.5\ \text{s}^{-1} = 0.5\ \text{Hz}$

64. A is correct.

$½\rho v^2 = \rho gh$

cancel ρ from both sides of the expression

$½v^2 = gh$

$v^2 = 2gh$

$v^2 = 2(9.8\ \text{m/s}^2)\cdot(0.8\ \text{m})$

$v^2 = 15.68\ \text{m}^2/\text{s}^2$

$v = 3.95\ \text{m/s} \approx 4\ \text{m/s}$

65. B is correct.

Faraday's Law:

$$V = N\Delta BA / \Delta t$$

where N = number of turns, B = magnetic field, A = area of coil and t = time

The *magnetic flux* is the product of the magnetic field strength and area of the coil exposed to the field:

$$\text{magnetic flux} = BA$$

If the magnetic flux and the number of turns are doubled, then the numerator increases by a factor of four, and the induced emf increases by a factor of four.

66. D is correct.

$$P = I^2R$$

$$I^2 = P / R$$

$$I^2 = 16\text{ W} / 18\ \Omega$$

$$I^2 = 0.89\text{ A}^2$$

$$I = 0.94\text{ A}$$

67. B is correct.

$$c = \lambda f$$

$$f = c / \lambda$$

Wavelength and frequency are inversely proportional.

68. C is correct.

Ionizing power and penetrating power are inversely related.

Alpha particles have the lowest penetrating power but highest ionizing power due to their high mass and charge.

69. B is correct.

Given the angular velocity, the linear velocity at a given radius is calculated by:

$$v = \omega r$$

where v is the linear velocity and ω is the angular velocity

Convert from revolutions/minute to radians/second to arrive at an answer in m/s.

Since 1 revolution = 2π radians and 1 min = 60 sec, angular velocity in radians/s is given by:

$$\omega = (8.3\text{ rev/min})\cdot(2\pi\text{ rad/rev})\cdot(1\text{ min}/60\text{ s})$$

$$\omega = 0.87\text{ rad/s}$$

continued…

Calculate linear velocity.

The problem gives the diameter, but the formula requires radius:

$v = (0.87\ \text{rad/s})\cdot(9\ \text{m})$

$v = 7.8\ \text{m/s}$

70. E is correct.

The magnitude of the instantaneous force a ball hits a bat if a batter hits a ball pitched with a 1,700 N instantaneous force equals 1,700 N.

71. D is correct.

In an elastic collision, KE and momentum are conserved:

KE before collision = KE after collision

$\text{Block}_1 + \text{Block}_2 = \text{Block}_1 + \text{Block}_2$

$\frac{1}{2}m_1v_1^2 + \frac{1}{2}m_2v_2^2 = \frac{1}{2}m_1u_1^2 + \frac{1}{2}m_2u_2^2$

Initially,

$v_2 = 0$

$\frac{1}{2}m_1v_1^2 = \frac{1}{2}m_1u_1^2 + \frac{1}{2}m_2u_2^2$

Solve for u_2:

$[m_1(v_1^2 - u_1^2) / m_2]^{\frac{1}{2}} = u_2$

Momentum:

$m_1v_1 = m_1u_1 + m_2u_2$

Solve for u_2:

$m_1(v_1 - u_1) / m_2 = u_2$

Set equal and isolate m_2:

$[m_1(v_1^2 - u_1^2) / m_2]^{\frac{1}{2}} = m_1(v_1 - u_1) / m_2$

$m_2 = m_1(v_1 - u_1)^2 / (v_1^2 - u_1^2)$

Solve for the m_2:

$m_2 = [(2.8\ \text{kg})\cdot(8.5\ \text{m/s} - (-1.1\ \text{m/s}))^2\,] / [(8.5\ \text{m/s})^2 - (-1.1\ \text{m/s})^2\,]$

$m_2 = [(2.8\ \text{kg})\cdot(92\ \text{m}^2/\text{s}^2\,)] / (71\ \text{m}^2/\text{s}^2\,)$

$m_2 = 3.6\ \text{kg}$

72. B is correct.

Work is defined as:

$$W = Fd \cos \theta$$

where d is the displacement in the direction of the force.

In this case, it is –h, where the minus sign indicates that the direction of the displacement is down

The force is the force of gravity:

$$F = -mg$$

where the minus sign indicates that the force is directed downward.

Therefore:

$$W = Fd$$

$$W = (-mg)\cdot(-h)$$

$$W = mgh$$

73. A is correct. Frequency = 1 / period

$$f = 1 / T$$

$$f = 1.6 \text{ kHz}$$

$$v = \lambda f$$

$$v = (0.25 \text{ m})\cdot(1.6 \text{ kHz})$$

$$v = (0.25 \text{ m})\cdot(1{,}600 \text{ s}^{-1})$$

$$v = 400 \text{ m/s}$$

74. E is correct.

60 mL = 60 g water

$$Q = (mc\Delta T)_{beaker} + (mc\Delta T)_{water}$$

Change in temperature is the same.

$$Q = \Delta T[(m_{beaker})\cdot(c_{beaker}) + (m_{water})\cdot(c_{water})]$$

$$2{,}200 \text{ cal} = (25 \text{ °C})\cdot[(m_{beaker})(0.18 \text{ cal/g·°C}) + (60 \text{ g})\cdot(1 \text{ cal/g·°C})]$$

$$2{,}200 \text{ cal} / 25 \text{ °C} = (m_{beaker})\cdot(0.18 \text{ cal/g·°C}) + 60 \text{ cal/°C}$$

$$88 \text{ cal/°C} = (m_{beaker})\cdot(0.18 \text{ cal/g·°C}) + 60 \text{ cal/°C}$$

$$m_{beaker} = (88 \text{ cal/°C} - 60 \text{ cal/°C}) / (0.18 \text{ cal/g·°C})$$

$$m_{beaker} = (28 \text{ cal/°C}) / (0.18 \text{ cal/g·°C})$$

$$m_{beaker} = 156 \text{ g}$$

75. C is correct.

Spherical mirror equation:

$$f = -r / 2$$

The negative is only included to imply convergence of the lens.

$$f = r / 2$$

$$r = 2f$$

$$r = (2)\cdot(20 \text{ cm})$$

$$r = 40 \text{ cm}$$

76. C is correct.

Find R_{eq} for entire circuit:

R_1:  in series:

2 Ω 1 Ω

$$R_1 = 2\ \Omega + 1\ \Omega$$

$$R_1 = 3\ \Omega$$

R_2:  in series:

5Ω 1 Ω

$$R_2 = 5\ \Omega + 1\ \Omega$$

$$R_2 = 6\ \Omega$$

R_1 and R_2 are in parallel:

$$1 / R_3 = 1 / R_1 + 1 / R_2$$

$$1 / R_3 = 1 / 3\ \Omega + 1 / 6\ \Omega$$

$$1 / R_3 = 2 / 6\ \Omega + 1 / 6\ \Omega$$

$$1 / R_3 = 1 / 2\ \Omega$$

$$R_3 = 2\ \Omega$$

R_3 is in series with the 4 Ω resistor:

$$R_{total} = 2\ \Omega + 4\ \Omega$$

$$R_{total} = 6\ \Omega$$

Calculate the current around the entire circuit:

$$V = IR$$

$$12 \text{ V} = I(6\ \Omega)$$

$$I = 2 \text{ A}$$

continued…

Calculate voltage drop across parallel section using R_3 as equivalent resistor:

$V = IR$

$V = (2\text{ A})\cdot(2\ \Omega)$

$V = 4\text{ V}$

*Note: voltage drop is equal across parallel branches.

Calculate current running through the branch the 2 Ω is on:

$V = IR$

$4\text{ V} = I(3\ \Omega)$

$I = 1.33\text{ A}$

Calculate power dissipated by the 2 Ω resistor:

$P = I^2R$

$P = (1.33\text{ A})^2\cdot(2\ \Omega)$

$P = 3.5\text{ W}$

77. A is correct.

$F = qv \times B$

The charge density ρ and the charge q are properties of the particle and do not change.

Because of the cross product, the force is always perpendicular to the velocity and thus perpendicular to the displacement of the charge over an infinitesimal interval of time.

Since the force is always perpendicular to the displacement, no work is done, so kinetic energy will not change.

Since the force is always perpendicular to the velocity, the force is centripetal, and the motion is a uniform circular motion, so the magnitude of the acceleration does not change.

The force does change the direction of the velocity, and thus the vector velocity itself.

78. C is correct.

gauge pressure $= P_{absolute} - P_{atm}$

$\rho gh = P_{absolute} - P_{atm}$

gauge pressure $= \rho gh$

gauge pressure $= (10^3\text{ kg/m}^3)\cdot(9.8\text{ m/s}^2)\cdot(10\text{ m})$

gauge pressure $= 9.8 \times 10^4$ Pa

79. E is correct.

The distance between two adjacent nodes is = $\frac{1}{2}\lambda$.

$\frac{1}{2}\lambda = 75$ cm

$\lambda = 150$ cm

80. C is correct.

$v = \lambda f$

$\lambda = v / f$

$\text{Hz} = \text{s}^{-1}$

$\lambda = (4{,}900\ \text{m/s}) / (640\ \text{s}^{-1})$

$\lambda = 7.7$ m

81. A is correct.

There is no force acting in the direction of motion (constant velocity), so no work is done.

$\text{W} = Fd \cos \theta$

Stacey performed no work because the mass experienced no acceleration due to its constant velocity, assuming no friction.

$F = ma$

$F = (15\ \text{kg})\cdot(0\ \text{m/s}^2)$

$F = 0$ N

$\text{W} = (0\ \text{N})\cdot(100\ \text{m})$

$\text{W} = 0$ J

82. E is correct.

The force delivered by the block to her hand and her hand to the block are equal by Newton's Third Law.

The time of impact is the same for the block and the hand.

Impulse is calculated by:

$J = F\Delta t$

Thus, if the force and time are equal for the block and the hand, they experience the same impulse.

83. E is correct.

The textbooks are following a straight path.

By turning the car's steering wheel, Karen pulls her car door into the path of the textbooks, giving the illusion that the textbooks have a force acting on them.

84. D is correct.

$d = (v_f^2 - v_i^2) / 2a$

$d = [(21\text{ m/s})^2 - (5\text{ m/s})^2] / 2(4\text{ m/s}^2)$

$d = [(441\text{ m}^2/\text{s}^2) - (25\text{ m}^2/\text{s}^2)] / (8\text{ m/s}^2)$

$d = (416\text{ m}^2/\text{s}^2) / (8\text{ m/s}^2)$

$d = 52\text{ m}$

85. C is correct.

$\Delta L / L_0 = \alpha\Delta T$

$\Delta T = \Delta L / \alpha L_0$

$\Delta T = 3 \times 10^{-3}\text{ m} / (30 \times 10^{-6}\text{ K}^{-1})\cdot(10\text{ m})$

$\Delta T = 10\text{ K}$

86. E is correct.

$F = ma$

$a = F / m$

$a = 10\text{ N} / 2\text{ kg}$

$a = 5\text{ m/s}^2$

$v_f = v_i + at$

$v_f = 0\text{ m/s} + (5\text{ m/s})\cdot(15\text{ s})$

$v_f = 75\text{ m/s}$

$\text{KE} = ½mv^2$

$\text{KE} = ½(2\text{ kg})\cdot(75\text{ m/s})^2$

$\text{KE} = 5{,}625\text{ J}$

87. D is correct.

The *beat frequency* is twice per second, or 2 Hz.

$f_{beat} = |f_1 - f_2|$

$2\text{ Hz} = |680\text{ Hz} - f_2|$

$f_2 = 678\text{ Hz or }682\text{ Hz}$

88. D is correct.

$F = ma$

$a = F / m$

The force is an inverse-square with respect to distance.

$F = (kqQ) / r^2$

F decreases as q moves away from Q, and F decreases asymptotically (never reaching zero).

The charges move away from each other because both are positive.

89. A is correct.

For a diverging lens, regardless of the object position, the image is:

- virtual
- upright
- reduced

90. D is correct.

The combined vectors A and B, when vector A has a length of 5 units and is directed to the North (N), and vector B has a length of 11 units and is directed to the South (S) is:

5 N + (–11) S = –6 S = 6 units, S

91. D is correct.

The energy uncertainty is given by the uncertainty principle:

$\Delta E \Delta t \geq h / 2\pi$

$\Delta E \geq (6.626 \times 10^{-34}\ \text{J·s} / 2\pi) / (30 \times 10^{-12}\ \text{s})$

$\Delta E \geq (3.5 \times 10^{-24}\ \text{J})$

Converting to eV (1.6×10^{-19} J /eV):

$\Delta E \geq 2.2 \times 10^{-5}\ \text{eV}$

92. B is correct.

Ignoring the dissipative forces due to friction, the total mechanical energy of a pendulum is conserved.

Gravitational PE is continuously converted into KE and vice versa.

The total mechanical energy (i.e., KE + PE) remains constant.

93. D is correct.

Volumetric flow rate is constant:

$v_1A_1 = v_2A_2$

$v_1 / v_2 = A_2 / A_1$

$v_1 / v_2 = (\pi / 4\ D_2^2) / (\pi / 4\ D_1^2)$

$v_1 / v_2 = (8\ cm)^2 / (3\ cm)^2$

$v_1 / v_2 = 7$

$v_1 = 7v_2$

If the diameter increases, then the speed decreases by a factor of 7.

94. E is correct.

Changing the current in a nearby wire changes the magnetic field around it and induces a voltage in the wire close to it by Faraday's Law.

Moving a magnet close to the wire or the wire close to a magnetic field changes the magnetic field around the wire and induces a voltage by Faraday's Law.

95. B is correct.

Ionizing radiation is dangerous to living tissue because it alters the chemical structure of atoms or molecules.

96. C is correct.

Isothermal process changes entropy:

$\Delta s = Q / T$

An isothermal process has work equal to the heat added.

$Q = W$

$\Delta s = W / T$

$W = T\Delta s$

$W = (273\ K)\cdot(2.6\ J/K)$

$W = 710\ J$

$W = 7.1 \times 10^2\ J$

97. A is correct.

The rock starts at the velocity Jack throws it, which is downward and, therefore, negative.

The velocity continues to increase negatively due to acceleration from gravity, which is constant.

Thus, the velocity is linear with a downward slope.

$$v_f = v_i + at$$

$$y = mx + b$$

98. E is correct.

By the definition of *impulse*:

$$F\Delta t = m\Delta v$$

Therefore:

$$F = m\Delta v/\Delta t$$

The problem is asking to find the force on the block.

Proceed by focusing on the block, first finding Δv, and then using the relation above to find the force.

To determine Δv, calculate velocity after the collision using conservation of momentum:

$$m_i v_i = (m_i + m_f)v_f$$

$$v_f = m_i v_i / (m_i + m_f)$$

$$v_f = (0.1\text{ kg})\cdot(50\text{ m/s}) / (0.1\text{ kg} + 0.9\text{ kg})$$

$$v_f = (5\text{ kg·m/s}) / 1\text{ kg}$$

$$v_f = 5\text{ m/s}$$

$$F\Delta t = m(v_f - v_i)$$

$$F = m(v_f - v_i) / \Delta t$$

$$F = (0.9\text{ kg})(5\text{ m/s} - 0\text{ m/s}) / 0.01\text{ s}$$

$$F = (4.5\text{ kg m/s}) / (0.01\text{ s})$$

$$F = 450\text{ N}$$

99. A is correct.

Use *conservation of momentum*:

$$m_A v_{Ai} + m_B v_{Bi} = m_A v_{Af} + m_B v_{Bf}$$

$$(2\text{ kg})\cdot(0.6\text{ m/s}) + 0 = 0 + (2.5\text{ kg})v_{Bf}$$

$$v_{Bf} = (1.2\text{ kg·m/s}) / (2.5\text{ kg})$$

$$v_{Bf} = 0.48\text{ m/s}$$

100. B is correct.

Radioactivity is when a nucleus of an unstable atom loses energy by emitting ionizing radiation (e.g., α particles, β particles, and γ rays).

A *rad* is a unit of absorbed radiation dose.

Roentgen (rem) is a related unit to the rad. It is used to quantify the number of rads deposited into a target after radiation exposure.

Becquerel (Bq) is the SI unit of radioactivity. One Bq is defined as the activity of a quantity of radioactive material in which one nucleus decays per second (s^{-1}).

Curie (Ci) is a non-SI unit of radioactivity. 1 Ci = 3.7×10^{10} decays per second.

Sievert (Sv) is the SI unit for a dosage of ionizing radiation and measures the health effect of low levels of ionizing radiation on the human body.

Notes for active learning

Glossary of Physics Terms

A

Absolute humidity (or saturation value) – the maximum amount of water vapor that could be present in 1 m^3 of air at a given temperature.

Absolute magnitude – a classification scheme that compensates for the differences in the distance to stars; calculates the brightness that stars would appear to have if they were at a defined, standard distance of 10 parsecs.

Absolute scale – temperature scale set so that zero is the theoretically lowest temperature possible (this would occur when random motion of molecules has ceased).

Absolute zero – the theoretically lowest temperature possible, at which temperature the molecular motion vanishes; –273.16 °C or 0 K.

Absorptance – the ratio of the total absorbed radiation to the total incident radiation.

Acceleration – the rate of change of velocity of a moving object with respect to time; the SI units are m/s^2; by definition, this change in velocity can result from a change in speed, a change in direction, or a combination of changes in speed and direction.

Acceleration due to gravity – the rate of change in velocity produced in a body due to the Earth's attraction; denoted by the letter g (SI unit – m/s^2); on the surface of the Earth, its average value is 9.8 m/s^2; increases when going towards the poles from the equator; decreases with altitude and with depth inside the Earth; the value of g at the center of the Earth is zero.

Achromatic – capable of transmitting light without decomposing it into its constituent colors.

Acid rain – rainwater with a pH of less than 5.7 is acid rain. It is caused by the gases NO_2 (from car exhaust fumes) and SO_2 (from the burning of fossil fuels) dissolving in the rain. Acid rain kills fish, kills trees, and destroys buildings and lakes.

Acoustics – the science of the production, transmission, and effects of sound.

Acoustic shielding – a sound barrier that prevents the transmission of acoustic energy.

Adiabatic – any change in which there is no gain or loss of heat.

Adiabatic cooling – the decrease in temperature of an expanding gas that involves no additional heat flowing out of the gas; the cooling from the energy lost by expansion.

Adiabatic heating – the increase in temperature of the compressed gas that involves no additional heat flowing into the gas; the heating from the energy gained by compression.

Afocal lens – a lens of zero convergent power whose focal points are infinitely distant.

Air mass – a large, uniform body of air with nearly identical temperature and moisture conditions throughout.

Albedo – the fraction of the total light incident on a reflecting surface, especially a celestial body, which is reflected in all directions.

Allotropic forms – elements with several structures with different physical properties (e.g., graphite and diamond).

Alpha (α) particle – the nucleus of a helium atom (i.e., two protons and two neutrons) emitted as radiation from a decaying heavy nucleus (α-decay).

Alternating current – the flow of charge (i.e., current) traveling in one direction for one-hundredth of a second but the opposite direction for the next hundredth of a second. An electric current that first moves in one direction, then in the opposite direction with a regular frequency.

Amorphous – term that describes solids that have neither definite form nor structure.

Amp – unit of electric current; equivalent to coulomb/second.

Ampere – the unit amp; the SI unit of electric current; one ampere is the flow of one coulomb of charge per second.

Amplitude – the maximum absolute value attained by the disturbance of a wave or by any quantity that varies periodically. The height of the wave crest above the average position.

Amplitude (of an oscillation) – the maximum displacement of a body from its mean position during an oscillatory motion.

Amplitude (of waves) – the maximum displacement of particles of the medium from their mean positions during the propagation of a wave.

Angle of contact – the angle between tangents to the liquid surface and the solid surface inside the liquid; both tangents are drawn at the point of contact.

Angle of incidence – the angle of an incident (arriving) ray or particle to a surface; measured from a line perpendicular to the surface (the normal).

Angle of reflection – the angle of a reflected ray or particle from a surface; measured from a line perpendicular to the surface (the normal).

Angle of refraction – the angle between the refracted ray and the normal.

Angle of repose – the degree of inclination of a plane with the horizontal such that a body placed on the plane is on the verge of sliding but does not.

Angstrom – a unit of length; 1 = 10^{-10} m.

Angular acceleration – the rate of change of angular velocity of a body moving along a circular path; denoted by a.

Angular displacement – the angle described at the center of the circle by a moving body along a circular path. It is measured in radians.

Angular momentum (or *moment of momentum*) – the cross-product of position vector and momentum.

Angular momentum quantum number – from the quantum mechanics model of the atom, one of four descriptions of the energy state of an electron wave; describes the energy sublevels of electrons within the principal energy levels of an atom.

Angular velocity – the rate of change of angular displacement per unit of time.

Annihilation – when a particle and an antiparticle combine and release their rest energies in other particles.

Antineutrino – the antiparticle of neutrino; has zero mass and spin ½.

Archimedes principle – a body immersed in a fluid experiences an apparent loss of weight equal to the weight of the fluid displaced by the body.

Area – the amount of surface enclosed within the boundary lines.

Astronomical unit – the radius of the Earth's orbit is defined as one astronomical unit (A.U.).

Atom – the smallest unit of an element that can exist alone or in combination with other elements.

Atomic mass unit – relative mass unit (amu) of an isotope based on the standard of the ^{12}carbon isotope; one atomic mass unit (1 amu) = 1/12 the mass of a ^{12}C atom = 1.66×10^{-27} Kg.

Atomic number – the number of protons in the nucleus of an atom.

Atomic weight – weighted average of the masses of stable isotopes of an element as they occur in nature; based on the abundance of each isotope of the element and the atomic mass of the isotope compared to ^{12}carbon.

Avogadro's number – the number of ^{12}carbon atoms in exactly 12.00 g of C is 6.02×10^{23} atoms or other chemical units; the number of chemical units in one mole of a substance.

Avogadro's Law – under the same temperature and pressure conditions, equal volumes of gases contain an equal number of molecules.

Axis – the imaginary line about which a planet or other object rotates.

B

Background radiation – ionizing radiation (e.g., alpha, beta, gamma rays) from natural sources.

Balanced forces – when some forces act on a body and the resultant force is zero; see *Resultant forces.*

Balmer lines – lines in the spectrum of the hydrogen atom in the visible range; produced by the transition between a high energy level to n = 2, with n being the principal quantum number.

Balmer series – a set of four-line spectra; narrow lines of color emitted by hydrogen atom electrons as they drop from excited states to the ground state.

Bar – a unit of pressure; equal to 10^5 Pascal.

Barometer – an instrument that measures atmospheric pressure; used in weather forecasting and determining elevation above sea level.

Baryon – subatomic particle composed of three quarks.

Beat – a phenomenon of the periodic variation in sound intensity due to the superposition of waves differing slightly in frequency; rhythmic increases and decreases of volume from constructive and destructive interference between two sound waves of slightly different frequencies.

Bernoulli's theorem – states that the total energy per unit volume of a non-viscous, incompressible fluid in a streamline flow will remain constant.

Beta (β) particle – high-energy electron emitted as ionizing radiation from a decaying nucleus (β-decay); also a *beta ray.*

Big bang theory – the current model of galactic evolution in which the universe is assumed to have been created by an intense and brilliant explosion from a primeval fireball.

Binding energy – the net energy required to break a nucleus into its constituent protons and neutrons; also, the energy equivalent released when a nucleus is formed.

Biomass – the chemical energy stored in fast growing plants.

Black body – an ideal body which would absorb all incident radiation and reflect none.

Black body radiation – electromagnetic radiation emitted by an ideal material (the black body) perfectly absorbs and perfectly emits radiation.

Black hole – the remaining theoretical core of a supernova that is so dense that even light cannot escape.

Bohr model – model of the structure of the atom that attempted to correct the deficiencies of the solar system model and account for the Balmer series.

Boiling point – the temperature at which a phase change of liquid to gas occurs through boiling; the same temperature as the condensation point.

Boundary – the division between two regions of differing physical properties.

Boyle's Law – for a given mass of a gas at a constant temperature, the volume of the gas is inversely proportional to the pressure.

Brewster's Law – states that the refractive index of a material is equal to the tangent of the polarizing angle for the material.

British thermal unit (Btu) – the amount of energy or heat needed to increase the temperature of one pound of water one-degree Fahrenheit.

Brownian motion – the continuous random motion of solid microscopic particles suspended in a fluid medium due to their ongoing bombardment by atoms and molecules.

Bulk modulus of elasticity – the ratio of normal stress to the volumetric strain produced in a body.

Buoyant force – the upward force on an object immersed in a fluid.

C

Calorie – a unit of heat; 1 Calorie = 4.186 joule.

Candela – the SI unit of luminous intensity defined as the luminous intensity in each direction of a source that emits monochromatic photons of frequency 540×10^{12} Hz and has a radiant intensity in that direction of 1/683 W/sr.

Capacitance – the ratio of the charge stored per increase in potential difference.

Capacitor – an electrical device used to store charge and energy in the electrical field.

Capillarity – the rise or fall of a liquid in a tube of a fine bore.

Carnot's theorem – no engine operating between two temperatures can be more efficient than a reversible engine working between the same two temperatures.

Cathode rays – negatively charged particles (electrons) emitted from a negative terminal in an evacuated glass tube.

Celsius scale of temperature – the ice-point is taken as the lower fixed point (0 °C), and the steam-point is taken as the upper fixed point (100 °C); the interval between the ice-point and the steam-point is divided into 100 equal divisions; the unit division on this scale is 1 °C; previously called the centigrade scale; the relationship relates the temperatures on the Celsius scale and the Fahrenheit scale, C/100 = (F – 32) / 180; the temperature of a healthy person is 37 °C or 98.6 °F.

Center of gravity – the point through which all the weight of an object appears to act.

Centrifugal force – an apparent outward force on an object in circular motion; a consequence of the third law of motion.

Centripetal force – the radial force required to keep an object moving in a circular path.

Chain reaction – a self-sustaining reaction where some of the products can produce more reactions of the same kind (e.g., in a nuclear chain reaction, neutrons are the products that produce more nuclear reactions in a self-sustaining series).

Charles' Law – for a given mass of a gas at constant pressure, the volume is directly proportional to the temperature.

Chromatic aberration – an optical lens defect causing color fringes due to the lens bringing different colors of light to focus at different points.

Circular motion – the motion of a body along a circular path.

Closed system – the system which cannot exchange heat or matter with the surroundings.

Coefficient of areal expansion – the fractional change in surface area per degree of temperature change; see *Coefficient of thermal expansion.*

Coefficient of linear expansion – the fractional change in length per degree of temperature change; see *Coefficient of thermal expansion.*

Coefficient of thermal expansion – the fractional change in the size of an object per degree of change in temperature at constant pressure; the SI unit is K^{-1}.

Coefficient of volumetric expansion – the fractional change in volume per degree of temperature change; see *Coefficient of thermal expansion.*

Coherent source – when there is a constant phase difference between waves emitted from different parts of the source.

Combustion (or *burning*) – the combining of a substance with oxygen.

Compass – a magnet, which is free to rotate and indicate direction.

Complementary colors – two colors which, when mixed, give white. For examples: blue + yellow = white; red + cyan = white and green + magenta = white.

Compound – a substance made up of two or more elements chemically combined.

Compression – a part of a longitudinal wave in which the density of the particles of the medium is higher than the typical density.

Compressive stress – a force that tends to compress the surface as the Earth's plates move into each other.

Concave lens – a lens that spreads out light rays.

Condensation – the changing of a gas to a liquid state.

Condensation (sound) – a compression of gas molecules; a pulse of increased density and pressure that moves through the air at the speed of sound.

Condensation (water vapor) – where more vapor or gas molecules are returning to the liquid state than are evaporating.

Condensation nuclei – tiny particles such as dust, soot, or salt crystals suspended in the air on which water condenses.

Condensation point – the temperature at which a gas or vapor changes back to liquid; see *Boiling point.*

Conduction – the transfer of heat from a region of higher temperature to a region of lower temperature by increased kinetic energy moving from molecule to molecule, without the movement of the solid.

Constructive interference – the condition in which two waves are arriving at the same place at the same time and in phase add amplitudes to create a new wave.

Control rods – material inserted between fuel rods in a nuclear reactor to absorb neutrons and control the rate of the nuclear chain reaction.

Convection – transfer of heat from a region of higher temperature to a region of lower temperature by the displacement of high-energy molecules. For example, the displacement of warmer, less dense air (higher kinetic energy) by cooler, denser air (lower kinetic energy).

Conventional current – the opposite of electron current; considers an electric current to consist of a drift of positive charges that flow from the positive terminal to the negative terminal of a battery.

Convex lens – a lens that brings light rays together.

Corrosion – an undesired process where a metal is converted to one of its compounds (e.g., rusting).

Coulomb – a unit used to measure the quantity of electric charge; equivalent to the charge resulting from the transfer of 6.24 billion particles such as the electron.

Coulomb's Law – the relationship between charge, distance, and magnitude of the electrical force between two bodies; the force between two charges is directly proportional to the product of charges and inversely proportional to the square of the distance between the charges.

Covalent bond – a chemical bond formed by the sharing of a pair of electrons.

Covalent compound – chemical compound held by covalent bonds.

Crest – the point of maximum positive displacement on a transverse wave.

Critical angle – the limit to the angle of incidence when all light rays are reflected internally.

Critical mass – the mass of fissionable material needed to sustain a chain reaction.

Current – a flow of charge. Unit is Ampere (A).

Curvilinear motion – the motion of a body along a curved path.

Cycle – a complete vibration.

Cyclotron – a device used to accelerate the charged particles.

D

De-acceleration – negative acceleration when the velocity of a body decreases with time.

Decibel – unit of sound level; if P_1 & P_2 are two amounts of power, the first is said to be n decibels greater, where n = 10 log10 (P_1/P_2).

Decibel scale – a nonlinear scale of loudness based on the ratio of the intensity level of a sound to the intensity at the hearing threshold.

Density – the mass of a substance per unit volume.

Destructive interference – the condition in which two waves arriving at the same point at the same time out of phase add amplitudes that cancel to create zero total disturbance; see *constructive interference*.

Dewpoint temperature – the temperature at which condensation begins.

Dew – condensation of water vapor into droplets of liquid on surfaces.

Diffraction – the bending of light around the edge of an opaque object.

Diffuse reflection – light rays reflected in many random directions, as opposed to the parallel rays reflected from a perfectly smooth surface such as a mirror.

Diopter – unit of measure of the refractive power of a lens.

Direct current – an electrical current that always flows in one direction only (i.e., from the positive terminal to the negative terminal).

Direct proportion – when two variables increase or decrease in the same ratio (at the same rate).

Dispersion – the splitting of white light into its component colors of the spectrum.

Displacement – a vector quantity for the change in the position of an object as it moves in a particular direction; also the shortest distance between the initial position and the final position of a moving body.

Distance – a scalar quantity for the length of the path traveled by a body irrespective of its direction.

Distillation – the vaporization of a liquid by heating and then the condensation of the vapor by cooling.

Doppler effect – an apparent change in the frequency of sound or light due to the relative motion between the source of the sound or light and the observer.

E

Echo – a reflected sound distinguished from the original sound, usually arriving 0.1 s or more after the original sound.

Einstein mass-energy relation – $E = mc^2$; E is the energy released, m is the mass defect, and c is the speed of light.

Elastic potential energy – the potential energy of a body by its configuration (i.e., shape).

Elastic strain – an adjustment to stress in which materials recover their original shape after stress is released.

Electric circuit – consists of a voltage source that maintains an electrical potential, a continuous conducting path for a current to follow, and a device where the electrical potential does work; a switch in the circuit is used to complete or interrupt the conducting path.

Electric current – the flow of electric charge; the electric force field produced by an electrical charge.

Electric field line – an imaginary curve tangent to which at any given point gives the direction of the electric field at that point.

Electric field lines – a map of an electric field representing the direction of the force that a test charge would experience; the direction of an electric field shown by lines of force.

Electric generator – a mechanical device that uses wire loops rotating in a magnetic field to produce electromagnetic induction to generate electricity.

Electric potential energy – potential energy due to the position of a charge near other charges.

Electrical conductors – materials that have electrons free to move throughout the material (e.g., metals); allows electric current to flow through the material.

Electrical energy – a form of energy from electromagnetic interactions.

Electric force – a fundamental force that results from the interaction of electrical charges; it is the most powerful force in the universe.

Electrical insulators – electrical nonconductors, or materials that obstruct the flow of electric current.

Electrical nonconductors – materials with electrons that do not move easily within the material (e.g., rubber); also called *electrical insulators.*

Electrical resistance – the property of opposing or reducing electric current.

Electrode – a conductor, which dips into an electrolyte and allows the electrons to flow to and from the electrolyte.

Electrolysis – the production of a chemical change using electricity. Electrolysis can be used to split up water into hydrogen and oxygen.

Electrolyte – water solution of ionic substances that will conduct an electric current.

Electromagnet – a magnet formed by a solenoid that can be turned on and off by turning the current on and off.

Electromagnetic force – one of four fundamental forces; the force of attraction or repulsion between two charged particles.

Electromagnetic induction – the process in which current is induced in a coil whenever there is a change in the magnetic flux linked with the coil.

Electromagnetic waves – the waves due to oscillating electrical and magnetic fields and do not need any material medium for their propagation; can travel through a material medium (e.g., light waves and radio waves); travel in a vacuum with a speed 3.0×10^8 m/s.

Electron – a subatomic particle that has the smallest negative charge possible; usually found in an orbital of an atom but is gained or lost when atoms become ions.

Electron configuration – the arrangement of electrons in orbits and sub-orbits about the nucleus of an atom.

Electron current – the opposite of conventional current; considers electric current to consist of a drift of negative charges that flows from the negative terminal to the positive terminal of a battery.

Electron pair – a pair of electrons with different spin quantum numbers that may occupy an orbital.

Electron volt – the energy gained by an electron moving across a potential difference of 1 V equals 1.60×10^{-19} Joules.

Electronegativity – the comparative ability of atoms of an element to attract bonding electrons.

Electroplating – where metal is covered with a layer of another metal using electricity.

Electrostatic charge – an accumulated electric charge on an object from a surplus of electrons or a deficiency of electrons.

Element – a pure chemical substance that cannot be broken down into anything simpler by chemical or physical means; there are over 100 known elements, the fundamental materials of which matter is made.

Endothermic process – the process in which heat is absorbed.

Energy – the capacity of a body to do work; a scalar quantity; the SI unit is the Joule; there are five forms: mechanical, chemical, radiant, electrical, and nuclear.

Equilibrium – the condition of a system when neither its state of motion nor its internal energy state tends to change with time; a balanced object is in equilibrium.

Escape velocity – the minimum velocity with which an object must be thrown upward to overcome the gravitational pull and escape into space; the escape velocity depends on the mass and radius of the planet/star, but not on the mass of the body being thrown upward.

Evaporation – a process of more molecules leaving a liquid for the gaseous state than returning from the gas to the liquid; can occur at any given temperature from the surface of a liquid; takes place from the surface of the liquid; causes cooling; faster if the surface of the liquid is large, the temperature is higher, and the surrounding atmosphere does not contain a large amount of vapor of the liquid.

Exothermic process – the process in which heat is evolved.

F

Fahrenheit scale of temperature – the ice-point (lower fixed point) is taken as 32 °F, and the steam-point (upper fixed point) is taken as 212 °F; the interval between these two points is divided into 180 equal divisions; the unit division on the Fahrenheit scale is 1 °F; the relationship relates the temperatures on the Celsius scale and the Fahrenheit scale, C/100 = (°F – 32) / 180; the temperature of a healthy person is 37 °C or 98.6 °F.

Farad – the SI unit of capacitance; the capacitance of a capacitor that, if charged to 1 C, has a potential difference of 1 V.

Faraday – the electric charge required to liberate a gram equivalent of a substance; 1 Faraday = 9,6485 coulomb/mole.

Fermat's principle – an electromagnetic wave takes a path involving the least time when propagating between two points.

First Law of Motion – every object remains at rest or in a state of uniform straight-line motion unless acted on by an unbalanced force.

Fluid – matter that can flow or be poured; the individual molecules of fluid can move, rolling by another.

Focus – the point to which rays that are initially parallel to the axis of a lens or mirror converge or from which they appear to diverge.

Force (F) – a push or pull which tends to change the state of rest or uniform motion, the direction of motion or the shape and size of a body; a vector quantity; the SI unit is a Newton, denoted by N; one N is the force which when acting on a body of mass 1 kg produces an acceleration of 1 m/s^2.

Force of gravitation – the force with which two objects attract by their masses; acts even if the two objects are not connected; an action-at-a-distance force.

Fossil fuels – formed from the remains of plants and animals that lived millions of years ago.

Fracture strain – an adjustment to stress in which materials crack or break because of the stress.

Fraunhofer lines – the dark lines in the spectrum of the sun or a star.

Freefall – the motion of a body falling to Earth with no other force except the force of gravity acting on it; free-falling bodies are weightless.

Freezing – the changing of a liquid to a solid state.

Freezing point – the temperature at which a phase change of liquid to solid occurs; the same temperature as the melting point for a given substance.

Frequency – the number of oscillations completed in 1 second by an oscillating body.

Frequency (of oscillations) – the number of oscillations made by an oscillating body per second.

Frequency (of waves) – the number of waves produced per second.

Friction – the force that resists the motion of one surface relative to another with which it is in contact; caused by the humps and crests of surfaces, even those on a microscopic scale; the area of contact is small, and the consequent high pressure leads to local pressure welding of the surface; in motion, the welds are broken and remade continually.

Fuel – any substance that burns in oxygen to produce heat.

Fuel rod – long zirconium alloy tubes are containing fissionable material for use in a nuclear reactor.

Fundamental charge – the smallest common charge known; the magnitude of the charge of an electron and a proton, which is 1.60×10^{-19} coulombs.

Fundamental frequency – the lowest frequency (longest wavelength) at which a system vibrates freely and can set up standing waves in an air column or on a string.

Fundamental properties – a property that cannot be defined in more straightforward terms other than to describe how it is measured; the fundamental properties are length, mass, time, and charge.

Fuse – a safety device in an electric circuit. If the current (i.e., the flow of charge) gets too high, the wire in the fuse melts, which breaks the circuit switching off the current.

G

g **–** a symbol representing the acceleration of an object in free fall due to the force of gravity; its magnitude is 9.80 m/s^2.

Gamma (γ) ray – a high energy photon of short wavelength electromagnetic radiation emitted by decaying nuclei (γ-decay).

Gases – a phase of matter composed of molecules that are relatively far apart moving freely in constant, random motion and have weak cohesive forces acting between them, resulting in the characteristic indefinite shape and indefinite volume of a gas.

Graham's Law of Diffusion – the diffusion rate of a gas is inversely proportional to the square root of its density.

Gram-atomic weight – the mass in grams of one mole of an element that is numerically equal to its atomic weight.

Gram-formula weight – the mass in grams of one mole of a compound that is numerically equal to its formula weight.

Gram-molecular weight – the gram-formula weight of a molecular compound.

Gravitational constant (G) – term in the equation for Newton's Law of Gravitation; numerically, equal to the force of gravitation, which acts between two bodies with a mass of 1 kg each separated by 1 m; the value of G is 6.67×10^{-11} Nm^2/kg^2.

Gravitational potential at a point – the amount of work done against the gravitational forces to move a particle of unit mass from infinity to that point.

Gravitational potential energy (PE) – the energy possessed by a body by its height from the ground; equals *mgh.*

Gravity – the gravitational attraction at the surface of a planet or other celestial body.

Greenhouse effect – the process of increasing the temperature of the lower parts of the atmosphere through redirecting energy back toward the surface; the absorption and re-emission of infrared radiation by carbon dioxide (CO_2), water vapor and a few other gases in the atmosphere.

Ground state – the energy state of an atom with its electrons at the lowest energy state possible for that atom.

H

Half-life – time required for one-half of the unstable nuclei in a radioactive substance to decay into a new element.

Halogens – the elements in group seven in the periodic table.

Hard water – water that finds it difficult to form lather with soap.

Heat – a form of energy that makes a body hot or cold; measured by the temperature-effect, it produces in any material body; the SI unit is the Joule (J).

Heisenberg uncertainty principle – states that there is a fundamental limit to the precision with which certain pairs of physical properties of a particle (i.e., complementary variables) can be known simultaneously (e.g., one cannot measure the exact momentum and position of a subatomic particle at the same time – the more certain is one, the less certain is the other).

Hertz (Hz) – unit of frequency; equivalent to one cycle per second.

Hooke's Law – within the elastic limit, stress is directly proportional to strain.

Horsepower – unit of power; 1 hp = 746 Watts.

Humidity – the ratio of water vapor in a sample of air to the volume of the sample.

Huygens' principle – each point on a light wavefront can be regarded as a source of secondary waves, the envelope of these secondary waves determining the position of the wavefront later.

Hypothesis – a tentative explanation of a phenomenon that is compatible with the data and provides a framework for understanding and describing that phenomenon.

I

Ice-point – the melting point of ice under 1 atm pressure; equal to 0 °C or 32 °F.

Ideal gas equation – PV = nRT.

Immiscible liquids – liquids that do not mix to form a solution, e.g. oil and water.

Impulse – equal to the product of the force acting on a body and the time for which it acts; if the force is variable, the impulse is the integral of Fd_t from t_0 to t_1; the impulse of a force acting for a given time interval is equal to change in momentum produced over that interval; $J = m(v - u)$, assuming that the mass m remains constant while the velocity changes from v to u; the SI units are kg m/s.

Impulsive force – acts on a body for a short time but produces a large change in the momentum of the body.

Incandescent – matter emitting visible light because of high temperature (e.g., a light bulb, a flame from any burning source, the Sun).

Incident ray – line representing the direction of motion of incoming light approaching a boundary.

Index of refraction – the ratio of the speed of light in a vacuum to the speed of light in a material.

Indicator – a substance, which shows using a color change if a substance is acidic or basic.

Inertia – the property of matter that causes it to resist any change in its state of rest or of uniform motion; there are three kinds of inertia: the inertia of rest, the inertia of motion and the inertia of direction; the mass of a body is a measure of its inertia.

Infrasonic – sound waves at a frequency below the range of human hearing (less than 20 Hz).

Insulators – materials that are poor conductors of heat or electricity (e.g., wood or glass); materials with air pockets slow down the movement of heat because the air molecules are far apart.

Intensity – a measure of the energy carried by a wave.

Interference – the redistribution of energy due to the superposition of waves with a phase difference from coherent sources, resulting in alternate light and dark bands.

Intermolecular forces – interaction between molecules.

Internal energy – the sum of the kinetic energy and potential energy of the molecules of an object.

Inverse proportion – the relationship in which the value of one variable increases while the value of a second variable decreases at the same rate (in the same ratio).

Ion exchange – a method of removing hardness from water, and it replaces the positive ions that cause the hardness with H^+ ions.

Ion – a charged atom or group of atoms, e.g. Na+.

Ionic bond – a force of attraction between oppositely charged ions in a compound, resulting from a transfer of electrons.

Ionization – a process of forming ions from molecules.

Ionized – an atom or a particle with a net charge because it has gained or lost electrons.

Isobaric process – pressure remains constant.

Isochoric process – volume remains constant.

Isostasy – a balance or equilibrium between adjacent blocks of Earth's crust.

Isothermal process – in which temperature remains constant.

Isotope – atoms of the same element with the same atomic number (i.e., number of protons) but with a different mass number (i.e., number of neutrons).

J

Joule (J) – the unit used to measure work and energy; can also measure heat; 1 J = 1N·m.

Joule's Law of Heating – states that the heat produced when current (I) flows through a resistor (R) for a given time (t) is given by $Q = I^2Rt$.

K

Kelvin scale of temperature (K) – the lower fixed point is taken as 273.15 K (0 °C), and the steam-point (the upper fixed point) is taken as 373.15 K (100 °C); the interval between these two points is divided into 100 equal parts; each division is equal to 1 K.

Kelvin's statement of Second Law of Thermodynamics – it is impossible that, at the end of a cycle of changes, heat has been extracted from a reservoir, and an equal amount of work has been produced without producing some other effect.

Kepler's Laws of Planetary Motion – the three laws describing the motion of the planets.

Kepler's First Law – in planetary motion, each planet moves in an elliptical orbit, with the Sun located at one focus.

Kepler's Second Law – a radius vector between the Sun and a planet moves over equal areas of the ellipse during equal time intervals.

Kepler's Third Law – the square of the period of an orbit is directly proportional to the cube of the radius of the major axis of the orbit.

Kilocalorie (Kcal) – the amount of energy required to raise the temperature of 1 kg of water by 1 °C; 1 Kcal = 1,000 calories.

Kilogram – the fundamental unit of mass in the metric system of measurement.

Kinetic energy (KE) – possessed by a body due to its motion; $KE = ½mv^2$, where m is mass and v is velocity.

L

Laser – a device that produces a coherent stream of light through stimulated emission of radiation.

Latent heat – energy released or absorbed by a body during a constant-temperature phase change.

Latent heat of vaporization – the heat absorbed when one gram of a substance changes from the liquid phase to the gaseous phase; also, the heat released when one gram of gas changes from the gaseous phase to the liquid phase.

Latent heat of fusion – the quantity of heat required to convert one unit mass of a substance from a solid state to a liquid state at its melting point without a change in its temperature; the SI unit is J kg^{-1}.

Latent heat of sublimation – the quantity of heat required to convert one unit of mass of a substance from a solid state to a gaseous state without a change in its temperature.

Law of Conservation of Energy – states that energy can neither be created nor destroyed but can be transformed.

Law of Conservation of Mass – mass (including single atoms) can neither be created nor destroyed in a chemical reaction.

Law of Conservation of Matter – matter can neither be created nor destroyed in a chemical reaction.

Law of Conservation of Momentum – the total momentum of a group of interacting objects remains constant in the absence of external forces.

Law of the lever – when a lever is balanced, the sum of the clockwise moments equals the sum of the anti-clockwise moments.

Lenz's Law – the induced current always flows in such a direction that it opposes the cause producing it.

Lever – a rigid body, which is free to turn about a fixed point called the fulcrum.

Light – a form of energy.

Light-year – the distance that light travels in a vacuum in one year (365.25 days); approximately 9.46×10^{15} m.

Line spectrum – an emission (of light, sound, or other radiation) spectrum consisting of separate isolated lines (discrete frequencies or energies); can be used to identify the elements in a matter of unknown composition.

Lines of force – lines drawn to make an electric field strength map, with each line originating on a positive charge and ending on a negative charge; each line represents a path on which a charge would experience a constant force; having the lines closer indicates a more energetic electric field.

Liquids – a phase of matter composed of molecules that have interactions stronger than those found in gas but not strong enough to keep the molecules near the equilibrium positions of a solid, resulting in the characteristic definite volume but the indefinite shape of a liquid.

Liter – a metric system unit of volume; usually used for liquids.

Longitudinal strain – the ratio of change in the length of a body to its initial length.

Longitudinal waves – the particles of the medium oscillate along the direction of propagation of a wave (e.g., sound waves).

Loudness – a subjective interpretation of a sound that is related to the energy of the vibrating source, related to the condition of the transmitting medium and the distance involved.

Lubricant – a substance capable of reducing friction (i.e., force that opposes the direction of motion).

Luminosity – the total amount of energy radiated into space each second from the surface of a star.

Luminous – objects that produce visible light (e.g., the Sun, stars, light bulbs, burning materials).

Lunar eclipse – when the Earth passes between the sun and the moon.

Lyman series – a group of lines in the ultraviolet region in the spectrum of hydrogen.

M

Magnetic domain – tiny physical regions in permanent magnets, approximately 0.01 to 1 mm, have magnetically aligned atoms, giving the domain an overall polarity.

Magnetic field – the region around a magnet where other magnetic objects experience its magnetic force; a model used to describe how magnetic forces on moving charges act at a distance.

Magnetic poles – the ends, or sides, of a magnet about which the force of magnetic attraction seems to be concentrated.

Magnetic quantum number – from the quantum mechanics model of the atom, one of four descriptions of the energy state of an electron wave; describes the energy of an electron orbital as the orbital is oriented in space by an external magnetic field, a kind of energy sub-sublevel.

Magnetic reversal – the changing of polarity of the Earth's magnetic field as the north magnetic pole and the south magnetic pole exchange positions.

Magnetic wave – the spread of magnetization from a small portion of a substance from an abrupt change in the magnetic field.

Magnification – the ratio of the size of the image to the size of the object.

Magnitude – the size of a measurement of a vector; scalar quantities that consist of a number and unit only.

Malleable – can be hammered into sheets (e.g., metals).

Malus Law – the intensity of the light transmitted from the analyzer varies directly as the square of the cosine of the angle between the plane of transmission of the analyzer and the polarizer.

Maser – microwave amplification by stimulated emission of radiation.

Mass (m) – the quantity of matter in a body; the SI unit is the kg; remains the same everywhere; a measure of inertia, which means resistance to a change of motion.

Mass defect – the difference between the sum of the masses of the individual nucleons of a nucleus and mass of nucleus.

Mass number – the sum of the number of protons and neutrons in a nucleus; used to identify isotopes (e.g., Uranium-238).

Matter – anything that occupies space and has mass.

Mean life – the average time during which a system (e.g., atom, nucleus) exists in a specified form.

Mechanical energy – the sum of the potential energy (PE) and the kinetic energy (KE) of a body; energy associated with the position of a body.

Mechanical waves – require a material medium for propagation (e.g., sound waves and water waves); also *elastic waves*.

Megahertz (MHz) – unit of frequency; equal to 106 Hertz.

Melting – the changing of a solid to a liquid state.

Melting point – the temperature at which a phase change of solid to liquid takes place.

Metal – matter having the physical properties of conductivity, malleability, ductility, and luster.

Meter – the fundamental metric unit of length.

MeV – a unit of energy; equal to 1.6×10^{-13} joules.

Millibar – a measure of atmospheric pressure equivalent to 1,000 dynes per cm^2.

Miscible fluids – fluids that can mix in any proportion.

Mixture – matter made of unlike parts with variable composition and separated into their parts by physical means.

Model – a mental or physical representation of something that cannot be observed directly; used as an aid to understanding.

Modulus of elasticity – the ratio of stress to the strain produced in a body.

Modulus of rigidity – the ratio of tangential stress to the shear strain produced in a body.

Mole – the amount of a substance that contains Avogadro's number of atoms, ions, molecules, or any other chemical unit; 6.02×10^{23} atoms, ions, or other chemical units.

Molecule – two or more atoms chemically combined.

Moment – a measure of the turning effect of a force. Moment of a force = force × perpendicular distance from the fulcrum.

Momentum – a measure of the quantity of motion; the product of the mass and the velocity of a body; SI units are kg·m /s.

Monochromatic light – consisting of a single wavelength.

N

Natural frequency – the frequency of oscillation of an elastic object in the absence of external forces; depends on the size, composition, and shape of the object.

Negative electric charge – one of two types of electric charge; repels other negative charges and attracts positive charges.

Negative ion – atom or particle that has a surplus or imbalance of electrons and a negative charge.

Net force – the resulting force after vector forces are summed; if a net force is zero, the vector forces have canceled, and there is no unbalanced force.

Neutralization – the reaction between an acid and a base to give salt and water.

Newton (N) – a unit of force defined as $kg{\cdot}m/s^2$; 1 Newton is needed to accelerate a 1 kg mass by 1 m/s^2.

Newton's First Law of Motion – a body continues in a state of rest or uniform motion in a straight line unless an external (unbalanced) force acts upon it.

Newton's Law of Gravitation – the gravitational force of attraction acting between two particles is directly proportional to the product of their masses and inversely proportional to the square of the distance between them; the force of attraction acts along the line joining the two particles; real bodies having spherical symmetry act as point masses with their mass assumed to be concentrated at their center of mass.

Newton's Second Law of Motion – the rate of change of momentum is equal to the force applied; the force acting on a body is directly proportional to the product of its mass and acceleration produced by force in the body.

Newton's Third Law of Motion – for every action, there is an equal and opposite reaction; the action and the reaction act on two different bodies simultaneously.

Noise – sounds made up of groups of waves of random frequency and intensity.

Non-uniform acceleration – when the velocity of a body increases by unequal amounts in equal intervals of time.

Non-uniform speed – when a body travels unequal distances in equal intervals of time.

Non-uniform velocity – when a body covers unequal distances in equal intervals of time in a direction, or when it covers equal distances in equal intervals but changes its direction.

Normal (N) – a line perpendicular to the surface of a boundary.

Nuclear energy – the form of energy from reactions involving the nucleus.

Nuclear fission – the splitting of a heavy nucleus into more stable, lighter nuclei with an accompanying release of energy.

Nuclear force – one of four fundamental forces; a strong force of attraction that operates over short distances between subatomic particles; overcomes the electric repulsion of protons in a nucleus and binds the nucleus.

Nuclear fusion – a nuclear reaction of low mass nuclei fusing to form a more stable and more massive nucleus with an accompanying release of energy.

Nuclear reactor – a steel vessel in which a controlled chain reaction of fissionable materials releases energy.

Nucleons – a collective name for protons and neutrons in the nucleus of an atom.

Nucleus – the central, positively charged, dense portion of an atom; contains protons and neutrons.

O

Octet rule – during bonding, atoms tend to reach an electron arrangement with eight electrons in the outermost shell.

Ohm – unit of resistance; 1 ohm = 1volt/ampere.

Ohm's Law – the current flowing through a conductor is directly proportional to the potential difference across the ends of the conductor. At a constant temperature, the voltage across a conductor is proportional to the current flowing through it. Voltage = current × resistance (V = IR).

Open system – a system across whose boundaries matter and energy can pass.

Optical fiber – a long, thin thread of fused silica; used to transmit light; based on total internal reflection.

Orbital – the region of space around the nucleus of an atom where an electron is likely to be found.

Origin – the point on a graph where the x and the y variables have a value of zero at the same time.

Oscillatory motion – the to and fro motion (periodic) of a body about its mean position; also, *vibratory motion.*

Oxidation – the addition of oxygen or the losing of electrons.

P

Pascal – a unit of pressure equal to the pressure resulting from a force of 1 N acting uniformly over an area of 1 m^2.

Pascal's Law – states that the pressure exerted on a liquid is transmitted equally in all directions.

Paschen series – a group of lines in the infrared region in the spectrum of hydrogen.

Pauli exclusion principle – no two electrons in an atom can have the same four quantum numbers; a maximum of two electrons can occupy a given orbital.

Peltier effect – the evolution or absorption of heat at the junction of two dissimilar metals carrying current.

Period (of a wave) – the time for a wave to travel through a distance equal to its wavelength; denoted by T; period of a wave = 1/frequency of the wave.

Period (of an oscillation) – the time to complete one oscillation; does not depend upon the mass of the bob and amplitude of oscillation; directly proportional to the square root of the length and inversely proportional to the square root of the acceleration due to gravity.

Periodic wave – a wave in which the particles of the medium oscillate continuously about their mean positions regularly at fixed intervals of time.

Periodic motion – a motion that repeats at regular time intervals.

Permeability – the ability to transmit fluids through openings, tiny passageways, or gaps.

pH scale – a scale from 0 to 14. If the pH of a solution is 7 it is neutral; if the pH of a solution is less than 7 it is acidic; if the pH of a solution is greater than 7 it is basic.

Permanent hardness – hardness in water that cannot be removed by boiling. It is caused by calcium sulfate.

Phase – when particles in a wave are in the same state of vibration (i.e., in the same position and the direction of motion).

Phase change – the action of a substance changing from one state of matter to another; always absorbs or releases internal potential energy that is not associated with a temperature change.

Photons – quanta of energy in the light wave; the particle associated with light.

Photoelectric effect – the emission of electrons in some materials when the light of a suitable frequency falls on them.

Physical change – a change of the state of a substance but not in the identity of the substance.

Pitch (sound) – depends on the frequency of the wave.

Planck's constant – proportionality constant in the ratio of the energy of vibrating molecules to their frequency of vibration; a value of 6.63×10^{-34} J·s.

Plasma – a phase of matter; a hot highly ionized gas consisting of electrons and atoms that have been stripped of their electrons because of high kinetic energies.

Plasticity – the property of a solid whereby it undergoes a permanent change in shape or size when subjected to a stress.

Plastic strain – an adjustment to stress in which materials become molded or bent out of shape under stress and do not return to their original shape after the stress is released.

Polarized light – light whose constituent transverse waves are vibrating in the same plane.

Polaroid – a film that transmits only polarized light.

Polaroid or polarizer – a device that produces polarized light.

Positive electric charge – one of the two types of electric charge; repels positive and attracts negative charges.

Positive ion – atom or particle with a net positive charge due to an electron or electrons being torn away.

Positron – an elementary particle having the same mass as an electron but an equal and positive charge.

Potential difference (or *voltage*) – the force, which moves the electrons around the circuit. Unit is Volt (V).

Potential energy (PE) – possessed by a body by its position or configuration; see the *Gravitational potential energy* and *Elastic potential energy*.

Power – scalar quantity for the rate of doing work; the SI unit is Watt; 1 W = 1 J/s. The rate at which energy is converted from one form to another. Unit is Watts (W). Power = voltage × current (P = VI).

Pressure – a measure of force per unit area (e.g., kilograms per square meter (kg/m^2). Unit is Pascal (Pa).

Primary coil – part of a transformer; a coil of wire connected to a source of alternating current.

Primary colors – three colors (red, yellow, and blue) combined in various proportions to produce any other color. When red, green, and blue are combined, it results in white. For example, red + green + blue = white.

Principal quantum number – from the quantum mechanics model of the atom, one of four descriptions of the energy state of an electron wave; describes the main energy level of an electron regarding its most probable distance from the nucleus.

Principle of calorimetry – states that if two bodies of different temperature are in thermal contact, and no heat is allowed to go out or enter the system; heat lost by the body with higher temperature is equal to the heat gained by the body of lower temperature (i.e., the heat lost = the heat gained).

Products – chemicals produced in a chemical reaction.

Progressive wave – a wave that transfers energy from one part of a medium to another.

Projectile – an object thrown into space horizontally or at an acute angle and under the action of gravity; the path followed by a projectile is its trajectory; the horizontal distance traveled by a projectile is its range; the time is taken from the moment it is thrown until the moment it hits the ground is its time of flight.

Proof – a measure of ethanol concentration of an alcoholic beverage; double the concentration by volume (e.g., 50% by volume is 100 proof).

Properties – qualities or attributes that, taken together, are usually unique to an object (e.g., color, texture, size).

Proportionality constant – a value applied to a proportionality statement that transforms the statement into an equation.

Pulse – a wave of short duration confined to a small portion of the medium at any given time; also a *wave pulse*.

Q

Quanta – fixed amounts; usually referring to fixed amounts of energy absorbed or emitted by matter.

Quantum limit – the shortest wavelength; present in a continuous x-ray spectrum.

Quantum mechanics – model of the atom based on the wave nature of subatomic particles and the mechanics of electron waves; also *wave mechanics*.

Quantum numbers – numbers that describe the energy states of an electron; in the Bohr model of the atom, the orbit quantum numbers could be any whole number (e.g., 1, 2, 3, etc.); in the quantum mechanics model of the atom, four quantum numbers are used to describe the energy state of an electron wave (n, m, l, and s).

Quark – one of the hypothetical fundamental particles; has a charge with magnitudes of one-third or two-thirds of the charge on an electron.

R

Rad – a measure of radiation received by a material (radiation-absorbed dose).

Radiant energy – the form of energy that can travel through space (e.g., visible light and other parts of the electromagnetic spectrum).

Radiation – the emission and propagation of waves transmitting energy through space or some medium. Heat transfer through invisible rays, which travel outwards from the hot object without a medium.

Radioactive decay – the natural, spontaneous disintegration or decomposition of a nucleus.

Radioactive decay constant – a specific constant for a isotope that is the ratio of the rate of nuclear disintegration per unit of time to the total number of radioactive nuclei.

Radioactive decay series – series of decay reactions that begins with one radioactive nucleus that decays to a second nucleus that decays to a third nucleus and so on, until a stable nucleus is reached.

Radioactive Decay Law – the rate of disintegration of a radioactive substance is directly proportional to the number of undecayed nuclei.

Radioactivity – spontaneous emission of particles or energy from an atomic nucleus as it disintegrates.

Rarefaction – a part of a longitudinal wave where the density of the particles of the medium is less than the typical density.

Real image – an image generated by a lens or mirror that can be projected onto a screen.

Reactants – chemicals that react together in a chemical reaction.

Rectifier – converts alternating current to direct current.

Rectilinear motion – the motion of a body in a straight line.

Reduction – the removal of oxygen or the gaining of electrons.

Reflected ray – a line representing the direction of motion of light reflected from a boundary.

Reflection – the bouncing back of light from a surface.

Refraction – the bending of a light wave, a sound wave, or another wave from its straight-line path as it travels from one medium to another.

Refractive index – the ratio of the speed of light in a vacuum to that in the medium.

Relative density – (i.e., *specific gravity*) is the ratio of the density (mass of a unit volume) of a substance to the density of given reference material. *Specific gravity* usually means relative density concerning water. The term *relative density* is more common in modern scientific usage.

Relative humidity – the percentage of the amount of water vapor present in a specific volume of the air to the amount of water vapor needed to saturate it.

Resistance (R) – the opposition of a conductor to current (i.e., the flow of charge). A good conductor has a low resistance, and a bad conductor has a high resistance.

Resolving power – the ability of an optical instrument to produce separable images of different points of an object.

Resonance – when the frequency of an external force matches the natural frequency of the body.

Restoring force – the force which tends to bring an oscillating body back to its mean position whenever it is displaced from the mean position.

Resultant force – a single force, which acts on a body to produce the same effect on it as done by other forces collectively; see *balanced forces*.

Reverberation – apparent increase in the volume of sound caused by reflections from the boundary surfaces, usually arriving within 0.1 seconds after the original sound.

Rigid body – an idealized extended body whose size and shape are fixed and unaltered when forces are applied.

S

Salt – when a metal replaces the hydrogen of an acid.

Saturated air – air in which an equilibrium exists between evaporation and condensation; the relative humidity is 100 percent.

Saturated solution – the apparent limit to dissolving a given solid in a specified amount of water at a given temperature; a state of equilibrium exists between dissolving solute and solute coming out of solution.

Scalar quantity – a physical quantity described entirely by its magnitude.

Scientific law – a relationship between quantities; usually described by an equation in the physical sciences; describes a broader range of phenomena and is more important than a scientific principle.

Scientific principle – a relationship between quantities within a specific range of observations and behavior.

Second – the standard unit of time in the metric and English systems of measurement.

Second Law of Motion – the acceleration of an object is directly proportional to the net force acting on that object and inversely proportional to the mass of the object.

Secondary coil – part of a transformer; a coil of wire in which the voltage of the original alternating current in the primary coil can be stepped up or down by electromagnetic induction.

Secondary colors – formed when two primary colors (i.e., red, green, blue) are mixed. The three secondary colors are yellow, magenta, and cyan. For example, red + green = yellow; red + blue = magenta; and blue + green = cyan.

Second's pendulum – a simple pendulum whose period on the surface of the Earth is 2 seconds.

Semiconductors – elements whose electrical conductivity is intermediate between that of a conductor and an insulator.

Shear strain – the ratio of the relative displacements of one plane to its distance from the fixed plane.

Shear stress – the restoring force developed per unit area when deforming force acts tangentially to the surface of a body, producing a change in the shape of the body without any volume change.

Siemens – the derived SI unit of electrical conductance; equal to the conductance of an element with a resistance of 1 ohm; also written as ohm^{-1}.

Simple harmonic motion – the vibratory motion occurs when the restoring force is proportional to the displacement from the mean position and is directed opposite to the displacement.

Simple pendulum – a heavy point mass (a small metallic ball) suspended by a light inextensible string from frictionless rigid support; a simple machine based on the effect of gravity.

Snell's Law – states that the ratio of *sin* i to *sin* r is a constant and is equal to the refractive index of the second medium concerning the first.

Solar eclipse – happens when the moon passes between the sun and the Earth.

Solenoid – a cylindrical coil of wire that becomes electromagnetic when current passes through it.

Solids – a phase of matter with molecules that remain close to fixed equilibrium positions due to strong interactions between the molecules, resulting in the characteristic definite shape and definite volume of a solid.

Solution – a mixture of a solute (usually a solid) and a solvent (usually a liquid).

Sonic boom – sound waves that pile up into a shock wave when a source is traveling at or faster than the speed of sound.

Sound – a form of energy.

Specific gravity – see *relative density*.

Specific heat – the amount of heat energy required to increase the temperature of 1 g of a substance by 1 °C; each substance has its specific heat value.

Speed – a scalar quantity for the distance traveled by a body per unit of time; if a body covers the distance in time, its speed is given by distance/time; SI units are m/s.

Spin quantum number – from the quantum mechanics model of the atom, one of four descriptions of the energy state of an electron wave; describes the spin orientation of an electron relative to an external magnetic field.

Stable equilibrium – a body is in stable equilibrium if, when slightly moved, its center of gravity rises.

Standing waves – the condition where two waves of equal frequency traveling in opposite directions meet and form stationary regions of maximum displacement due to constructive interference and stationary regions of zero displacement due to destructive interference.

State of motion – when a body changes its position concerning a fixed point in its surroundings; the states of rest and motion are relative to the frame of reference.

State of rest – when a body does not change its position concerning a fixed point in its surrounding; the states of rest and motion are relative to the frame of reference.

Steam-point – the temperature of steam over pure boiling water under 1 atm pressure; taken as the upper fixed point (100 °C or 212 °F) for temperature scales.

Stefan-Boltzmann Law – the amount of energy radiated per second per unit area of a perfectly black body is directly proportional to the fourth power of the absolute temperature of the surface of the body.

Sublimation – the changing of a solid directly to a gas. (Iodine is an example of a substance that sublimes).

Superconductors – some materials in which, under certain conditions, the electrical resistance approaches zero.

Super-cooled – water in the liquid phase when the temperature is below the freezing point.

Supersaturated – containing more than the average saturation amount of a solute at a given temperature.

Surface tension – the property of a liquid due to which its surface behaves like a stretched membrane.

Suspension – a mixture of a liquid and a finely divided insoluble solid.

T

Temperature – a measure of the hotness or coldness of a body; according to the molecular model, measures the average kinetic energy of the molecules; heat flows from a body at a higher temperature to a body at a lower temperature.

Temporary hardness – hardness in water that can be removed by boiling. It is caused by calcium hydrogen carbonate.

Tensional stress – the opposite of compressional stress; when one part of a plate moves away from another part that does not move.

Tesla (T) – SI unit of magnetic flux density; the magnetic flux density of a magnetic flux of 1 Wb through an area of 1 m^2.

Thermal capacity – the quantity of heat required to raise the temperature of the body by one degree (1 K or 1 °C).

Thermal equilibrium – when two bodies in contact are at the same temperature, and there is no heat flow between them; also, the average temperature of the bodies in thermal equilibrium.

Thermal expansion – the increase in the size of an object when heated.

Thermometer – a device used for the measurement of temperature; the mercury thermometer is commonly used.

Third Law of Motion – when two objects interact, the force exerted on one object is equal in size and opposite in direction to the force exerted on the other object; forces always occur in matched pairs that are equal and opposite.

Titration – the process of adding one solution from a burette to a measured amount of another solution to find out exactly how much of each is required to react.

Total internal reflection – a condition where all light is reflected from a boundary between materials; occurs when light travels from a denser to a rarer medium, and the angle of incidence is greater than the critical angle.

Transformation of energy – converting one form of energy into another (e.g., when a body falls, its potential energy is converted to kinetic energy).

Transverse wave – a wave in which the particles of the medium oscillate in a direction perpendicular to the direction of propagation of the wave (e.g., water waves, light waves, radio waves).

Trough – the point of maximum negative displacement on a transverse wave.

U

Ultrasonic – sound waves too high in frequency (above 20,000 Hz) to be heard by the human ear.

Unbalanced forces – when some forces act on a body and the resultant force is not zero.

Uniform acceleration – when the velocity of a body increases by equal amounts in equal intervals of time.

Uniform circular motion – the motion of an object in a circular path with uniform speed; accelerated motion.

Uniform speed – when a body travels equal distances in equal intervals of time.

Uniform velocity – when a body travels along a straight line in a direction with equal distance in equal time intervals.

Universal Law of Gravitation – every object is attracted to every other object with force directly proportional to the product of their masses and inversely proportional to the square of the distance between the centers of the two masses.

Unpolarized light – light consisting of transverse waves vibrating in all possible random directions.

Unstable equilibrium – a body is in unstable equilibrium if, when slightly moved, its center of gravity falls.

V

Van der Waals force – general term for weak attractive intermolecular forces.

Valency – the number of electrons an atom wants to gain, lose, or share to have a full outer shell.

Vapor – the gaseous state of a substance that is generally in a liquid state.

Vector quantity – a quantity that needs magnitude and direction to describe it.

Velocity (v) – distance traveled by a body in a direction per unit time; the displacement of the body per unit time; a vector quantity; the SI units are m/s.

Vibration – a back and forth motion that repeats itself.

Virtual image – an image formed when the reflected or refracted light rays appear to meet; this image cannot be projected on a screen.

Volt (V) – a unit of potential difference equivalent to joules/coulomb.

Voltage drop – the difference in electric potential across a resistor or other part of a circuit that consumes power.

Volume – the amount of space an object occupies.

W

Watt (W) – SI unit for power; equivalent to joule/s.

Wave – a disturbance or oscillation that moves through a medium.

Wavelength (λ) – the distance between the two nearest points on a wave in the same phase; the distance between two adjacent crests or two adjacent troughs.

Wave (mechanical) – a periodic disturbance produced in a material medium due to the vibratory motion of the particles of the medium.

Wave mechanics – alternate name for quantum mechanics derived from the wavelike properties of subatomic particles.

Wave motion – the movement of a disturbance from one part of a medium to another involving the transfer of energy but not the transfer of matter.

Wave period – the time required for two successive crests (or successive regions) of the wave to pass a given point.

Wave velocity – the distance traveled by a wave in one second; depends on the medium through which it passes.

Weight – the force with which a body is attracted towards the center of the Earth; the SI unit is N; the gravitational units are kg·wt and g·wt; the weight of a body is given by *mg*. Weight = mass × acceleration due to gravity.

Weightlessness – the state when the apparent weight of a body becomes zero; objects while falling freely under the action of gravity are seemingly weightless.

Wien's Displacement Law – states that for a black body, the product of the wavelength corresponding to its maximum radiance and its absolute temperature is constant.

Work – work is done when a force acting on a body displaces it; work = force × displacement (W = Fd) in the direction of the force; work is a scalar quantity; the SI unit is Joule.

Y

Young's modulus of elasticity – the ratio of normal stress to the longitudinal strain produced in a body.

Z

Zeeman effect – the splitting of the spectral lines in a spectrum when the source is exposed to a magnetic field.

Zeroth Law of Thermodynamics – states that if body A is in thermal equilibrium with body B, and B is also in thermal equilibrium with C, then A is necessarily in thermal equilibrium with C.

College study aids by Sterling Test Prep

Cell and Molecular Biology Review

Organismal Biology Review

Cell and Molecular Biology Practice Questions

Organismal Biology Practice Questions

Physics Review (Paert 1 and 2)

Physics Practice Questions (Vol. 1 and 2)

Organic Chemistry Practice Questions

United States History 101

American Government and Politics 101

Environmental Science 101

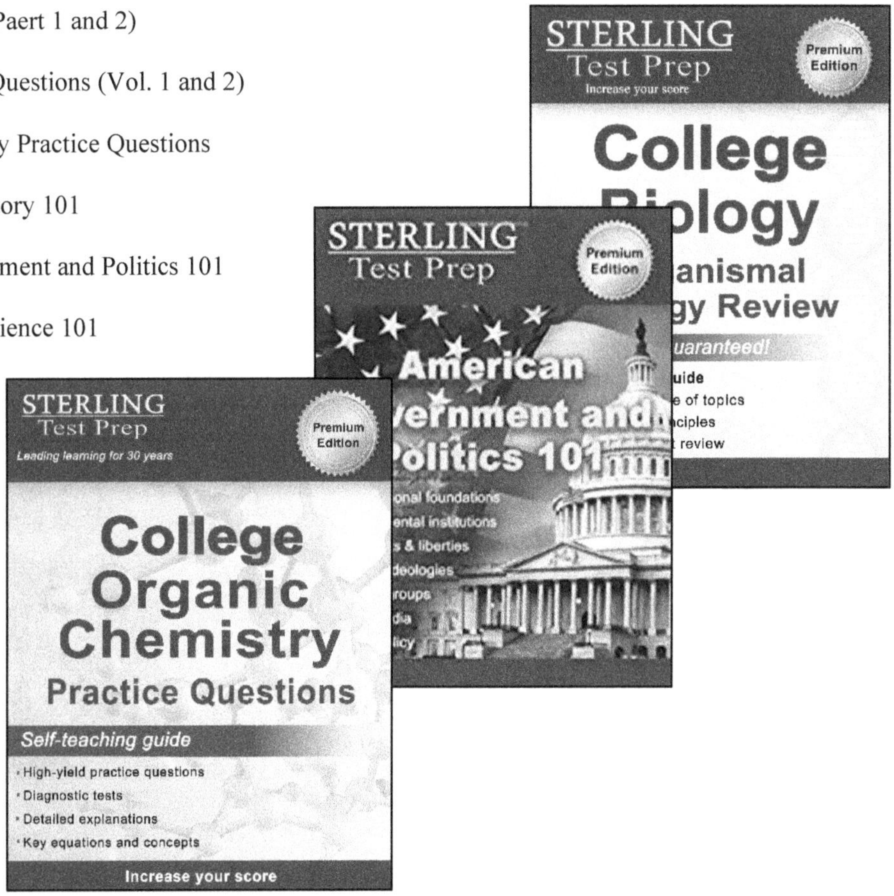

Visit our Amazon store